TIMELESS REALITY

TIMELESS
REALITY

Symmetry, Simplicity, and Multiple Universes

VICTOR J.
STENGER, PH.D.

Prometheus Books

59 John Glenn Drive
Amherst, New York 14228-2119

Published 2000 by Prometheus Books

Inquiries should be addressed to
Prometheus Books
59 John Glenn Drive
Amherst, New York 14228–2119
VOICE: 716–691–0133, ext. 210
FAX: 716–691–0137
WWW.PROMETHEUSBOOKS.COM

12 11 10 09   5 4 3

Library of Congress Cataloging-in-Publication Data

Stenger, Victor J., 1935–
   Timeless reality : symmetry, simplicity, and multiple universes / Victor J. Stenger.
      p. cm.
   Includes bibliographical references and index.
   ISBN 1–57392–859–3 (alk. paper)
   1. Time reversal. 2. Quantum theory. I. Title.

QC173.59.T53 .S74 2000
530.12—dc21   00–059860

Printed in the United States of America on acid-free paper

CONTENTS

TIMELESS REALITY

[6]

PREFACE

All great discoveries in experimental physics have been due to the intuition of men who made free use of models, which were for them not products of the imagination but representatives of real things.

Max Born (1953)

ATOMS AND THE VOID

Most people would agree that science must tell us something about reality. However, no one has stated what exactly that may be. Scientists do not speak with one mind on the matter. They, and the philosophers who study science, have not reached anything approaching consensus on the nature of reality revealed by science, or even if any has been revealed. Still, despite this collective uncertainty, something must be out there in the real world. And, based on its track record, science is still the best tool we have at our disposal to help us find it.

TIMELESS REALITY

Since the seventeenth century, science has occupied first place among the various approaches that humans have taken in their attempts to understand and control their environment. This special status did not come about as a consequence of a jeweled crown being placed on its head by some higher authority. Rather, science proved itself by results. Its instruments have greatly extended the range of vision provided by the human senses, and the theories of science have profoundly altered the way humanity thinks about itself and its place in the overall scheme of things.

Observation and theory constrained by an uncompromising methodology have worked together to present us with a picture of a universe beyond the imagining of the most talented poet or pious mystic. No other human intellectual or creative endeavor, whether philosophy, theology, art, or religious experience, from East or West, has come close to fantasizing the universe revealed by modern physics and astronomy. Reality is out there telling us that this is the way it is, whether we like it or not, and that reality is far beyond our simple, earthbound imaginations.

In recent years, the privileged position for science has been challenged. Some sociologists and other scholars who have examined science within a cultural context have concluded that statements made within Western science are simply narratives that have no more claim on the truth than the myths of any other culture. So far they have convinced no one but themselves.

Science is not fiction. Although it involves creativity, it is not the sole product of unbridled imagination. Scientists build equipment and mathematical theories, gather and analyze data, and come to an always-tentative consensus on what should be added to or subtracted from the library of scientific knowledge. That library is then utilized by technologists to build the many devices that mark the dramatic difference between the lives of humans today and those of the not-too-distant past.

To be sure, much imagination went into the development of the computer that sits on my desk. But that imagination was forced to act within a framework of constraints such as energy conservation and gravity. These constraints are codified as the "laws" of physics. Surely they represent some aspect of reality and are not pure fantasy.

Scientists themselves, including great numbers of non-Western persuasion, continue to maintain confidence in the exceptional power and value of their trade. They are sure they are dealing with reality, and most people outside of a few departments in academia agree. But, we must still ask, what is the reality that scientists are uncovering?

In this book I suggest that the underlying reality being accessed by the instruments of science is far simpler than most people, including many scientists and philosophers, realize. The portion of reality that responds to the probing of scientific experiment and theory is not terribly mysterious. For those portions remaining unresponsive pending further discovery, we have no basis to believe that they fall outside the naturalist tradition that has developed over millennia. No one need think, after this time, that any phenomena currently lacking full scientific explanations can only be revealed by nonscientific or supernatural means.

Based on all we know today, the complete library of data from across the full spectrum of the sciences is fully consistent with a surprisingly simple model: the natural universe is composed, at the elementary level, of localized material bodies that interact by colliding with one another. All these bodies move around in an otherwise empty void. No continuous, etheric medium, material or immaterial, need be postulated to occupy the space between bodies. Applying an insight more ancient than Plato and Aristotle, but continually ignored because of human propensities to wish otherwise, *atoms and the void* are sufficient to account for observations with the human eye and the most powerful telescopes, microscopes, or particle accelerators of today.

The four-dimensional space-time framework introduced by Einstein and Minkowski, along with the associated rules of relativity and all the rest of physics, are adequate to describe the motion of these primal bodies. Furthermore, we find that the great foundational "laws" of physics—the principles of energy, linear momentum, and angular momentum conservation—are not rules imposed on the universe from outside. Rather, they represent physicists' way of theoretically describing the high degree of symmetry and simplicity that the universe, on the whole, exhibits to their instruments.

Four centuries ago, Galileo observed that an object falls with an acceleration that is independent of its mass. We find that the same is true when the experiment is done today (with the usual caveats that we neglect air friction), and we measure the same acceleration of gravity he did. When we look with our telescopes at the farthest galaxies, where the light left billions of years ago, we find that the properties of that light, such as the relative positions of spectral lines, are exactly the same as we observe in the laboratory today.

The fact that the same behavior is found over such an enormous time scale implies that the basic principles of physics do not change over time. No moment, what the Greeks called *kairos* and philosopher Martin Heidegger translated to the German as *augenblick*, stands apart from any other (despite the recent millennial fever). When we proceed to incorporate this fact in our theoretical descriptions, lo and behold we find that energy is conserved. That is, the total energy of any isolated system within the universe is a constant. Energy conservation is simply another way of saying that the universe exhibits no special moment in time.

Galaxies are distant in space as well as time; some are billions of light years away. The fact that the same physical phenomena are observed at all distances in space tells us that the principles of physics are the same at all places. No special position in space can be found where the physics is different. This is what Copernicus discovered when he realized that the earth was not absolutely at rest with the rest of the universe circling about it. When the absence of any special place in space is incorporated into our theoretical descriptions, we find that the physical quantity of momentum is conserved. That is, the total momentum of any isolated system within the universe is a constant. And so, momentum conservation is simply another way of saying that no special place in space exists.

When we look in several directions with our telescopes, we find again that the basic behavioral patterns of the observed light are the same. This absence of any special direction in space is represented in our theories as conservation of *angular momentum*. That is, the total angular momentum of any isolated system within the universe is a constant. Angular momentum conservation is simply another way of saying that the universe exhibits no special orientation in space.

These conservation "laws" are global, applying throughout our universe. Extending rotational symmetry to the full four dimensions of space-time, the principles of Einstein's special theory of relativity join the conservation principles already mentioned. In other words, the most fundamental notions of physics hardly need explanation. Any other form of these laws would be so astounding as to force us to look for some more complex explanation. They eloquently testify to the lack of design to the universe.

While the idea that many of the most important principles of physics follow from space-time symmetries may not strike a familiar chord, this connection has been known for a century or more. You will find it described in advanced physics textbooks in both classical and quantum mechanics. So my assertion represents nothing new, merely a public exposition of well-established physics. Indeed, nothing I will say in these pages should be taken as a proposal to change a single fact or equation in the existing body of physics—or any other science for that matter. I am merely reporting what that science seems to be telling us about reality.

The model of reality I propose is basically the one strongly implied by modern particle physics theory, when the esoteric mathematics of that field is recast in the admittedly less precise medium of words and images. This model cannot be proved correct by any process of deductive logic or mathematics. It is probably not verifiable by additional observations or experiments beyond what has already been done, although more experiments will yield more details and could, in principle, falsify the picture. Nevertheless, the proposed model is based on observation and experimental data, and the theories that currently describe all currently existing data without anomaly. The primary alternative models of reality are likewise not capable of being proved by logic, but I will argue that they are less reasonable, less rational, and less convincing.

Most physicists will object that only the empirically testable merits our consideration. I will not adopt that view, since it leaves us with nothing we can then say about the nature of the reality behind bald statements of fact about observations. I believe we have every right to talk about nontestable ideas, so long as we do so in a logically consistent (that is, non-self-contradictory) fashion that does not disagree with the data. And criteria other than testability must be available to allow us to make a rational choice among alternatives and make our speculations worthwhile.

The reader will not be asked to believe the proposed picture on the basis of the author's or any more famous physicist's authority. The model I will present is simple, economical, and possibly even useful, and these are rational criteria for making a choice. At the very least, I hope to demonstrate that nothing we currently know from our best sources of knowledge requires anyone to buy into one or more of the many extravagant claims that are made by those who would try to use science to promote their own particular mystical or supernatural worldview. Since these promoters introduce extraneous elements of reality not required by the data, their proposals fail the test of parsimony. It then follows that they have the burden of proving their schemes, not I the burden of disproving them.

Of course, the universe we see with eyes and instruments is not "simple" by our normal understanding of the term. The details we observe are very complex, with many layers of structure and other physical laws besides conservation principles that follow from the global symmetries of the universe. However, I will try to show that these complex structures and laws can still be grossly understood in surprisingly simple terms, where the details are unimportant. I will describe a scenario, consistent with current knowledge, where complex order arises from the spontaneous, that is, uncaused and accidental, breaking of symmetries that themselves were uncaused.

Just as the structure of living organisms is the result of spontaneous events acting within the global constraints of energy conservation and other limiting factors like gravity and friction, so, too, could the structural properties of elementary matter have evolved spontaneously in the early universe. At least nothing we currently know rules this out. In the proposed scenario, what emerged in terms of particle properties and force laws during the early evolution of the universe was not pre-determined by either natural or supernatural law. Rather, it arose by chance. Start the universe up again and it will turn out different. Thus, much of the detailed structure of the universe, so important to us as earthbound humans, is not of great importance to our basic understanding of reality. This structure could be wildly different, and that basic understanding would be unchanged.

I will discuss the possibility of other universes besides our own. These might be imagined to have different structures, different laws. While no one can demonstrate that other universes exist, current cosmological theories allow, and even suggest, that they do. Again, no known principle rules them out. To assume ours is the only universe is to take the narrow view of humans before Copernicus that the earth is the only world beneath the heavens. It seems very likely that the sum of reality includes a vastness of possibilities in which our universe is but a speck, even as our earth is a but speck within that universe. The so-called anthropic coincidences, in which our universe appears, to some, incredibly fine-tuned for the formation of carbon-based life, are readily accounted for in a universe of universes. And those who think these coincidences provide evidence for some special design, with humans in mind, exhibit the same lack of imagination as those who once thought that only one world existed and all else revolved about it.

Moving from the vast to the tiny, one place where a model of a reality containing only localized bodies may be reasonably questioned is at the level of quantum phenomena. Quantum events have been widely interpreted as providing a basis for any number of strange or even mystical and holistic effects.

For seventy years, the Copenhagen interpretation of quantum mechanics has presided as the consensus view of physics, a position that has only recently begun to erode. The way in which the observer and observed are intertwined in this interpretation has suggested to some that human consciousness has a controlling role in determining material behavior. I discussed this issue thoroughly in my previous book, *The Unconscious Quantum: Metaphysics in Modern Physics and Cosmology* (Stenger 1995), and have tried not to be too repetitive here. In some ways, this is a sequel to that book; however, *Timeless Reality* should be self-contained.

As I described in some detail in *The Unconscious Quantum*, and will only briefly summarize here, David Bohm postulated the existence of a mysterious holistic field that acts instantaneously throughout the universe to bring everything together into one irreducible whole. Bohm's model could not be more diametrically opposed to the one I will describe here. I will not disprove the Bohm model, but argue against it on the basis of parsimony.

The *many worlds interpretation*, which envisages our universe as an array of parallel worlds that exist in ghostly connection to one another, is less in conflict with the ideas I will present. Many *worlds* is not to be confused with the many *universes*, mentioned above, that go their own separate ways, presumably never coming into contact after they are formed. Many worlds might be found in each of many universes. However, the many worlds interpretation is not required in the proposed scheme and other alternative views will be presented. It may be possible to retain the ideas of the many worlds interpretation within in single world.

Other interpretations of quantum mechanics exist, but none have the dramatic implications of the big three: Copenhagen, Bohmian, and many worlds. No consensus has developed as to which, if any, is to be preferred, though each has a list of distinguished supporters. As we will see, there are many ways to skin Schrödinger's cat.

All attempts to come to grips with the observed outcomes of a wide range of quantum experiments, by applying familiar notions based on Newtonian classical physics, have conclusively failed. Still, classical physics remains highly successful when carefully applied to its own still very wide domain, which encompasses most familiar physical phenomena. Since common sense is based on our normal experience of these phenomena, something of common sense must give in trying to understand the quantum world.

One commonsense notion that may be expendable is that time changes in only one direction. As we will see, quantum events proceed equally well in either time direction, that is, they appear to be "tenseless." This is what I mean in the title: *Timeless Reality*. By allowing time to change in either direction, many of the most puzzling features of quantum mechanics can be explained within the framework of a reality of atoms and the void.

It may be a matter of taste whether you find timeless quanta more palatable than conscious quanta, holistic fields, or ghost worlds. In any case, I am not proposing an alternative interpretation of quantum mechanics, and various ideas from the many proposals in the literature may still be necessary to provide a complete picture. I merely urge that time symmetry be considered part of any interpretive scheme. As we will see, it provides for a particularly simple and elegant model of reality.

I am not the first to suggest that time symmetry might help explain some of the interpretive problems of quantum mechanics. In fact, this has been long recognized but discarded because of the implied time-travel paradoxes. I will show, again not originally, that the time-travel paradoxes do not exist in the quantum world. I feel that the possibility of time-reversal has been widely neglected for the wrong reason—a deep prejudice that time can only change from past to future. Evidence for this cannot be found in physics. The only justification for a belief in directed time is human experience, and human experience once said that the world was flat.

I ask you to open your mind to the possibility that time can also operate from future to past. The symmetry between past and future is consistent with all known physical theories, and, furthermore, is strongly suggested by quantum phenomena themselves. The indisputable asymmetry of time in human experience arises, as Ludwig Boltzmann proposed over a century ago, from the fact that macroscopic phenomena involve so many bodies that certain events are simply far more likely to happen in one direction rather than the reverse. Thus, aging is more likely than growing younger. We do not see a dead man rising, not because it is impossible but because it is so highly unlikely. But time asymmetry is no more fundamental than the left-right asymmetry of the face you see in the mirror—a simple matter of chance.

Classical physics is well-known to be time symmetric. Although the second law of thermodynamics is asymmetric by its very nature, it simply codifies the observed fact of everyday life that many macroscopic physical processes seem to be irreversible. However, the second law does not demand that they be so. In fact, every physical process is, in principle, reversible. Many simply have a low probability of happening in reverse. As Boltzmann showed, the second law amounts to a definition of the arrow of time.

Time symmetry is commonly observed in chemistry, where all individual chemical reactions can occur in either direction. The same is true in nuclear and elementary particle physics. In only a few very rare particle processes do we find the probability for one time direction very different from the other, and even then to just one part in a thousand. While this exception requires us to strictly reverse the spatial as well as time axes and change particles to antiparticles when we reverse the time direction, this will not negate our conclusions about time symmetry. In fact, these small complications will give us an even deeper understanding of the principles involved, which are that natural, global symmetries lead to the great conservation laws of physics.

With all this taken into account, we can state that the microscopic world is quite time symmetric. The equations that describe phenomena at that level operate equally well in either time direction. But, more importantly, the experiments themselves seem to be telling us not to make an artificial distinction between past and future. Those quantum phenomena that strike most people as weird are precisely the ones where the future seems to have some effect on the past. Weirdness results only when we insist on maintaining the familiar arrow of time and defining as weird anything that is not familiar.

Experiments demonstrate unequivocally that quantum phenomena are *contextual*. That is, the results one obtains from a measurement depend on the precise experimental setup. When that setup is changed, the results of the experiment generally change. This may not sound surprising, but what people do find surprising is that the results change even in cases where common sense would deny that any change was possible without a superluminal signal.

In Einstein's theory of relativity, and modern relativistic quantum field theory, no physical body or signal can travel faster than the speed of light. In experiments over the past three decades, two parts of a quantum system well-separated in space have been found to remain correlated with one another even after any signal between them would have to travel faster than the speed of light. While some correlation is expected classically, after this correlation is subtracted an additional connection remains that many authors have labeled mysterious—even mystical. While these observations are exactly as predicted by quantum theory, they seem to imply an inseparability of quantum states over spatial distances that cannot be connected by any known physical means.

As has been known for years, time reversibility can be used to help explain these experiments. We will see that their puzzling results can be understood, without mystical or holistic processes, by the simple expedient of viewing the experiment in the reverse time direction. Filming the experiment and viewing it by running a film backwards through the projector, we can see that no superluminal signalling takes place.

Let us consider another example that leaves people scratching their heads: A photon (particle of light) that left a galaxy hundreds of millions of years ago may be bent one way or another around an intervening black hole. Suppose the photon arrives on earth today, and triggers one of two small photon detectors, separated in space, that tell us which path the photon took.

A special arrangement of mirrors can be installed in the apparatus so that the light beams from both paths around the black hole are brought together and made to constructively interfere in the direction of one detector and destructively interfere in the direction of the other. The first detector then always registers a hit and the other registers none. This is as expected from the wave theory of light.

Now, the puzzle is this: the decision whether or not to include the mirrors is made today. Somehow it reaches back to the time of the dinosaurs to tell the photon whether to pass one side of the black hole, like a good particle should, or pass both sides and interfere, like a good wave should.

While this particular astronomical experiment has not, to my knowledge, actually been conducted, a large class of (much) smaller-scale laboratory experiments imply this result. These experiments, I must continually emphasize, give results that agree precisely with the predictions of quantum mechanics, a theory that has remained basically unchanged for almost seventy years. So any discussions and disputes I may report are strictly over the philosophical or metaphysical interpretation of the observations, not any inconsistency with calculations of the theory.

In the traditional methods of classical mechanics, a system is initially prepared in some state and equations of motion are then used to predict the final state that will then be observed in some detection apparatus. However, quantum mechanics does not proceed in this manner. In the most commonly applied procedure, the initial state of the system is defined by a quantity called the **wave function**. This wave function evolves with time in a manner specified by the **time-dependent Schrödinger equation** to give the state at some later time. The wave function, however, does not allow one to predict the exact outcome of a measurement but only the statistical distribution of an ensemble of similar measurements.

Although not exhibiting any preference for one time direction or another, this series of operations still seems to imply a time-directed, causal process from initial to final state quite analogous to Newtonian physics. By removing our classical blinkers, however, we can see that quantum mechanics basically tells us how to calculate the probability for a physical system to go from one state to another. These states may be labeled "initial" and "final" to agree with common usage, but such a designation is arbitrary as far as the calculation is concerned. Nothing in the theory distinguishes between initial and final.

Furthermore, whereas classical physics would predict a single path between the two states, quantum physics allows for many different paths, like the two paths of the photon around the black hole. The interference between these paths leads to many of the special quantum effects that are observed. It is as if all paths actually occur, and what we observe is some combination of them all.

When we try to think in terms of particles following definite paths, however, we run into conceptual difficulties. In our cosmic experiment, for example, the photon somehow has to pass on both sides of the black hole to interfere in our apparatus a hundred million years later. We can arrange our detector to count a single photon at a time, so we can't think of it as two different photons. The same photon must be in two places at once.

The conventional wisdom has held that physicists should not speak of anything they cannot directly measure or test against measurements. So, according to this rule, we are not allowed to regard the photon as following a particular path unless we actually measure it. When we try to do that, however, the interference effect goes away. The party line for many years has been to leave this as it is. The equations give the right answer—what is observed. However, this policy has never provided a satisfactory response to the question of what is "really" happening.

I suggest that at least part of the solution has been there all along in the time symmetry of quantum theory and the apparent backward causality evident in quantum experiments. As long as they remain in a pure quantum state, photons can reach just as far back in time as they can forward, as can electrons and other subatomic particles. These bodies can appear in two or more places at once, because a backward moving particle can turn around and go forward again, passing a different place at the same time it was somewhere else. A particle going one way in time can be accompanied by its antiparticle going backward in time in a single (coherent) state of one particle. Together they constitute the timeless quantum.

Every author must gauge his or her audience and write with that audience in mind. This book, like *The Unconscious Quantum*, is written for a science-literate audience. That is, the reader is not expected to be a scientist or other scholar highly trained in science or the philosophy of science. I write for the much larger group of generally educated people who enjoy reading about science at the popular and semipopular level, in books and magazines. They also follow the science media, and are interested in the grand scientific issues of the day, especially as they interact with other areas of thought in philosophy, religion, and culture. Scientists and philosophers who may themselves not be experts in these specific issues are also kept in mind as potential readers. Hopefully, the experts will not object too much to what they read, for I have undoubtedly oversimplified in places and provided insufficient caveats.

As we will see, many of the matters being heatedly debated today are ancient, even eternal. They will not be settled by me, any more than they have been settled by the thousands who have discoursed on the nature of reality from the time words were first used as a medium for that discourse. I am not trying to finalize these matters but make some of their latest manifestations more accessible to the reader and to perhaps open up a few neglected lines of thinking for the professional.

I will regard my goal as satisfied if I succeed in slightly deflecting thinking in directions that have not been, in my view, adequately explored. Most philosophical, theological, cultural, and historical discourses implicitly assume directed time. Most models of physical reality implicitly assume the existence of material continua. Most attempts to account for the order of nature have suggested a Platonic reality which theists call God and nontheists call the "theory of everything." I suggest that, based on current knowledge, all of these approaches are at best weakly founded. As I will attempt to show, no basis exists for assuming that the detailed structure of the universe is the product of either logical necessity or supernatural design.

In adopting the medium of a semipopular but still scholarly book to suggest new ideas, I am catching a ride on a recent trend. For most of the twentieth century, scientific and philosophical discussions were largely confined to the professional journals and a few, highly priced technical monographs of limited circulation and even more limited comprehensibility. In more recent years, biologists Stephen Jay Gould, Richard Dawkins, Francis Crick, and E. O. Wilson have used the medium of the popular book to promote original ideas in that field that differ from the mainstream. In physics and cosmology, Stephen Hawking, Roger Penrose, David Deutsch, Lee Smolin, and many others have done the same. Other fields, such as neuroscience, complexity theory, and artificial intelligence, have seen similar use of this medium to promote new and controversial ideas. The excellent sales by this array of authors testifies to the market for new and challenging thoughts about fundamental issues of life, mind, and the universe. In some cases, the proposed ideas have begun to trickle down into the heavily conservative, formal disciplines in which their authors have usually made substantial contributions.

Of course, the less formal presentations and freer flow of ideas of popular literature force the reader to search for the pony in a huge mountain of horse manure. In the case of the Internet, this mountain is of Everest proportions. But the lesson of the Internet so far seems to be that the shoveling is well worth the effort, once that beautiful pony is found.

I hope that this book will not require a large shovel. In fact, it is the result of many years of spade work on my own part. In the four decades before its publication I have taught physics at every level and participated in research that helped elucidate the properties of almost every type of elementary particle, from strange mesons and charmed quarks to gluons and neutrinos. I have looked for gamma rays and neutrinos from the cosmos whose energies exceed anything yet produced on earth. This was with many collaborators to whom I owe a great debt of gratitude. While not a trained theorist or philosopher, I have also published a few theoretical and philosophical papers.

Over these four decades I have constantly tried to understand and explain the basis of the physical world in simple terms. As much as possible, I have supplemented or replaced the equations and abstract symbols of physics with words and visual concepts. We humans seem not to "understand" an idea until it is expressed in terms of words and pictures, although equations can, more compactly and precisely, say the same thing. In this book I present these words and ideas. In a few places, I use a symbol or a simple equation, but these are nothing more than shorthand. For completeness, the endnotes contain a few short derivations so that the mathematical reader can understand somewhat more precisely what I am saying.

A few caveats are also included in the notes, in the interest of accuracy, but these are minimal. Technical terms are boldfaced in the text the first time they appear, or when they haven't appeared for a while, and are defined in the glossary. I have tried to keep the discussion in the main text as complete as possible. In some places, technical aspects of a subject are described in some detail that may make for rough going for those not already familiar with the ideas. If the thread is lost in going through these sections, it should be possible to pick it up again at some point a paragraph or two later. The reader is encouraged to simply plunge ahead in those cases where he or she finds the going rough. The basic ideas are summarized and repeated many times.

I have been helped enormously in this work by the availability of the Internet and its unprecedented communication power. I formed an electronic mail discussion list (avoid-l@hawaii.edu), whose membership at times exceeded fifty, and placed the drafts and figures for this manuscript on a World Wide Web page (http://www.phys.hawaii.edu/vjs/www/void.html). Members of the list could then read the latest drafts and post comments for myself and others to read, all in a completely open fashion. No one was excluded or censored, and the discussion often ranged far and wide.

While not everyone on the list joined in the discussions, I must mention those who have directly helped me prepare the work you see before you by providing comments, suggestions, and corrections: Gary Allan, Perry Bruce, Richard Carrier, Jonathan Colvin, Scott Dalton, Keith Douglas, Peter Fimmel,Taner Edis, Eric Hardison, Carlton Hindman, Todd Heywood, James Higgo, Jim Humphries, Bill Jefferys, Norm Levitt, Chris Maloney, John H. Mazetier Jr., David Meieran, Ricardo Aler Mur, Arnold Neumaier, Huw Price, Steven Price, Jorma Raety, Wayne Spencer, Zeno Toffano, Ed Weinmann, Jim Wyman, and David Zachmann. That is not to say that any or all of these individuals subscribe to the views expressed in this book. Indeed, several hold strong opposing views and helped me considerably as friendly but firm devil's advocates. Many thanks to all, and to the many others who also helped by their interest and encouragement. As always, I am grateful to Paul Kurtz and Steven Mitchell of Prometheus Books for their continued support for my work. Thanks also go to those at Prometheus who helped produce the book, especially Grace Zilsberger in the art department and the copy editor, Art Merchant. And, as always, I have been advised, supported, and sustained by the wonderful companion of my life, my wife Phylliss, and our two adult offspring, Noelle and Andy.

1

ATOMS
AND
FORMS

By convention sweet, by convention bitter, by convention hot, by convention cold, by convention colour, but in reality atoms and void.

Democritus (Kirk 1995)

THE PRESOCRATICS

In sixth-century B.C.E. Greece, a diverse group of philosophers, collectively known as the "presocratics," began to explore unchartered territory. In place of the anthropocentric myths that had dominated human thinking until then, the presocratics visualized an impersonal reality external to the world of thoughts and dreams. This reality, called *physis*, or *nature*, constituted the world of sensory experience. Furthermore, all the objects of that experience were composed of the same basic stuff, inanimate, material bodies from which all else was assembled by forces acting within that world alone. The supernatural did not exist. Even the gods and human souls were "natural."

Thales of Miletus (d. 550 B.C.E.) proposed that all of this was simply water. He probably did not come to this conclusion by way of internal, philosophical meditations. Rather, he likely got the idea of a single, basic stuff from observation. With his own eyes, he saw that liquid water freezes into ice and evaporates into vapor—that one substance appears in three widely different forms. Aristotle says of Thales' principle of water:

> Presumably he derived this assumption from seeing that the nutriment of everything is moist, and that heat itself is generated from moisture and depends upon it for its existence (and that from which a thing is generated is always its first principle). He derived his assumption, then, from this; and also from the fact that the seeds of everything have a moist nature, whereas water is the first principle of the nature of moist things. (Aristotle, Metaphysics 984a)

Unlike the popular myth about eskimos, Thales did not invent a separate word for every slightly different appearance of snow. Snow was water. Ice was water. Steam was water. He saw the unity of each. Only one word, one concept, was necessary to specify it, not a multitude.

We should not interpret Thales too literally here. The revolutionary idea was not so much that everything is water, but that the universe is simple, natural, composed of familiar materials, and within the capacity of human understanding.

By similar arguments, Anaximenes (c. 545 B.C.E.) concluded that the fundamental stuff was air. Heraclitus (c. 484 B.C.E.) said it was fire. Empedocles (d. 424 B.C.E.) combined all three ideas, suggesting that material objects were a mixture of fire, air, earth, and water in varying proportions. Each of these "elements" are characterized by certain observable properties: fire is bright, light, and dry; air transparent, light, and wet; earth dark, heavy, and dry; water transparent, heavy, and wet. In Empedocles' picture, the elements combine in different proportion to produce the materials of experience. And so, gold is a composite of fire and earth: bright, heavy, and dry. In the Middle Ages, alchemists would try to manufacture precious metals by adding fire to baser substances. They failed, of course, because gold was itself a chemical element and its manufacture by transmuting other elements had to await the development of nuclear physics in the 1940s. The alchemists' fires were simply not hot enough.

Included in the natural, material universe of the Milesians as the source of life was the *soul*. No distinction was made between soul and matter. Soul was as material as the wind.

To the followers of Thales, matter and motion were "natural." Matter behaved the way it did because nature just happened to be that way. The heavy elements earth and water naturally moved downward, while the light elements air and fire naturally moved upward. The heavenly bodies, composed of a fifth, celestial element not found on earth, moved naturally in circles about the earth. Nothing was needed to "cause" a body to behave naturally. Only when we think something is behaving unnaturally are we prompted to seek causal explanations. What must be explained is any deviation from naturality. One of the themes of this book is that the common perception of unnaturality is wrong. What is often thought to be unnatural, and thus requiring explanation, is in fact perfectly natural and only if it had occurred in a different way would we be compelled to seek causes. This will include much of the observed world, including the most universal of the so-called laws of physics. They will be seen to be natural and uncaused.

While these ancient ideas about nature are naive when viewed in terms of what we know today, they nevertheless represented a remarkable feat of intellect. Imagine the immense chasm that must have existed at the time between mythological thinking and this new idea that the world is so simple as to be composed of a few basic ingredients that behave the way they do because it is in their nature to do so.

The idea that the universe could be understood in terms of a small number of easily visualized ideas reached its most profound ancient elaboration in the theory of atoms. In a remarkable intuitive leap, Leucippus (c. 440 B.C.E.) and Democritus (c. 370 B.C.E.) proposed that matter was composed of atoms—uncuttable particles (*atomos* = not to cut) that moved around in an otherwise empty void.

The atomic theory of matter would take almost 2,500 years to be fully justified, theoretically and empirically. Despite all the caveats we must make about the imperfections of human conceptions and our inability to ever be certain of any notion of fundamental "truth," a universe composed of nothing more than irreducible elementary objects and the void continues to represent the most economical, useful, and "natural" model for ultimate reality.

BEING, MIND, AND NUMBER

Not all presocratic philosophers were atomists, or even materialists. Nevertheless, they can be characterized collectively by their common quest for a natural simplicity to the universe. Parmenides (c. 450 B.C.E.) argued that only a unitary, unchanging *being* can constitute ultimate reality. The observed world was rife with change and so, Parmenides concluded, matter cannot be the prime ingredient of eternal nature. An unchanging ingredient must exist that is more fundamental than the experiences of finite, temporal humans. I do not know if Parmenides considered the possibility that material atoms, uncuttable particles, might very well constitute exactly the unchanging ingredient he demanded. In what is undoubtedly an incomplete historical record of his views, he seems to have not thought of this obvious solution.

Anaxagoras (c. 428 B.C.E.) took another tack. He contended that mind was superior to matter and so must constitute the ordering force of the universe. He asserted that mental concepts were not simply tools used by humans to describe the observed order of an objective reality, but reality itself. Mind was the source of all motion. Like the many who have professed the same doctrine throughout the ages to the present, Anaxagoras provided no serious model or mechanism for mind or how it does these wondrous things. Aristotle complained that Anaxagoras introduced mind as a cause only when he knew of no other.

A century after Thales, Pythagoras (c. 530 B.C.E.) and his disciples discovered the power of mathematics and geometry in describing the world, including music, and decided that *number* was the basic stuff of reality. In this they previewed an idea that was later developed by Plato and will play an important role in this book—the notion that abstract ideals are more real than the observed objects of experience. As Aristotle puts it:

> At the same time [i.e., as the Milesians and atomists], however, and even earlier the so-called Pythagoreans applied themselves to mathematics, and were the first to develop this science; and through studying it they came to believe that its principles are the principles of everything. And since numbers are by nature first among these principles, and they fancied that they could detect in numbers, to a greater extent than in fire and earth and water, many analogues of what is and comes into being—such and such a property of number being justice . . . and since they saw further that the properties and ratios of the musical scales are based on numbers, and since it seemed clear that all other things have their whole nature modelled upon numbers, and that numbers are the ultimate things in the whole physical universe, they assumed the elements of numbers to be the elements of everything, and the whole universe to be a proportion or number. (Metaphysics 985b)

Actually, the Pythagorean claim was not as abstract then as it may appear to us today. They thought of numbers as geometrical shapes, which were observed in nature, and did not make a distinction between arithmetic and geometry.

However, their own greatest discovery, the Pythagorean theorem, led to a conundrum in interpreting geometry in terms of numbers. The hypotenuse of a right triangle is not in general a natural number, that is, a number defined in terms of a counting procedure that is easily related to common experience. Say you define a unit of length as the width of your thumb and lay out a right triangle on the ground with sides that you measure as ten thumbs each. You then proceed to measure the hypotenuse and obtain a result that is not an integer number of thumbs but rather something between fourteen and fifteen thumbs. The Pythagorean theorem allows you to calculate the hypotenuse as

10 √2, but this cannot be expressed as so many thumbs. Even if you further refine your scale by introducing fractions of thumbs—half a thumb, a quarter thumb, $\frac{1}{64}$ of a thumb, and so on—your measured hypotenuse will never be exactly 10 √2, at least "in theory." In short, you cannot solve the problem by using such *rational* numbers because √2 is an *irrational* number, inexpressible as either an integer or ratio of integers.

Until the more abstract mathematics of irrational numbers like √2 was invented centuries later, geometry was the primary mathematical tool for describing observations. The irrational numbers were, in a sense, already built in. Problems could be solved geometrically that were unsolvable arithmetically. Influential thinkers such as Plato (c. 347 B.C.E.) made a distinction between geometry and arithmetic. When Euclid (c. 300 B.C.E.) laid out geometry as a logical, deductive process he avoided the use of arithmetic proofs. Even Isaac Newton (d. 1727) used Euclidian methods in his published *Principia*, desiring not to hand competitors the more powerful methods of the infinitesimal calculus he had invented and had actually used to provide the solutions to the problems he was reporting.

The logical beauty of geometry, and its ability to accommodate unmeasurable quantities such as the *exact* hypotenuse of a right triangle or the *exact* ratio of the circumference of a circle to its diameter, would take on great mystical proportions in the Platonic cosmology. Eventually, when the abstract differential equations of calculus were applied with dramatic results in physics, even usually pragmatic physicists would be led to surmise that these equations and the symbols they contain are "more real" than the observations they describe.

SOPHISTRY, SOCRATES, AND SOUL

In the fifth century B.C.E., a group of wandering street philosophers called *sophists* shifted learned discourse away from inanimate nature and back toward more popular human concerns, particularly politics, law, and worldly happiness. The most notable sophist was Protagoras (d. 420 B.C.E.), who is famous for his statement: "Man is the measure of all things." This appears to affirm the popular delusion that the universe revolves about us humans. But a fuller quotation suggests Protagoras's possible intent: "Man is the measure of all things, of those that are that they are, and of those that are not that they are not."

This seems to mean that all knowledge is relative, varying from person to person. According to Protagoras, we all decide for ourselves what exists and what does not. Protagoras may have been making an epistemological statement rather than an ontological one. That is, knowledge is molded by human thought processes. We scientists make measurements using tools we invent. We describe these measurements in words, symbols, and mathematics that we invent.

Protagoras was a religious skeptic: "Concerning the gods I cannot say whether that they exist or that they do not, or what they are like in form; for there are many hindrances to knowledge." But Protagoras was also skeptical of the notion that humans can uncover irrefutable truths about reality.

The view that neither science nor religion is capable of producing indisputable knowledge, because all knowledge is operated on by human thinking processes, resonates down to the present day. Every new generation of intellectuals seems to unearth this ancient homily and claim it as its own. The relativists and postmodernists of our current period are the latest to make this discovery. Like the legions who have appeared on the historical stage since Protagoras, all saying more or less the same thing, they are greatly impressed by their own profundity.

Socrates (d. 399 B.C.E.) had no patience for trivial platitudes. The "Socratic method" of questioning the assertions of others is still widely used as a rhetorical tool. Socrates found the sophists too worldly, too amoral. Although he agreed that the practical matters of law and politics were important, more abstract concepts such as virtue and morality must be probed to attain the goal of human happiness. He discarded the relativism of the sophists and the materialism of the natural philosophers. "Know thyself," he said. That was all that mattered. Look inward rather than outward.

Looking inward, Socrates defined the human soul as the essence of an individual's existence. What's more, Socrates asserted, this essence continues beyond bodily death. Of course, by the time of Socrates, soul was already an ancient concept, in one form or another, and most cultures had notions of survival after death. However, soul was typically associated with the vital substance or force believed to give life to living things, something that distinguished the living from the nonliving and departed the body upon death. We have already seen how the Milesian notion of soul, which was called *psyche*, envisaged a fully material substance—specifically, air or breath. Ancient Hebrew thinkers had a similar idea, which they called *ruach*. The more subtle concept of an incorporeal soul, separate from matter, occurred in India. In ancient Hindu philosophy, spirit-soul was the fundamental stuff of reality and matter-body an illusion.

In both its material and spiritual renditions, in ancient times and today, the human soul is regarded as part of a greater cosmic soul. It follows that some aspect of soul lives on as part of the eternal, unchanging whole. The Greeks called this cosmic soul *pneuma*, the breath of the gods. The Latin translation *spiritus* leaves no doubt to the direction in which this concept evolved. In related fashion, the Greek root *psyche* seen today in *psychic* and *psychology* is clearly connected with the notion promoted by Descartes in the seventeenth century, that the soul is the seat of the mind.

Athens. The Lyceum had far more resources than the Academy of Plato that Aristotle had attended a half-century earlier. Plato's school was mainly concerned with ideas anyway, and its master felt no need to make measurements or collect observational data. Besides funding the Lyceum, Alexander ordered an expedition up the Nile that was most likely prompted by Aristotle (they discovered the Abyssinian source), and saw that the Macedonian armies brought back biological specimens for the Lyceum from their distant conquests. And so, *observational,* if not *experimental,* science had its beginnings. The concept of controlled experimentation did not occur to Aristotle—nor to anyone else for centuries. But observational science as a complement to pure theorizing was a good start.

Aristotle did far more than gather data. He codified the rules of deductive logic in a fashion that we still use today (although other logics are now recognized), and then applied his logic to the understanding of the world of observations. He proposed theories of physics, meteorology, biology, and psychology. He was way off base on most of his physics, violating his own rules of logic and epistemology, according to Galileo. However, the bulk of the remainder of Aristotelian science, especially his biology, was of lasting value. Aristotle also wrote on history, politics, and ethics. His later editors coined the term "metaphysics" (*metaphysikos,* meaning "after the physics") to identify the writings that followed Aristotle's *Physics* and included speculations that went beyond physics.

Rejecting the relativism of Protagoras, Aristotle argued for the existence of invariable, universal truths. As I have mentioned, he also rejected Plato's notion that universals existed in a separate realm of transcendent Forms. To Aristotle, the universals were inseparable parts of substance, coexisting with matter. Substance was a combination of matter and form.

Still, the invariable and permanent nature of universals called for an explanation of change and impermanence in the observed world. Although he did accept the Milesian view that "some beginnings are originally inherent in things, while others are not" (Metaphysics 1012b), Aristotle proposed that four "causes," or, more accurately, "beginning principles," act to create, move, or change the nature of an object:

(1) *Material cause:* the elements that compose an object;
(2) *Efficient cause:* the means by which the object is changed;
(3) *Formal cause:* the concepts used to describe the process;
(4) *Final cause:* the purpose toward which an object moves or progresses.

Since millions of words have already been written attempting to explain what Aristotle meant by his four causes, let me do something different and try to put them in the context of modern physics.

Within that context, the changes that take place at the fundamental level include the motion of physical bodies, their interaction with one another, and their transformation from one type to another. We can quickly identify *material cause* with the elementary objects themselves—the quarks, leptons, and bosons of the current standard model that will be discussed in detail later in this book. The basic forces—gravity, electromagnetism, the weak and strong nuclear forces—manifest *efficient cause*. These forces act to create matter out of energy, accelerate bodies toward or away from one another, and transform one form of matter to another.

Formal cause is readily associated with the mathematical symbols and equations embedded in the theories of modern physics.

But what about *final cause*? Let us consider for a moment the classical, Newtonian physics that predated twentieth century physics. As with modern physics, Newtonian physics is characterized in terms of particles, forces, and mathematical laws. So, it likewise accommodates Aristotle's first three causes. However, final cause does not exist as an independent attribute in Newtonian physics. The properties of an object, the forces on it, and the mathematics that describe the two are sufficient to determine how the object behaves. In the deterministic universe implied by Newtonian mechanics, the outcome is already built in at the start. God might have brought it all into existence, but she has no need to further fix things up as time goes by.

Apparently this was not what Aristotle had in mind when he recorded his four causes. The metaphor of the mechanical universe as a vast clock-work had to await the age of clocks and other machines. Scientific metaphors are not pulled out of thin air, but chosen from the array of concepts that characterize the thinking of the times. Virtually all ancient thinking, including materialism, utilized the metaphor of the living organism, with a strong distinction made between the animate and inanimate, living and dead. In the Enlightenment and the following Industrial Revolution, inanimate machines became the dominant metaphor. Today we commonly hear the computer used as the metaphor for the operation of the universe and we are beginning to debate whether computers may someday evolve into living beings.

When the mechanical universe first appeared in the seventeenth century, its primary creators were not quite ready to abandon final cause, nor to accept the full implications of the machine metaphor. René Descartes (d. 1650), who laid the conceptual foundation for mechanics, went to great pains to find a continuing role for God. No doubt prompted by his political needs (to keep his head), he divided the substance of the world into matter and mind independent of matter.

Newton (whose head was in less danger in protestant England), after formulating the exact mathematical principles of mechanics, pursued in his studies of alchemy the divine principles he believed were necessary for the animation and variety of life. Newtonian mechanics, in its inventor's view, was not enough. Indeed it wasn't. It would take modern quantum mechanics to provide a coherent scheme for a purely material model of life.

Later interpreters of Newtonian mechanics, especially those in revolutionary France, removed God from the mechanical universe. According to a possibly apocryphal story, Pierre Simon Laplace (d. 1827) horrified Napoleon by telling him that he had "no need for the hypothesis" of God. In a similar vein, Julien de La Mettrie (d. 1751) published *Man a Machine* which left no room for either spirit or God (De la Mettrie 1778).

But many still respected the wisdom of the ancients. Surely Aristotle, the inventor of deductive logic, would not have hypothesized a fourth cause unless he believed it could not be derived from the other three. That is to say, we can reasonably assume Aristotle intended that final cause is *independently* needed to explain the world. The world, according to Aristotle, has purpose as well as design; the agent of this purpose actively and continually participates in the processes of change.

As Bertrand Russell has noted, when we ask "why?" concerning an event, we are looking for one or the other of two answers: either a *mechanical* one, that is, "what earlier circumstances caused this event?" or a *teleological* one, that is, "what purpose did this event serve?" (Russell 1995, 67). Aristotle believed both answers must be provided for a complete description—or at least that one answer did not automatically imply the other.

In the twentieth century, Newtonian mechanics was revised by quantum mechanics. As conventionally interpreted, quantum mechanics implies that we do not live in a clockwork universe, with everything that happens set down at the creation. When doing quantum mechanics, a physicist calculates the probability that a quantum system will move from some given initial state to some desired final state, but makes no certain prediction for where the system will specifically end up. As I will elaborate on later in great detail, the designations "initial" and "final" are simply conventional and no this-cause-always-produces-this-effect relationship is involved.

This may simply be, as Einstein thought, an incompleteness in an otherwise useful and correct theory. However, no measurement or observation with the finest instruments of our technology has so far forced us to conclude that, deep-down, the universe is "really" deterministic or causal. And so, while design may still be read into the Newtonian universe without grievous assault on common sense, purpose takes on mystical and indeed supernatural proportions in the quantum universe. In a deistic perspective of Newtonian mechanics, purpose is built into the very fabric of the universe. God has his place as the weaver of this fabric, rather than a force that regularly intervenes to keep the world on course. In the quantum universe, God is not necessary and purpose is not evident, although, as we will see later, the new *Intelligent Design* movement in theology claims the opposite.

Aristotle's final cause thus marks a clear dividing line between scientific materialism and those systems of thought that assert a purposeful universe. Proponents of the latter usually accept physics as far as it goes, but insist that a fourth cause beyond physics, divine plan, is needed to explain the world. The modern materialist needs only physics, neither predetermined nor postdetermined.

Belief that some teleological principle acts in the universe is not limited to those who hold traditional religious beliefs. Some imaginative theorists have recently claimed that the laws of nature must have been fine-tuned to produce life and that this argues for a universe with the evolution of intelligence as the goal (Barrow 1986). This claim will be discussed in some detail in chapter 13.

THE STOICS

Although Aristotle rejected Platonic spiritualism and viewed the soul as substance, his insistence on a prime-mover God transcending the material world certainly removed any possible stigma of his being labelled an atheist by subsequent historians. St. Thomas Aquinas (d. 1274) hardly would have constructed Christian theology on the teachings of a wicked materialist!

Stoicism arose shortly after Aristotle, in the third century B.C.E. Mainly an ethical philosophy of civic duty and resignation to one's fate, stoicism was essentially materialist in its metaphysical outlook, viewing the soul as coextensive with the body. The stoic universe operated rationally according to the principle of *logos*, with all information entering the human mind through the senses. Following Heraclitus, the stoics viewed fire as the basic substance. God and soul were substantial, composed of the primal fire, which was also associated with reason and harmony, beauty and design.

Stoicism had lasting impact, remaining influential until the Middle Ages. It attained its pinnacle with the Roman emperor Marcus Aurelius (d. 80 C.E.), whose *Meditations* record the private thoughts of history's nearest personification of the philosopher king. However, the stoics, like the sophists, were more concerned with human affairs than science and did little to advance either physics or metaphysics.

The atomic picture resurfaced with Epicurus (d. 270 B.C.E.) and his disciples, most notably the Roman poet Lucretius (c. 55 B.C.E.). In his remarkable poem *De Rerum Natura* (The Nature of Things), Lucretius maintained that the laws of nature, rather than divinity, create the universe. He also argued that humans, not gods, were responsible for civilization and morality. The gods were human inventions.

In the sixteenth century, a copy of *De Rerum Natura* was discovered in Germany and brought to Italy by Poggio Bracciolini. Despite attempts at suppression by the church, atoms soon become part of the new cosmology that Nicolaus Copernicus (d. 1543), Galileo Galilei (d. 1642), and Isaac Newton introduced to the world.

NEOPLATONISM AND NOMINALISM

Even today, the Platonist, whether in theology, philosophy, or science, carries on the tradition of belief in a deeper, truer reality beyond the material world exposed to our external senses. As we have seen, this idea probably originated in ancient India, drifted to Greece, and then became integrated into Christian thinking as St. Augustine (d. 430) and other early theologians incorporated it into Church teaching. While the later version is still referred to as *neo-Platonism*, it is no longer "new."

Although Aquinas preferred Aristotle to Plato, Plato's Form of the One, the Good, the idea from which all other ideas derive in a "Great Chain of Being," evolved into the God of Christianity who bears little resemblance to either Zeus or Yahweh—except on the ceiling of the Sistine Chapel in the Vatican. Aquinas, whose theology soon became Catholic dogma, gave God the Aristotelian qualities of Prime Mover and Creator that went beyond Plato, but was more consistent with Genesis. Thus, the two greatest Christian theologians succeeded in weaving traditional beliefs together with the teachings of the two greatest pagan philosophers into the blanket of faith that so many find comforting to this very day.

The medieval theologians who followed Aquinas found room for both Plato and Aristotle. The one who stands out as perhaps the most original thinker of the age is William of Occam (d. 1349). Occam (or Ockham) founded a school of thought called *nominalism* that distinguished science (that is, natural philosophy) from metaphysics in an important way. Science, according to Occam, deals not with what is but with what is known. He asserted that the so-called universals are mental images of sense impressions and have no objective reality. This is essentially the distinction I will be making between ontology and physics. Scientific theories, like those of physics, deal with abstractions that themselves need not correspond to reality but serve as problem-solving tools embedded in theories or what more recently have been termed "paradigms."

Historian Richard Carrier informs me that Occam was simply rediscovering what Claudius Ptolemy (c. 70) had earlier synthesized from the Epicureans and Stoics. In *Kriterion and Hegemonikon* (Huby 1989) Ptolemy says: "The intellect . . . could not begin to think without the transmission from sense perception (unless anyone wanted to indulge in pure fantasy)" (8.5). And, in a blast against Platonism he says it only "sometimes seem[s]" that the intellect acquires "the first concept" of things "by its own agency," so good is our memory at retaining and presenting recorded perceptions, for the intellect apprehends the Forms of things only "through memory that is an extension from sense objects" (12.3).

Nominalism was not the sole intellectual contribution of the Franciscan monk Occam, who, while deeply embroiled in the turmoil of Church politics of his day, somehow still found time to think about more important things. Occam is of course famous for *Occam's razor*, or the principle of parsimony that is usually rendered as "Entities are not to be multiplied beyond necessity," though there is no record of Occam using these exact words. Applied to a scientific theory, Occam's razor shaves away unneeded hypotheses, namely, those assumptions which are neither logically necessary nor called for in the data the theory seeks to describe. Thus, the most economical theory, namely, the one with the fewest assumptions, rules over any alternative that is less parsimonious. Certainly, a less parsimonious theory may eventually turn out to be correct. But we have no right to assume its truth until evidence is found that cannot be explained by a more parsimonious alternative.

MECHANICS AND MONADS

The influence Occam may have had on the development of the new science in the sixteenth century, if any, is not clear. The traditional metaphysical teachings of the Church, now supplemented and somewhat modified by Protestantism, continued to dominate European thinking.

Copernicus and Galileo showed that a more economical explanation of astronomical observations could be envisaged when one assumes that the planets revolve around the sun. The earth, from the Copernican perspective, is just another planet—no more special than Venus, Mars, or Jupiter.[1] This realization struck a great blow to human pride. Most cultural traditions have held that the earth, that is to say, humankind, is the center of the universe. As centuries passed, and astronomers probed farther beyond the solar system to galaxies billions of light years away, the traditional view has become increasingly difficult to sustain.

This was, however, only temporarily unsettling. As human knowledge expanded and revealed a universe immense and ancient far beyond previous estimates and totally unimagined in any holy scripture, these results were soon reinterpreted so as to further enhance the glory of God. God ruled over vast domains, not just the tiny earth. Furthermore, Newtonian mechanics seemed to support the argument from design. Indeed, Newton believed he had succeeded better than anyone before in reading the "mind of God," a far more wondrous world revealed not in the Bible but in the book of nature.[2]

Descartes, who laid the conceptual foundation for the mechanical universe that Newton codified with his laws of motion and gravity, saw matter as the basest component in a universe of matter, mind, and God. Only after Descartes had "proved" to his—or at least his Catholic superiors'—satisfaction that mind and God exist, did matter enter into his metaphysics.

First Descartes famously "proved" the existence of his own mind by the argument *cogito ergo sum*. He then expanded upon the contorted logic of St. Anselm (d. 1109), using a variation on the so-called *ontological argument* to "prove" the existence of God. This argument basically said that since the mind can conceive of a being greater than anything else that can be conceived, that being must exist. Descartes added that the concept of God is innate. Once God's existence was thus "proven," Descartes argued that external bodies are real because God would surely not deceive us by making them an illusion.

According to Descartes, knowledge is not exclusively prompted by sensory observations of material bodies but can arise internally from within our minds as well. God, in his goodness, has given us the mental capacity to reason and infer the truth of propositions by thought alone. This was the Age of Reason, and Descartes reasoned, or more accurately rationalized, his way to mind, God, and matter.

Gottfried Wilhelm von Leibniz (d. 1716) was as brilliant a rationalizer as ever lived. He invented calculus, independent of Newton (we still use Leibniz's notation), and symbolic logic. He devised a calculating machine that did square roots. Leibniz was a major figure of the Enlightenment, whose genius rivaled Newton's. But Leibniz was also the precursor of a fantastical German idealism that would one day replace scientifically motivated Enlightenment thinking as materialist philosophers reasoned themselves into a corner.

Leibniz's law of optimism stated that we live in the best of all possible worlds:

> God has chosen (to create) that world which is the most perfect, that is to say, which is at the same time the simplest in its hypotheses and the richest in phenomena. (*Discourse on Metaphysics* as quoted in Rescher 1967, 19)

Although this idea was satirized by Voltaire in *Candide*, where Leibniz is modelled as "Dr. Pangloss," we can see from the quotation that Leibniz had a notion of simplicity consistent with richness that, when stripped of its theological baggage, represents a significant proposal for a guiding metaphysical principle.

Leibniz started from the premise that an infinity of worlds was possible and that God had a wide, but still limited, range of choices in creating the universe. Even God could not break the rules of logic, but otherwise, in his goodness, selected the most perfect world consistent with logic. This may seem inconsistent with God's supposed omnipotence, but you would not expect an omnipotent being to be inconsistent, and logic is little more than consistency.

Leibniz explained that suffering in the world is a necessary evil, required in the logically consistent best of all worlds in order to make possible greater goods such as free will. Leibniz's term *theodicy* has since been used to identify the task of apologizing for God.

Leibniz also proposed a specific metaphysics in which the basic substance of the universe is composed of structureless points he later called *monads*.

> There are only atoms of substance, that is to say, real unities, that are absolutely devoid of parts, which are the sources of action and the absolute first principles of the composition of all things and, as it were, the ultimate elements in the analysis of substantial things. One could call them metaphysical points. They have something vital, a kind of perception; and mathematical points are their points of view, from which they express the universe. (Rescher 1967, 12)

Leibniz's monads were of substance, but living, vital substance—little pointlike souls that contained "representations" of what is outside. God is the supreme monad. Leibniz disputes Descartes notion of physical substance as pure extension. Suggestive of Einsteinian relativity, Leibniz points out that size, figure, and so on are relative to our perceptions. The basic metaphysical stuff must be independent of human perception (Rescher 1967, 13).

While Leibniz's monads might sound superficially like today's atoms or elementary particles, clearly he was not proposing a materialist metaphysics. Indeed the monads were spiritual entities that did not interact with one another but were kept in perfect harmony by God. If anything, Leibniz's monads illustrate how far metaphysical speculations can take you from the reality of physical observations, once you allow your imagination to run free of any constraints provided by those observations. Nevertheless, we can find some interesting insights in Leibniz's ideas that hint at modern notions, in particular, the relativity of time and the monad as a kind of clock.

NON-PLATONIC MATERIALISM

Despite the attempts by early Enlightenment figures to maintain the Platonic Christian metaphysics, and even enhance it by application of the new tools of reason, a non-Platonic materialism subsisted as the scientific revolution progressed. Thomas Hobbes (d. 1679) argued that the concept of "incorporeal substance" such as spirit or soul was a logical contradiction and that all substance must be material by definition. John Locke (d. 1704) founded what we call *empiricism*, the doctrine that all of our knowledge, with the possible exceptions of logic and mathematics, is derived from experience. As usual, this was not a completely new idea but Locke moved it to the front burner. He disagreed with Plato and Descartes, arguing that we have no innate ideas, calling the mind a *tabula rasa* upon which experience is inscribed (Russell 1945, 609).

Bishop George Berkeley (d. 1753) agreed with Locke on the primacy of experiential knowledge, but disputed that it provided a reliable picture of the real world outside our minds. Since sensory knowledge is still in our heads, only thoughts and mind can possibly be real. However, Berkeley still was a clergyman and could not allow his argument to proceed to its logical conclusion: *solipsism*, the doctrine that only one's own mind was real. He decided instead that only God's mind is real.

Regardless of the merit of this doctrine, Berkeley was the first to draw a clear line between material monism, in which all is matter, and ideal monism, in which all is mind. Even Plato had not gone that far, and Descartes had concluded that substance existed in the dual phases of matter and mind.

David Hume (d. 1776) agreed with Locke that all our ideas are sensory in origin. But he also agreed with Berkeley that we have no way of knowing that sense data are true matters of fact. Thus, he was willing to tread where Berkeley would not, asserting a doctrine of complete skepticism concerning the ability of the human being to determine truth by any means, empirical or mental. Shattering tradition, Hume rejected intuition and induction as sources of knowledge.

As part of his skeptical doctrine, Hume issued the first real challenge to cause and effect and, indeed, to the very notion of natural law. He asserted that all we actually observe is the spatio-temporal conjunction between events. Just because two events occur close to one another in space and time, we cannot conclude that one was the cause of the other, that is, in the absence of one event the other would not have happened. Similarly, we cannot assume that events will *always* happen in a certain sequence and pattern just because they have always happened that way so far.

Applying this to science, we can never be certain that the "laws of physics" are true, even if we were to attain the fantastical goal of a theory of everything that agrees with all the data that may have been gathered by that time. The notion that the mathematical equations of such a theory would encompass a Platonic reality of which we would then have certain knowledge is a pipe dream from the Humean perspective.

However, Hume's rejection of induction went too far. If we cannot predict events with complete certainty from the principles we establish from the conjunction of observations, we can certainly do better than a simple toss of the dice. Science is highly successful in predicting the future occurrence of many events with sufficient probability for most practical purposes. And while we can never be certain that any given ontological interpretation of a scientific theory is true, we can still explore the relative merits, logical foundations, and possible consequences of various proposals. Indeed, we have the intellectual freedom to talk about anything we want, as long as we do so in a coherent and consistent fashion.

INNATE KNOWLEDGE

Immanuel Kant (d. 1804) reported that when he read Hume he awoke from his "dogmatic slumber." Kant proceeded to analyze what could and could not be known about the universe, God, and the human mind. He decided that the mind does in fact possess innate knowledge of "things in themselves," what he called the *synthetic a priori*, that are not derived from experience. In yet another variation on Platonic forms, the mind is not a tabula rasa at birth but has a priori knowledge of the forms of phenomena. The purest of these, the *transcendental aesthetic*, are space and time. The notions of space and time are not simply abstractions from experience but inner knowledge about what lies beyond the world of experience.

Kant gave Euclidean geometry as an example of a priori knowledge. Such knowledge is applied to the real world rather than inferred from observations, at least according to Kant, yet it successfully describes those observations. However, this example collapsed, along with the rest of Kant's proposition, when Carl Friedrich Gauss (d. 1855) and others showed that non-Euclidean geometries were mathematically possible. In 1916, Einstein used non-Euclidean geometry in the theory of general relativity, striking the final blow against the transcendental aesthetic.

Euclidean geometry turns out to be an approximation that we use in everyday life because it happens to work well in describing most of the observations in our neighborhood. We infer it by observation, not pure thought. By the early twentieth century, scientific instrumentation had greatly extended the range of our observations and revealed that both Euclidean geometry and classical Newtonian physics were approximations to the new physics of relativity and quantum mechanics.

How can we ever be sure that relativity, quantum mechanics, or any future theories we might invent are not similar approximations? Based on the evidence from scientific history, you would be foolish indeed to think you ever had finally achieved the ultimate theory of everything. Nevertheless, this does not mean we should throw up our hands and stop doing science. Approximations can still be applied to the appropriate range of phenomena where they give sufficient accuracy to be useful. In the view of science as a problem-solving activity, no inconsistency arises and the progress of science provides strong support for the contention that it converges on reality.

GEIST AND FEELINGS

It was awhile before Kant's version of Platonic forms was definitively repudiated. In the meantime, he inspired the weird period in Western philosophical history called *German idealism* that culminated in the baffling abstractions of Georg Wilhelm Hegel (d. 1831).

Hegel saw the universe as a collective spirit, mind, or soul he called *geist*. Humans are an intrinsic part of this Absolute, which is continually evolving and gradually gaining self-knowledge through the *dialectic* method of question and answer utilized in Plato's *Dialogues*. Kant had also seen the dialectic as the process by which we gain knowledge of ultimate reality independent of sense experience. Hegel's idealism thus appears as yet another attempt to find reality by the inner process of pure thought rather than the outer process of objective observation.

With no little irony, Karl Marx (d. 1883) claimed to use Hegelian thinking in his *dialectic materialism*, which comes to the directly opposite conclusion from Hegel. To Marx, ultimate reality is fully material and independent of the human mind. Nothing better demonstrates the worthlessness of a theory than having it produce mutually contradictory conclusions. Think of all the human suffering that might have been avoided in the twentieth century had the poverty of the dialectic been earlier recognized!

Another notable German idealist, Arthur Schopenhauer (d. 1860), pictured the world as "will and idea." He is reported to have defended his legendary obscurity, which matched even Hegel's, by noting that, after reading Kant, "the public was compelled to see that what is obscure is not always without significance" (Durant, 1953, 221). Physicist Ludwig Boltzmann was not so impressed by obscurity, however. Reputedly, he once made his views on Schopenhauer unambiguous by proposing to give a lecture to philosophers with the title "Proof that Schopenhauer Was a Degenerate, Unthinking, Unknowing, Nonsense Scribbling Philosopher, Whose Understanding Consisted Solely of Empty Verbal Trash" (Bernstein 1993, 33).

Nevertheless, Schopenhauer made some sense, arguing that the human mind is incapable of knowing about things in themselves and simply engages in a struggle to find meaning where none exists. In this he became the first major Western philosopher to directly incorporate aspects of Buddhist thinking into his philosophy.

Even before the German idealists took center stage in humanity's ongoing intellectual drama, the Romantic movement led by Jean Jacques Rousseau (d. 1778) had reacted against the scientistic universe of the

Enlightenment. The romantics emphasized "feeling" over reason. This point of view has dominated thinking in the arts to the present day. However, it should not be taken to imply any great take-no-prisoners war between opposing ideologies comparable to that between realism and idealism. Few scientists deny the value of painting, poetry, music, and the other fine arts, and indeed many have highly developed senses of appreciation for the aesthetic. No conflict between feeling and thinking exists, at least in the feeling and thinking minds of scientists. Most just happen to think and feel that feeling and thinking are being done by gray matter, not some imaginary spirit.

MATERIALISM HANGS ON

Materialism often gets blamed for the decline of "spiritual values" in the modern world. Hegel and the bewildering twentieth-century philosopher Martin Heidegger amply illustrate the direction to which the application of "pure thinking," unfettered by external reality, can lead.

Materialistic metaphysics was kept alive in the eighteenth and nineteenth centuries by authors such a Paul d'Holbach (d. 1789) and Ludwig Buechner (d. 1899), who are largely forgotten today. D'Holbach, in *The System of Nature; or the Laws of the Moral and Physical World*, attempted to develop a completely materialistic picture of the world based on Enlightenment science, with no room for spirit or religion (D'Holbach 1853).

Buechner's *Force and Matter* was a complete statement of materialist philosophy that was one of the most widely read German books of the nineteenth century, with twenty-one German editions and translations in seventeen other languages (Buechner 1870, 1891). Needless to say it was also widely opposed, with diatribes against it by theologians and priests rolling off the presses. But *Force and Matter* continued to be printed and had some small influence on people such as Einstein. By his death, however, Buechner had lost much of the hope he had with his first edition in 1855 of converting most of humankind to materialism.[3]

Perhaps one of the reasons materialism did not catch on was a general disenchantment among those who (shades of Socrates and the sophists) saw the improvement of the human social and economic condition as the highest goal and viewed the speculations of both idealists and atomists rather remote from everyday life. Nevertheless, the industrial revolution, fueled by materialist science, had enormous if uneven impact on human material welfare. This great success story suggested that scientific method should be applied as well to human society.

Auguste Comte (d. 1857) coined the term *sociology* to refer to the scientific study of social systems. He also founded the philosophical school of *positivism* in which metaphysics was viewed as an obsolete stage of human intellectual development. In the first stage, according to Comte, humanity looked to gods and other mythological beings to explain events. In the second stage, these were replaced by abstractions such as Plato's perfect Forms or Hegel's *geist*. Finally, human thinking has now advanced to the scientific stage of experiment, hypothesis, and mathematical theory that should be applied to every phase of human life.

While Comte was primarily a political theorist, positivism found a place in the philosophy of science largely through the influence of the Austrian physicist and philosopher Ernst Mach. Mach was highly skeptical of the atomic theory and many of the ideas that physicists were tossing about around the turn of the century. He was chair of philosophy at the University of Vienna at the same time that Boltzmann was chair of physics, and the two disagreed vehemently. Boltzmann had used the atomic model to derive the laws of thermodynamics. During one of Boltzmann's lectures at the academy of science, Mach spoke up and said, "I don't believe atoms exist." He often would say about atoms, "Have you seen one?" Of course no one had (yet), and Mach was justified in being skeptical. But he remained so, even after Einstein had convinced most other skeptics by showing that Brownian motion, in which suspended particles are observed to bounce about randomly, could be explained as the effect of atomic collisions. No doubt, if he were alive today he would say about quarks, "Have you seen one?" (Bernstein 1993, 28–37).

Shortly after Mach's death in 1916, the University of Vienna become the center for the philosophical school of *logical positivism*, which was partially based on the notion that only statements that have testable observable consequences can be meaningful.

Positivists of both centuries found an ally in Herbert Spencer, who agreed that metaphysics was a waste of time and philosophy should concern itself with unifying the results of science. He held the view that evolution was evidence for progress and applied this idea to psychology, sociology, and morality. Questions like the existence of God or the origin of the universe, according to Spencer, were unknowable.

While the power of science as applied to the machines of industry was universally acknowledged in the nineteenth century, and Charles Darwin (d. 1882) had shown how life, including *homo sapiens*, had evolved by the mechanism of natural selection, few humans were ready to accept de La Mettrie's notion that "man" is a machine. The modern antimechanical crusade is exemplified by the writing of Henri Bergson, who argued that human creativity cannot be explained by any Darwinian mechanism.

However, this remains wishful thinking. No evidence exists for any component to the physical universe other than matter. No observation of the physical world requires any processes other than natural ones. A picture of uncuttable atoms in the void remains the simplest and best ontology for explaining the data of modern science. If a mystical, Platonic world beyond the senses cannot be ruled out, such a world is not required by the data.

NOTES

1. Actually, to fit the data, Copernicus had the earth revolving not about the center of the sun but a point that itself revolved about the sun.

2. The catchy phase, "the mind of God," was used in Stephen Hawking's best seller *The Brief History of Time* (Hawking 1988, 175). However, Hawking was speaking metaphorically. Many physicists use "God" to refer to the mathematical order of the universe, which unfortunately can be misinterpreted in sectarian terms that were not intended.

3. For a complete discussion of d'Holbach, Buechner, and the history of materialism in general, see Vitzthum 1995.

2

THE WHOLE
IS EQUAL TO
THE SUM OF
ITS PARTICLES

The extension, hardness, impenetrability, mobility and inertia of the whole, results from the extension, hardness, impenetrability, mobility and inertia of the parts.

Isaac Newton, *Principia*

THE LAST PARADIGM SHIFT

L ike most of his contemporaries, Isaac Newton did not regard the mechanical universe as a complete description of reality. After explaining the motions of apples and planets in terms of the laws of mechanics and gravity, he spent many years in fruitless alchemic studies that he hoped would lead him to the source of the "vegetative forces" that he, like most of his contemporaries, assumed must animate living things. And God was not left out of Newton's picture, for where else could the discovered laws of mechanics and the still-undiscovered laws of life and soul have come about but by the act of the Creator?

Newton never found evidence for his vegetative forces. Nor has anyone else in the centuries since, although the notion of a vital force—"bioenergetic field" is the current designation—remains a common belief (Stenger 1999c). This belief is especially prevalent today in various forms of alternative medicine that claim connections to Eastern and other "spiritual" healing methods. Because of the lack of any evidence for such a force, these ideas never have become mainstream Western science. In the meantime, Newtonian mechanics and the developments it engendered grew steadily in power and utility to dominate the mainstream. As the twentieth century approached, mechanistic physics (which we now call *classical physics*) seemed sufficient to explain the behavior of all matter, living or dead.[1] The clockwork universe had succeeded far beyond Newton's expectations or, it seems, desires.

Then, in rapid succession during the early years of the twentieth century, two monkey wrenches were tossed into the mechanism: the twin "revolutions" of relativity and quantum mechanics. Newtonian mechanics was found to give slightly incorrect results when applied to tiny but, by then, observable effects in planetary motion. Classical electrodynamics was unable to account for the observed spectra of light emitted by heated bodies and high voltage gas discharges, and for a few other phenomena such as the photoelectric effect.

Relativity and quantum mechanics demanded new principles, equations, and methods. In particular, quantum mechanics was characterized by uncertainty and indeterminacy, in contrast to classical physics which implied a universe where everything that happens follows predictably from preceding events.

Nevertheless, in the present day, classical physics still plays a dominant role in our high-tech society. The language of classical physics is spoken in virtually every marketplace of modern life, including the shops that sell our most sophisticated technology. We build aircraft, automobiles, and bridges with classical physics. We explore the planets with classical physics. Although some quantum physics is needed to design the microchips and lasers that are used as components in thousands of computers and other high-tech devices, the bulk of these systems are largely fabricated on the basis of principles of classical mechanics and electrodynamics.

While quantum mechanics is needed to understand the structure of matter, the physics that comes into play once atoms and molecules have assembled into biological and other familiar macroscopic systems is almost exclusively classical. Even the human brain appears to be an electrochemical device that operates largely, if not exclusively, according to classical rules. Several authors have recently asserted a strong role for coherent quantum processes in the brain (Penrose 1989, 1995; Squires, 1990; Stapp 1993). The issue remains a contentious one. I have argued elsewhere that no convincing basis exists for such claims (Stenger 1995, 268–93). Indeed, as I have suggested, natural selection may have acted to ensure that life evolved at or near the edge of the classical domain where predictability, perhaps supplemented by a touch of quantum randomness, can be found.

You often hear that science is always making new discoveries that prove previous theories to be wrong. This seems to be one of those myths that most people accept without looking carefully at the historical facts. While tentative proposals can prove incorrect, I cannot think of a single case in recent times where a major physical theory that for many years has successfully described all the data within a wide domain was later found to be incorrect in the limited circumstances of that domain. Old, standby theories are generally modified, extended, often simplified, and always clarified. Rarely, if ever, are they shown to be "wrong."

No doubt, dramatic changes have occurred in the progress of physics. What Thomas Kuhn, in his famous book *The Structures of Scientific Revolutions*, labeled as "paradigm shifts" account for much of the progress in science (Kuhn 1970). For example, consider the transition from classical to modern physics, just mentioned, that occurred early in the twentieth century. Methods and concepts used in modern physics, especially quantum mechanics, are quite different from those in classical physics. This is exemplified in quantum mechanics by the calculation of probabilities for particles following certain paths, rather than calculating the exact paths themselves. But, these paradigm shifts did not negate the applicability of the old paradigms, as Kuhn seemed to imply. Classical mechanics and electrodynamics, in particular, still find widespread utility.

This point has been made by physicist Steven Weinberg in a recent retrospective of Kuhn's work:

> It is not true that scientists are unable to "switch back and forth between ways of seeing," and that after a scientific revolution they become incapable of understanding the science that went before it. One of the paradigm shifts to which Kuhn gives much attention in *Structure* is the replacement at the beginning of this century of Newtonian mechanics by the relativistic mechanics of Einstein. But in fact in educating new physicists the first thing that we teach them is still good old Newtonian mechanics, and they never forget how to think in Newtonian terms, even after they learn about Einstein's theory of relativity. Kuhn himself as an instructor at Harvard must have taught Newtonian mechanics to undergraduates.

Weinberg points out that the last "mega-paradigm shift" in physics occurred with the transition from Aristotle to Newton, which actually took several hundred years: "Nothing that has happened in our understanding of motion since the transition from Newtonian to Einsteinian mechanics, or from classical to quantum physics fits Kuhn's description of a paradigm shift" (Weinberg 1998). We will take another look at Kuhn and some of the unfortunate consequences of his influential work in chapter 14.

ATOMS REDISCOVERED

Newton reportedly said, "If I have seen farther than others, it is because I have stood on the shoulders of giants." He owed a great debt to Copernicus and, of course, Galileo, who died the year Newton was born. Other giants supporting Newton included the French philosophers René Descartes and Pierre Gassendi (d. 1655), and the English philosophers Walter Charleton (d. 1707), Robert Boyle (d. 1691), and Thomas Hobbes, among others.

Descartes had created a mechanical world system, the first since Aristotle. While it was largely qualitative, it set the stage for the more quantitative developments that followed. In the Cartesian system, matter could be indefinitely ground into finer and finer pieces. The universe was a *plenum*, or what we commonly call today a *continuum*. The void, as Aristotle had tried to prove logically, does not exist in this plenum. This system contrasted with that of Gassendi, which was based on the principle of atoms and the void carried down from Democritus and Leucippus by way of Epicurus and Lucretius. In the atomic picture, as you divide matter into ever finer pieces you eventually reach a place where further division is impossible. Here is the crux of the issue: is the universe continuous or discrete? The answer to that question remains the primary disagreement between the competing views of ultimate reality among physicists.

Charleton had published an English version of Gassendi's atomic system, making it widely available in Britain. However, the mechanical philosophies being developed at the time in England and the continent were not strictly based on the atomic model. Boyle used ideas of both Descartes and Gassendi in his studies of gases and other material systems, asserting that his mechanical hypotheses required no metaphysical assumptions. He argued for economy of thought and reduced everything to matter and motion. If immaterial substances existed, Boyle argued, then they were unintelligible to us and, furthermore, had to act through the medium of matter (Hesse 1961, 115).

Hobbes also argued for a completely materialistic world system. However, he was severely criticized as being atheistic by Henry More (d. 1687). Even Descartes was too materialistic for More and, as I have already noted, Newton himself did not promote a fully materialistic world view (Dobbs 1995, 6).

Descartes clung to some of the elements of Aristotle's physics which his contemporary, Galileo, the first modern empiricist, rejected because of inconsistencies and disagreement with experiment. Newton followed Galileo and discarded most of the physics of Aristotle, while respecting the ancient master's thinking on biology, logic, and other matters. Reading the mechanistic thinkers as a student, Newton became a "corpuscularian," holding that matter was composed of small material particles or "corpuscles." His student notebooks indicate that he early adopted the notion that the corpuscles were ultimately indivisible, that is, composed of atoms (Dobbs 1995, 12).

Today, classical mechanics is routinely taught to students from the corpuscular perspective, although nothing in this approach requires ultimate indivisibility. That does not come until quantum mechanics. As far as classical mechanics is concerned, particles can be as small as you want to make them, and their properties can be measured with precisions limited only by the quality of the measuring apparatus.

Of course, few of the bodies of everyday experience look at all like point-like particles. In the rigid body approximation, the constituent particles of a solid object remain in an approximately fixed position relative to one another as the body, a system of particles, moves through space. From the basic *particle* laws of motion, we can prove that a composite body, whether rigid or not, moves as if all its mass were concentrated at a certain point called the center-of-mass. This justifies the treatment of obviously multi-particle objects, like automobiles, planets, and galaxies, as if they were single particles. Indeed, they look like particles when viewed from far enough away and obey particle mechanics on that scale. As you look closer, however, you see internal motion, such as rotation or vibration. These can be treated with particulate mechanics as well, with all the particles in the body moving with respect to the center-of-mass of the whole.

However, not all bodies are rigid (actually, none are). In fluids, such as water and air, the constituent particles move relative to one another. A volume of water or air is composed of particulate molecules, but the number is usually so huge that calculating the paths of each is impossible—and uninteresting anyway. Rather, the fluid is treated as the continuous medium it appears to be to the naked eye. Quantities such as pressure and density are introduced to describe this apparent continuum. These are called **fields**, which are defined as quantities that have values at all points within the continuum, not just at the position of a particle.

Field quantities represent averages over the many particles in a macroscopic system. For example, the density field of a fluid is the average mass per unit volume at each spatial point in the fluid. The pressure applied by a gas on the wall of its container is the average force per unit area produced by all the molecules bouncing off the wall. Since the number of particles is usually of the order of 10^{24} or more, statistical fluctuations are small (one part in a trillion in the typical case); therefore, we do not normally notice deviations in instantaneous density or pressure fields away from their averages.

Infinitesimal fluid elements are introduced that occupy particular positions at particular times, just like particles. These fluid elements follow paths exactly as predicted by the corpuscular laws of motion, with "external" forces applied by neighboring fluid elements or the environment.

Calculus is then used to sum up the contributions of all these component elements to provide equations that describe the motion of the whole system. As they get deeper into increasingly complex applications of continuum mechanics, both the students and their instructors alike tend to lose sight of the fact that their field equations were derived from particle mechanics. The new fields tend to take on a life of their own.

In fact, classical fluid mechanics is still particle mechanics. No independent laws of motion exist for physical continua. The fluid laws are all derivable from particle mechanics. The same is true for thermodynamics, the science of heat phenomena. Many of the principles of thermodynamics were developed in the nineteenth century from direct observations of the behavior of what appears to the naked eye as continua, such as the gases in a heat engine. Yet by the end of the century, these principles were fully understood in particulate terms.

At the most advanced level of classical field theory, it can be shown that the equations of motion for a field can be rewritten in terms of coordinates and momenta that obey the normal particle equations of motion. As we will see, this notion is exploited in quantum field theory where a particle is associated with every field. The ontological question then becomes, which is more real—the particle or the field? And, as we will see, an unacceptable answer is "some of both."

The classical paradigm provides us with the means for predicting the motion of all material systems in the classical domain. Whether continuous or discrete, these systems of bodies are treated as composed of constituents that obey Newton's laws of particle motion and the various principles derived from them. Given the initial position and velocity of the constituent, and knowing the net force on it, you can predict its future position and velocity. Doing this for all the constituents in the system, and adding them up, you obtain the motion of the system as a whole. The methods of calculus enable you to sum the infinite number of terms that are assumed when an ultimately discrete many particle system is approximated as a continuum. While nothing requires that the particles of matter be ultimately indivisible, let there be no doubt about it: in classical physics, *the whole is equal to the sum of its particles.*

STICKING TOGETHER

With the laws of motion alone, you can predict the future behavior of some systems, such as rigid bodies in collision with one another. In most cases, however, you need to know something about the forces on each particle or element. The means needed to calculate the forces external to a particle, which act as the agents of changes in its motion, are not included in the laws of motion. Additional *laws of force* are required to supplement the laws of motion.

Repulsive forces are easy to visualize in an atomic impact picture. If our generic atoms are visualized as impenetrable spheres, then we can imagine them bouncing off one another like billiard balls. Even where the bodies are ultimately penetrable, this model can be used to account for phenomena such as gas pressure, as described above, and the transfer of heat. The temperature of a body is proportional to the average kinetic energy of its constituent particles. It rises when the body is placed in contact with another of higher temperature, as energy is transferred by collisions at the interface between the two bodies. Friction and viscosity result when the molecules in one system collide with those in a neighboring one, such as when a fluid passes through a pipe. All this is well accounted for in the atomic model, with Newton's laws applied to the motion of particulate constituents.

From its beginning, however, the model of atoms colliding with one another in an otherwise empty void has had trouble accounting for attraction and cohesion. Atomic collisions provide no obvious picture for how atoms stick together to form composite objects. Lucretius had imagined them as having little hooks; Gassendi followed him in suggesting a similar kind of Velcro system. Still, even this unlikely model could not explain the instantaneous "action at a distance" that seemed to be observed with gravity, electricity, and magnetism, where the interacting bodies did not come into contact. How can forces be transmitted in a void? A void is by definition absent of substance. The traditional answer, going back to Aristotle, has been that no void can occur, that something must exist in the space between visible bodies.

If the void does not exist, then instantaneous action at a distance poses no problem. Instead a continuum of matter pervades the universe. Bodies experience forces as they move through this medium, invisible to the naked eye. Even so, explaining attraction still is not very easy. Greek mechanists had suggested a *vortex* picture that Descartes included in his mechanical system. They noted that when solid matter is caught in a rotating eddy, the denser matter tends to go to the center and the light matter moves outward, according to Archimedes' principle. Descartes's system contained vortices of invisible matter rotating around the earth and other bodies. This idea was developed into a theory of gravity by Christiaan Huygens (d. 1695); but this was rejected by Newton because it disagreed with the data (Hesse 1961, 52–53).

Many other explanations, too vague to be useful, were made for the nature of attraction and cohesion during the Middle Ages and Enlightenment. For example, Francis Bacon (d. 1626) suggested that certain immaterial spirits or "pneumaticals" shaped the structure of matter, an idea similar to one of the Stoics (Dobbs 1995, 13). The term "spirit" often appears in the context of these discussions. But this should not be interpreted as necessarily referring to supernatural forces. The natural could still be immaterial, as distinguished from normal, visible matter. Newton used the term "subtle spirit" in his speculations, but generally as a natural medium. As we will see, the idea of an underlying **aether** as a continuous medium pervading all the universe became prominent in the later nineteenth century, only to be discarded in the twentieth.

Newton became acquainted with mechanistic and atomistic ideas as a student. His notebooks from the time show sketches of how a rain of fine aetheric particles (note his use of particles here, even to describe the aether) might be harnessed to produce a perpetual motion machine somewhat like a water wheel. Much later, in a 1675 letter written during the period when he had largely developed, but not yet published, his principles of mechanics, he speculates that a subtle, elastic aether might contain electric and magnetic effluvia and the "gravitating principle" (Hesse 1961, 151). However, in a 1679 letter to Boyle, Newton emphasizes that his explanation of planetary motion rested on "mathematical demonstration grounded upon experiments without knowing the cause of gravity" (Jaki 1966, 62).

"HYPOTHESES NON FINGO"

After the publication of the first edition of *Principia* in 1687, Newton would speculate further about the causes of gravity, electricity, magnetism and their possible connection between gravity and the "vegetative" forces responsible for animating life that he sought in his alchemic studies. However, in *Principia* he steered clear of metaphysics, famously claiming in the second edition of 1713 to "frame no hypotheses." He explains that he has

> not been able to discover the cause of those properties of gravity from phenomena. I frame no hypotheses [hypotheses non fingo]; for whatever is not deduced from the phenomena is to be called an hypothesis; and hypothesis, whether metaphysical or physical, whether of occult qualities or mechanical, have no place in experimental philosophy. In this philosophy particular propositions are inferred from the phenomena, and afterwards rendered general by induction.

In the *General Scholium* added to the 1713 edition, Newton still broached the subject of the fundamental nature of force. In particular, he rejects Descartes's vortex theory outright, since it cannot explain the highly eccentric paths of comets. Here, Newton is being quite consistent with his position of framing no hypotheses. Descartes's theory was wrong because it disagreed with observations. Thus, we see in the Newton of that period clear premonitions of positivism, empiricism, and, indeed, falsificationism.

Strictly speaking, by today's conventions Newton's laws of motion and gravity were still hypotheses. But we cannot hold that against him. They were inferred from observations, and testable by experiment, and this was what Newton most likely was trying to say—that he made no *empty* hypotheses. In any case, this has been the message of science ever since: you can speculate all you want about causes, but science deals with observations and the description of those observations in ways that can be usefully applied to the prediction and description of further observations. Causal language is often used in these descriptions, and, as we will see, just as often misused.

Newton's law of gravity stated, with great simplicity, that the gravitational force of attraction between two bodies is proportional to the product of their masses, and inverse to the square of the distance between their centers. Note that in no place does this law make any reference to the means by which bodies produce this force or how it reaches out over vast distances. Newton does not say *why* the gravitational force is proportional to mass, or *why* it falls off as the inverse square of distance. He simply provides us with an equation, inferred from observations, that enables us to calculate the gravitational force in a wide range of applications—with sufficient accuracy to fly to the moon

Laypeople are accustomed to hearing that the force of gravity "causes" a pin to drop to the ground and that this force is "produced" by the enormous mass of the earth. But Newton's **third law of motion** ("for every action there is an equal and opposite reaction") tells us that the tiny pin applies a force of equal magnitude on the earth! Did the earth cause the pin to drop, or did the pin cause the earth to rise up to it? Of course, the acceleration of the earth is infinitesimal, but you might still ask how a pin could produce a force on the earth. Newton and the physicists who followed after recognized the difficulty of answering such metaphysical questions and contented themselves with the accurate description of past measurements and the prediction of future ones. Despite this, even today causal language is commonly used when describing physical phenomena in the vernacular. And it still leads people to paradoxical, nonsensical, and often bizarre conclusions.

Avoiding such distractions by sticking to the observations and their mathematical description, Newton used the methods of calculus he had invented to prove that the gravitational force between two extended bodies was the same as that for two point particles of the same mass located at their respective centers of gravity. This explained why the force law applies to you standing on the earth; the force on you is the same as if all the earth's mass were concentrated at a point at the center of the earth. And the force you apply to the earth is the same as if all your mass is concentrated at your center of gravity.

As I mentioned earlier, Newton used only geometrical proofs in *Prin-*

cipia, in order to avoid uncovering the unique mathematical tools of calculus that gave him a leg up on his competitors. He made them public when he became concerned about credit after learning that Leibniz had independently invented calculus.

The successful application, by Galileo and Newton, of a methodology deeply based on experiment and observation established the empiricist position that has since been the hallmark of science. After *Principia*, "natural philosophers," now called physicists, chemists, or biologists, did science. They made measurements and constructed theories that sought agreement with these measurements, while metaphysics was relegated to the metaphysical philosophers whose writings scientists largely ignored.

THE PRINCIPLE OF LEAST ACTION

Newton had attempted to derive the properties of light, such as its rectilinear motion and the laws of reflection and refraction from mechanical principles, assuming light was corpuscular in nature. Looking back from our modern perspective we can see that he essentially derived these properties from conservation of momentum and energy—in spite of the fact that the concept of energy had not been formulated yet. In this he followed Descartes, who had invented the notion of momentum (although he did not use the term) as the quantity of motion that Newton exploited so effectively in *Principia*.

The reflection of light can be seen to follow easily from conservation of momentum, by way of the elastic (kinetic energy-conserving) scattering of light from a mirror. For example, a billiard ball will bounce off the cushion of a pool table at an angle of reflection equal to the angle of incidence, conserving momentum parallel to the cushion while the table applies an equal and opposite impulse in the perpendicular direction.

In the case of refraction, as light travels from one medium to another, such as from air to water or glass, it bends closer to the perpendicular in the denser medium. Descartes and Newton also described this in terms of conservation of momentum. Momentum is the product of mass and speed, and the observed relationship, known as **Snell's law**, implied that the light moves faster in the denser medium.

The speed of light in various transparent materials was not measured until 1846, by Armand Fizeau (d. 1896), who found that light moves slower, not faster, in the denser substances. For example, the speed of light in glass is two-thirds of the speed in air. Conservation of momentum, then, would seem to imply that light should bend *away* from the axis in dense media, opposite to what is observed. Descartes and Newton appear to have been wrong.

However, as things eventually turned out, both the electromagnetic wave theory of light and the modern photon theory of light give the observed law of refraction. What's more, that law can be explained in terms of conservation of momentum—exactly as Descartes and Newton had intended. They had no way of knowing at the time that the momentum of an electromagnetic wave, or a photon, is its energy divided by its speed. Consequently, light of a given energy has a greater momentum in the denser medium where its speed is lower, and a light ray must move closer to the perpendicular as it passes through the interface in order to conserve momentum. This process is to a good approximation elastic, meaning that negligible energy is lost in the medium.

In 1658, well before Newton's publications on light, Pierre Fermat (d. 1665) had proposed that the path a light ray takes as it traverses various media is that path which minimizes the total time it takes to make the trip. Assuming, correctly as it turned out, that light moves slower in a denser medium, Fermat was able to derive Snell's law. However, **Fermat's principle** was not immediately accepted. The Cartesians objected because of the disagreement on the speed of light in dense media, and for other, metaphysical reasons, which took precedence in those days. Leibniz, who also opposed the principle, wrote in 1682 that light takes the easiest path, which is not to be confused with the shortest time. Instead he proposed a principle in which the path is the one of least "resistance" (Dugas 1955, 260).

In 1744, Pierre de Maupertuis (d. 1759) proposed a generalization of Fermat's principle to mechanical systems called the **principle of least action**. De Maupertuis argued that "Nature, in the production of her effects, always acts in the most simple way." The path taken by bodies was, according to de Maupertuis, the one for which a quantity called the **action** was the least. This was similar to what Leibniz had said, but de Maupertuis made it quantitative. He defined the action this way: "When a body is carried from one point to another a certain action is necessary. This action depends on the velocity that the body has and the distance that it travels, but it is neither the velocity nor the distance taken separately. The quantity of action is the greater as the velocity is the greater and the path which it travels is longer. It is proportional to the sum of the distances, each one multiplied by the velocity with which the body travels along it."

De Maupertuis applied his principle of least action to refraction and got the same result as Descartes and Newton. That is, he got the wrong result that implies a greater speed for light in a denser medium. But, as with Descartes and Newton, de Maupertuis did not know he was wrong, since the speed of light had not yet been measured in any medium. Likewise, de Maupertuis was not conceptually wrong in using the principle of last action. He simply used the wrong expression for the action of light, one that applied only for bodies moving at speeds low compared to the speed of light.

Despite these initial confusions, Fermat and de Maupertuis had in fact glimpsed what is probably the most important principle in classical mechanics. Indeed, this principle can be generalized to describe a large class of problems called *isoperimetrical* problems. In this class of problems, you seek a curve of fixed length that maximizes or minimizes some given quantity.

Consider the legend of Dido, the founder of Carthage and lover of Aeneas in Virgil's epic poem. She is said to have negotiated a transaction with a North African chieftain in which she would receive the amount of land that could be enclosed with a certain ox hide. She proceeded to cut the hide into thin strips and tied them together to form the perimeter of a large territory. This is perhaps the earliest reference to anyone solving an isoperimetrical problem (Gossick 1967, 1).

Objects, even people and animals, try to get from place to place with the least effort. The problem, then, is to calculate that optimum path. Indeed, we might suppose that this represents the basic problem that all living organisms must deal with every day: how to get from some initial condition A to some final condition B, while maximizing or minimizing various implications of the journey that takes them from A to B. We might want to simply expend the least energy in walking from one building to another, or, if it is raining, find the most sheltered route. Or, we might be conducting a more complex task, such as manufacturing an automobile with minimum cost in minimum time and maximum reliability.

The shortest distance between two points in Euclidean space is a straight line. This is the solution to the isoperimetrical problem that seeks to minimize the distance. On the non-Euclidean surface of a sphere, the shortest distance is a great circle. In other circumstances, you might be more interested in minimizing some other parameter, such as time, rather than distance.

For example, imagine a Bay Watch life guard on a beach hearing a call for help from a swimmer in the water off some point up the beach. The lifeguard needs to get to the swimmer in minimum time, and since she can run faster than she can swim, the direct, straight-line route is not the optimum path she should follow. Instead, she heads up the beach at an angle directed away from the swimmer, and then turns back toward the swimmer when she enters the water. The optimum solution to this problem is exactly that given by Snell's law for the refraction of light, derived from Fermat's principle of minimum time.

The principle of least action was eventually generalized by the great Irish mathematician William Rowan Hamilton (d. 1865) and is now called **Hamilton's principle** in text books. Hamilton himself called it the principle of constant action. I will stick to the "least action" designation for reasons of familiarity and also to avoid confusion with an alternative formulation of classical mechanics that is also associated with the name of Hamilton.

When the same guy has invented the two best theories, you can't refer to them both with his name. That's a peril of being so much smarter than everyone else!

In the years after *Principia*, a brilliant new breed of scientists developed the forms of calculus and classical mechanics that we still use today. Hamilton was the last of this remarkable group, the one who put most of the finishing touches on the subject. In 1788, Joseph-Louis, Comte de Lagrange (d. 1813) had developed generalized methods for doing mechanics that enabled the equations of motion for complex mechanical systems with many degrees of freedom, and angular as well as linear coordinates, to be written down. Lagrange's equations enable you to do this in terms of a quantity called the **Lagrangian** that, in most cases is simply the kinetic energy *minus* the potential energy of the system.

In 1834, Hamilton developed an equivalent set of generalized equations from which the equations of motion could be determined. These equations were expressed in terms of a quantity now called the **Hamiltonian** that in most cases is simply the kinetic energy *plus* the potential energy of the system, that is, the total energy. Both Lagrangian and Hamiltonian formulations can be derived from the principle of least action, or Hamilton's principle, where the action is simply the average Lagrangian over the path from point A to point B times the time interval for the trip.

We will see later that the Hamiltonian formulation became the starting point for the conventional quantization procedure that takes you from classical to quantum mechanics. As we will find, this quantization approach led to difficulties that were most efficiently solved by a return to the principle of least action.

CORPUSCULAR MECHANICS APPLIED

In the decades and centuries following Newton, the mechanics of fluids was also developed as a generalization of Newtonian mechanics. Fluids, such as air and water, appear continuous to the naked eye and yet can be described in terms of concepts that are directly derivable from Newtonian corpuscular mechanics. Historically, Johann Bernoulli (d. 1748) was the first to write the equations of motion for the elements of a continuous fluid by treating them as point particles and using particle mechanics to describe the motion of these elements. Leonhard Euler (d. 1783), a student of Bernoulli, who made many important contributions to both mathematics and physics, further developed Bernoulli's ideas and laid the foundation of fluid mechanics and field theory as they are still implemented.

More progress on fluids was made by Daniel Bernoulli (d. 1782), who showed how discontinuous pulses of pressure and density in a fluid could be handled, anticipating Fourier analysis. He also developed the first kinetic theory of gases based on the atomic model, although this was not accepted at the time because of its disagreement with the *caloric* theory then in vogue which held that heat is a material substance that flows between bodies, a theory that proved to be completely incorrect.

Newton had adopted Descartes's definition of the quantity of motion, what we now call momentum. However, the related concept of energy, now recognized as equally profound and indeed very closely related to momentum, was not introduced into physics until much later. In 1797, Benjamin Thompson, Count Rumford, (d. 1814) made the connection between heat and motion that was experimentally quantified by James Joule (d. 1889). This refuted the caloric theory by acknowledging that heat is a form of energy. Just rub your hands together and you will experience motional energy converted to heat. In 1847, Hermann Helmholtz (d. 1894) conceived of the principle of **conservation of energy**, which recognized that mechanical energy appears in different forms: kinetic (motional), gravitational, electrical, and so on. Energy can be transformed back and forth among forms, and used to do work, but is gradually dissipated by friction as heat, a disordered form of energy that is difficult, but not impossible, to reuse.

The process of using heat to do work was implemented in the heat engines that powered the industrial revolution. William Thomson, Lord Kelvin, synthesized earlier ideas of Joule and Sadi Carnot (d. 1832) and proposed the principle that heat always flows from hot to cold. Rudolph Clausius (d. 1888) drew the implication that you can never build a perfect heat engine, or a perpetual motion machine in which energy is taken directly from the environment and used to do work. He introduced the important concept of **entropy**, a measure of disorder, and elucidated the **second law of thermodynamics**.

The second law described the fact that certain physical processes, such as the flow of heat in an isolated system, occur irreversibly. This is expressed by demanding that the total entropy or disorder of the system and its environment increase with time. A system can lose entropy and become more orderly, but only at the expense of increasing the entropy of its environment by at least the same amount.

Thus, heat does not flow from a colder body to a hotter one, in an otherwise isolated system, because the reverse process would lower the total entropy. In another familiar example, a freely expanding gas has an increasing entropy that cannot be reversed unless work is done on the gas from the outside to recompress it. A gas will not recompress by itself. We will revisit the second law later, when we discuss how physics has, so far, been unable to find a convincing microscopic basis for the time irreversibility that is implied by these macroscopic observations. Indeed, our every experience as human beings cries out for time to flow in a singular direction. Even the most prominent scientists have difficulty relinquishing time irreversibility, and we will see how this has led to a serious misunderstanding of the fundamental simplicity of reality.

THE CHEMICAL ATOMS

By the eighteenth century, the successful applications of the scientific method in physics had encouraged the evolution of the medieval, occult art of alchemy into the science of chemistry. In 1772, experiments by Antoine Lavoisier (d. 1791) demonstrated that combustion resulted from the interaction of the burning substance with some component of air he later named oxygen. In 1774, Joseph Priestly (d. 1804) succeeded in isolating oxygen from air. In 1783, Henry Cavendish (d. 1810), the physicist who first measured Newton's gravitational constant, G, thereby "weighing the earth," determined that water was composed of hydrogen and oxygen. It was considered a wonder that these two highly reactive substances could unite into the benign substance so common and essential on earth.

Two volumes of hydrogen gas were needed for each volume of oxygen for the combination to be completely converted to water. Careful, quantitative experiments on other substances and their mutual chemical reactions led to the discovery of many such rules, which John Dalton (d. 1844) codified as his *New System of Chemical Philosophy*, first published in 1808.

Without acknowledging Democritus, Lucretius, or any earlier atomist except Newton, Dalton used as the basis of his New System the notion that all matter was composed of indivisible elementary substances he called atoms. He made a list of atoms and atomic weights that contained many errors and deficiencies, but he still was able to deduce from the atomic theory three "laws of proportions" that explained many of the measurements being made in chemical laboratories.

Dalton believed his laws of proportions were not just mathematical tricks but a consequence of the reality of atoms. Indeed, he first called his basic units "particles," and they only later came be known as "atoms" (McDonnell 1991, 96–98). However, many saw the hypothesis that atoms are actual, discrete particles as unnecessary, or at least not proven. No one had seen an atom then, although we have pictures of them today, taken with the Scanning Tunneling Microscope (see the frontispiece in Stenger 1990a). None of the observations of chemical reactions offered any idea of the scale of atoms, that is, their individual masses or sizes. Matter could still be a continuum and just happen to obey Dalton's laws by some underlying continuum principle yet to be discovered.

A huge step in the ultimate confirmation of the discreteness of matter and the atomic model, and determining the scale of atoms, was made in

1811 by Amedeo Avogadro (d. 1856). Avogadro proposed that familiar gases such as hydrogen, nitrogen, and oxygen were not fundamental units, as Dalton had assumed, but themselves composed of two atoms. That is, they are what we now call diatomic molecules. Avogadro's name is familiar to all students of science today in **Avogadro's number**, 6.022×10^{23}, the number of molecules in a mole of a substance. Avogadro did not know the value of this number, but had hypothesized that it was a constant. In what is called Avogradro's law, an equal number of moles of different gases contain the same number of molecules.

Avogadro's conjecture can be readily understood today in terms of the nuclear model of the atom. It is essentially the number of nucleons (protons or neutrons) in gram of matter, independent of substance. However, his number is so huge, or, equivalently, atomic masses are so small (the reciprocal of Avogadro's number is the mass of the nucleon in grams), that no direct determination of its value was possible given the technology of the early nineteenth century. As the century progressed several estimates were made that turned out to be in the right ballpark. Nevertheless, the value of Avogadro's number, and thus the scale of atoms, was not established with confidence until the twentieth century.

In his 1905 Ph.D. thesis, Einstein showed how Avogadro's number could be measured from observations of Brownian motion, the random movements of macroscopic particles suspended in liquids and gasses. Einstein hypothesized that Brownian motion was the result of invisible molecular bombardment, and that by measuring the size of the fluctuations Avogadro's number could be determined. Measurements by Jean-Baptist Perrin applied Einstein's method to determine Avogadro's number with good precision. He confirmed the earlier, cruder estimates and established that, indeed, atoms are very small and very many of them are contained in the bodies of everyday experience.

It was not until these developments that the atomic picture of matter became firmly and finally established. Previously, evidence had gradually accumulated as the atomic theory explained increasingly more observations in chemistry and physics. James Clark Maxwell (d. 1879) developed the kinetic theory of gases, first proposed a century earlier by Daniel Bernoulli. This accounted for all the observed thermodynamic properties of gases in terms of the motions of molecules.

In the kinetic theory, the internal energy of a gas equates to its total molecular energy. Temperature is proportional to the average molecular kinetic energy. The first law of thermodynamics is equivalent to the principle of energy conservation, with the total heat going into a system plus any external work done on the system equal to the change in internal energy. The second law of thermodynamics is associated with the increasing disorder of the molecular motions of system and its environment with time.

Entropy is interpreted as the amount of disorder in the molecular motion. These ideas were extended to liquids and solids.

If a substance in its vapor phase is made of particles, as implied by the kinetic theory of gases, then surely it must be so in its liquid and solid phases. When this was added to the fact that the properties of the chemical elements followed from the atomic model, the conclusion was ultimately inescapable: the continuity of matter as it appears to the naked eye is an illusion brought about by the great numbers of particles involved and the inability of our eyes to resolve them.

Maxwell's attitude about the reality of scientific concepts is instructive at this juncture. His views are significant in regard to both atoms and the discovery electromagnetic fields, discussed in the next section. In both arenas, Maxwell played a central position. He rejected both the use of purely mathematical formulas, which cause us to "entirely lose sight of the phenomena to be explained," and the use of hypotheses that postulate the existence of unobservable entities and causes, which lead us to a "blindness to facts and a rashness in assumption." Rather, Maxwell preferred the method of physical analogies that avoids speculations about unobservables (Achinstein 1991, 159).

Maxwell permitted what he called "physical speculation," where your hypotheses involve known phenomena and you can demonstrate independent, empirical justification. Thus, although the molecules of a gas in kinetic theory were not directly observable (at the time), they were reasonably expected to possess properties such as mass, momentum, and energy that are already known to be possessed by observable bodies, and to be bound by the same laws. Furthermore, bodies are clearly made of parts; phenomena such as the diffusion of liquids and gases through each other illustrate that these systems possess parts not directly perceptible to us (Achinstein 1991, 161).

It may seem that Maxwell was making a fine distinction between two invisible realities, but it was an important one. He said that speculation is acceptable, but not wild speculation that has no rational basis, no connection to what is in fact observed in the real world. Maxwell was evidently not a positivist, like physicist and philosopher Ernst Mach, who would not believe atoms existed until he saw one. Maxwell said that if atoms behave in your theory in ways that are consistent with known principles, and if the theory produces results that agree with the data and make successful predictions, then the assumptions of the theory can be treated as containing meaningful concepts that relate to reality.

Today, no conflict exists on the corpuscular nature of matter since we have instruments that allow us to "see" chemical atoms and molecules. Without a doubt, rocks, trees, planets, and people are made up of discrete particles with mostly empty space in between. But we who are alive at the turn of the millennium still want to know: Can the elementary particles as now specified be further divided and if so, what are those final, uncuttable atoms? Before we address this further, we must take another look at a component of the universe that is clearly not made of the same stuff as the chemical elements—light.

PARTICLES OF LIGHT

When Newton was a young man, he possessed exceptional eyesight and mechanical skills that enabled him to conduct remarkably accurate laboratory measurements. He was as great an experimentalist as mathematical theorist, and many of his experiments concerned properties of light. He developed theories to describe these observations, but delayed publishing them until 1704. At that time they appeared in *Opticks* as a series of "Queries." In Query 29, Newton advanced the notion that light was particulate, or "corpuscular," in nature. The particles of light were material and obeyed his three equations of motion, published earlier in *Principia*. Newton felt no need to introduce a special force, like gravity, to describe the behavior of light. He sought to explain light in mechanical terms alone. We have already discussed his use of Cartesian ideas in explaining reflection and refraction. He also sought to explain why a beam of white light spreads out into a spectrum of colors when refracted by a glass prism.

Newton associated color with the mass of the particles of light, red being the heaviest and violet the lightest. This turned out to be wrong in specifics, but we will see later that a connection between color and energy exists, in our modern view, in which light is composed of photons. In any case, Newton explained the spreading of white light into various colors by a prism as the consequence of the different refrangibilities, that is, refraction properties, of the different mass particles. Put simply, this again can be understood in terms of particle motion. Heavier particles are more difficult to deflect from their paths than lighter ones. Newton also offered explanations for other phenomena, such as *Newton's rings* in which the alternate bright and dark rings are seen when light travels through a thin film (Achinstein 1991, 14).

In *Opticks*, Newton rejected the alternative theory, prevalent at the time, that light is an impulse phenomenon in a continuous medium, that is, a wave. Descartes and Newton's bitter adversary Robert Hooke (d. 1703) had promoted this idea, which was developed into a sophisticated model by Huygens in 1678, twenty-five years before the publication of *Opticks*. Huygens's principle of summing wavelets is still used today to offer a simple, visual explanation of wave effects such as interference and diffraction, and can also be used to demonstrate the laws of reflection and refraction. Newton wrongly argued that light did not diffract around corners, like water waves: "Light is never known to follow crooked Passages nor to bend into the Shadow." But here one of the smartest human beings of all time was wrong. Light does indeed bend into the shadow, and it is surprising that Newton missed this since the effect can be seen with the naked eye and, otherwise, his optical observations were impeccable. Diffraction cannot be readily understood in terms of particulate behavior, at least not without quantum mechanics.

WAVES OF LIGHT

Largely by virtue of his prodigious reputation, Newton's corpuscular view of light prevailed as the consensus until early in the nineteenth century. By then, quantitative experiments had strongly supported the wavelike behavior of light. In 1801, Thomas Young (d. 1829) performed his famous **double slit interference experiment**. Young reintroduced the wave theory, with light understood as an undulation of the aether. Euler had been one of the few to hold on to the notion of aether-waves, and now his field theory of fluids was applied to the aether whose vibrations were assumed to constitute light.

Euler's theory had been used to describe the propagation of sound in fluids. In this case, the oscillations are longitudinal, that is, back and forth along the direction of propagation. Polarization effects observed in experiments with light suggested instead a transverse oscillation at right angles to the direction of propagation, what are called shear waves. Euler had only considered shearless fluids that are incapable of transverse oscillation. Nevertheless, his basic equations of fluid motion, derived as we have seen from Newton's laws, proved sufficiently general and were successfully applied to transverse oscillations in elastic solids by George Stokes (d. 1903).

In the meantime, the fruitful partnership between mathematical theory and laboratory experimentation, in which minimal attention is given to metaphysical speculation, had also resulted in great advances in the knowledge of electricity and magnetism. Like gravity, these forces seem to operate over empty space. As we have seen, fields are used to describe how a force applied at one place in a material medium is transmitted through that medium to another place without the actual movement of any matter between the two points. For example, a stretched sheet of rubber acts as a medium for the fields of stress and strain that produce forces that act everywhere in the sheet. Vibrations induced at one end of the sheet will transmit throughout the sheet in the form of acoustic waves. Air, water, and solid materials from our everyday experience provide the media for sound propagation. Sound phenomena are well-described by the field equations of acoustic theory.

Michael Faraday (d. 1867) discovered the phenomenon of **electromagnetic induction** in which a time-varying magnetic field induces an electric field. This implied that the electrical and magnetic forces are deeply related. He imagined invisible **lines of force** transmitting electromagnetic forces through space by means of "contiguous particles" propagating along the lines of force. Faraday seemed to be thinking in terms of discrete particles rather than a continuum, which still required instantaneous action at a distance in the microscopic void between atoms (Hesse 1961, 199–201).

In his 1873 *Treatise on Electricity and Magnetism,* Maxwell modified the field equations describing electricity and magnetism to allow also for a time-varying electric field to produce a magnetic field, although such an effect had not been observed at that time. Then he showed that this led to waves that propagated at exactly the speed of light. Around 1886, Heinrich Hertz (d. 1894) generated electromagnetic waves in the laboratory and measured their speed to be that of light.

Hertz's detection of electromagnetic waves had been made possible by the development of high voltage technology. A spark is an electrical discharge, a miniature lightning bolt, that can be generated with high voltage in air or other gasses. Hertz generated a high voltage spark that produced another spark in a glass chamber well out of contact some distance away.

Other observed properties of Hertzian waves, such as polarization, strengthened their connection to light. Evidently, the only difference between light and Hertzian waves was the lower frequency or greater wavelength of the latter. This implied that the electromagnetic spectrum extended in both directions, well beyond the capacity of the human eye to detect. We have now measured electromagnetic wavelengths smaller than the nuclei of atoms and as large as a planet. Visible light occupies only a tiny band in this spectrum.

With the confirmation of electromagnetic waves, it seemed to most physicists that everything was now clear. Light was the propagation of vibrational electromagnetic waves in an invisible medium that extends throughout the universe—the aether.

But is the aether real? Recall that Maxwell was very careful when talking about the reality of atoms. He rejected both a strictly mathematical approach and a strictly empirical one. Atoms could not be seen, but they could be believed if the theory that described them was logically consistent and supported by independent empirical evidence. On the reality of the aether, Maxwell noted that you can find many possible mechanisms for electromagnetic waves and the aether is just one of them. His use of the aether, he says in the *Treatise,* was only intended to show that a mechanism can be imagined for electromagnetic waves (Hesse 1961, 5). It seems that he was more favorably disposed towards the reality of atoms than to the reality of the aether.

THE ELECTRON:
STILL ELEMENTARY, MY DEAR WATSON

In the late nineteenth century, experiments with gas discharges were providing many other clues to the nature of matter and light. These experiments were done in metal and glass chambers that contained various gases at different pressures. As vacuum pump technology advanced, the gas pressure could be decreased to unprecedented levels. The glass chambers were penetrated by electrodes that enabled the application of electric fields at various positions.

It was noticed that a bluish glow often appeared near the negative electrode. This emission was called **cathode rays**. The glow was observed to change in the presence of a externally applied magnetic field, which suggested that the cathode rays were charged particles. However, in that case they should have been affected by an externally applied electric field as well, and Hertz found that this did not seem to be the case. Furthermore, cathode rays were able to penetrate thin layers of metal. At the time, no atoms or molecules were known that could do this, even the lightest, hydrogen. Electromagnetic waves, on the other hand, were known to be able to penetrate thin gold foils and so many thought that cathode rays were waves of some sort.

In 1895, Perrin demonstrated that cathode rays were negatively charged, although this interpretation was not completely convincing since the effect could have been an artifact of accompanying negative ions from the cathode. Joseph Thomson improved on Perrin's experiment, avoiding the objection by blocking out all but the cathode rays, and confirmed the result. He then repeated Hertz's experiment at a much lower gas pressure and saw the deflection of the cathode rays by an electric field, as was to be expected if they were charged particles. The conductivity of the higher pressure gas in the Hertz experiment had shielded the cathode rays from the external electric field he had applied to check if they were charged.

Continuing to improve his techniques, Thomson measured the mass-to-charge ratio of the cathode particles and found it to be much lower than that for hydrogen ions, which was the smallest measured mass at that time. This result was believed to be due to the lower mass of the cathode particles, but also could be explained by them having an exceptionally high charge. In 1899 Thomson used the cloud chamber, newly invented by his student Charles Wilson, to make an independent determination of the charge. He found it to be comparable to the unit of electric charge first measured in 1874 by George Stoney.

Thomson thus determined that the cathode ray particle had a mass at least a thousand times less than the hydrogen ion. He called his discovery the "corpuscle," and continued to insist on this designation. However, the cathode ray particle became known as the **electron**, the name Stoney had given in 1891 to the quantum of electric charge, e. The electron carries one negative unit of charge, –e. An ionized hydrogen atom, now recognized as its bare nucleus and called the proton, carries one positive unit, +e.

In the century since the discovery of the electron, many new particles have been discovered and later found to be composed of even more elementary constituents. For example, the proton is now known to be made up of **quarks**. However, the electron remains one of the elementary particles in the **standard model** of quarks and **leptons**. The electron, along with the **muon, tauon**, and three kinds of **neutrino** constitute the leptons of the standard model. But, of the leptons, only the electron is present in ordinary matter.

Thousands of experiments have led to the determination of the properties of the electron to precisions unmatched by any other object in the natural world. Nothing we know, including the earth we walk upon, is more real than the electron. Except, perhaps, the photon.

THE PHOTON: THE ELEMENTARY PARTICLE OF LIGHT

The wave theory of light was expressed in terms that were already familiar for acoustic waves. Sound propagates as pressure and density waves in any continuous material medium. This is understood fully within the classical, Newtonian scheme. Euler's equations, from which acoustic waves are derived, are themselves derived from the application of Newton's laws of particulate motion to fluid elements. In the case of sound, the fluid medium is air, water, or any other known material.

The same concept was assumed to apply in the wave theory of light, except that the medium of propagation, the aether, was not visible or otherwise detectable. The equations of electrodynamics were not obtained from a direct application of mechanical principles to a medium with known properties, as are the equations of acoustics. Being undetected, none of the properties of the aether were measured. The laws of electrodynamics were inferred from experiments on electricity and magnetism and can be viewed as simply mathematical representations of those observations. The aether exists in electromagnetic theory, as Maxwell made very clear, only as an analogy. He allowed that it was one of many possible analogies. It follows that the fields and waves in this aether can have little standing other than that of analogies.

As we will see, searches for direct evidence for the aether failed. At the same time, we will also discover that **Maxwell's equations** for electrodynamics contained the seeds for the development of the theory of relativity. So it would be wrong to think that classical electrodynamics turned out to be "wrong" just because the aether does not exist. Maxwell did not roll over in his grave when the aether failed to be confirmed. Indeed, classical electrodynamics is still immensely useful. It remains as correct as the rest of classical physics, in the pragmatic sense. Classical electrodynamics is correct as a theory applicable within its proper domain, but not suitable as a model of reality. The fields in Maxwell's equations are mathematical objects, not tensions in a continuum pervading all of space. Furthermore, long before these equations were first written down, electromagnetic phenomena had been observed that could not be understood in terms of continuous fields and waves.

As early as 1802, William Hyde Wollaston (d. 1828) had observed dark lines in the spectrum of light from the sun. These were studied in detail by Joseph von Fraunhofer (d. 1826), who mapped 576 of them. These **Fraunhofer lines** are now recognized as the absorption of very well-defined wavelengths from what appears, at first glance, to be an otherwise continuous spectrum of light from the sun. Later in the nineteenth century, the spectrum of the emitted light in gas discharge experiments was observed to contain discrete sharp, bright lines, indicating the emission of well-defined wavelengths. Eventually the sun and other stars were found to exhibit such emission spectra, and indeed the element helium was first observed in the solar spectrum, before being isolated on earth.

The wave theory of light gave no explanation for emission or absorption lines, instead predicting a smooth, continuous spectrum. Explaining these line spectra would be a major triumph of the quantum theory and the ultimate nail in the coffin of continuity. What's more, even when the apparently continuous part of the spectrum of radiation from hot bodies, like the sun, was studied it was found to disagree with the wave theory. In 1900, the theoretical spectra of hot bodies (so-called black bodies) was calculated from the wave theory by John Strutt, Baron Rayleigh, and James Jeans. The calculated spectrum exhibited an "ultraviolet catastrophe," going to infinity at short wavelengths. This disagreed with the data, which went to zero. If the wave theory were correct, we would all be bathed in intense short wavelength, or high frequency, radiation.

Fortunately, we are not, and Max Planck explained why. In 1900, Planck proposed that light was not a continuous phenomenon but occurs in discrete packets of energy or **quanta**. By postulating that the energy of a quantum was proportional to the frequency of the corresponding (theoretical) electromagnetic wave, he was able to explain the observed spectrum of radiation from heated bodies. A body contains only so much energy, and so cannot radiate high frequency quanta without violating energy conservation.

In 1905, Einstein built on Planck's idea and proposed that the quanta of light were actual particles, later dubbed photons. He related the energy, E, of a photon to its frequency, f, using Planck's relation E = hf, where h is Planck's constant. With photons, Einstein was able to explain a phenomenon, also observed in nineteenth-century vacuum chamber experiments, called the **photoelectric effect**.

In the photoelectric effect, cathode rays are emitted from metallic electrodes irradiated by light. Thomson had observed that these rays had the same mass-to-charge ratio as normal cathode rays, and so were very likely electrons.

This phenomenon was also inconsistent with the wave theory of light. Depending on the particular metal, the electron emission shut off below a certain frequency independent of the intensity of the light. The wave theory offered no explanation, but in the photon picture the answer is simple: a photon collides with the electron and kicks it out of the metal. To do so, it needs a certain minimum energy or frequency.

As the twentieth century advanced, other experiments that refuted the electromagnetic wave theory while supporting the photon theory followed. Today we have no doubt. Individual photons are now routinely detected. Photons are the "atoms" of light. And so, with the appearance of Einstein's 1905 papers, the first on Brownian motion, which we recall enabled the determination of the scale of chemical atoms, and the second on the photoelectric effect, we could say that both matter and light are composed of discrete bodies. But where does that leave the gravitational and electromagnetic fields? What happened to the aether? For these questions we must discuss Einstein's most famous 1905 paper in which he introduced the world to the theory of relativity.

NOTE

1. My use of the term *classical physics* should be clarified. Physics textbooks usually use "classical" to represent any theory in physics that is not a quantum theory. This places both special and general relativity under the classical umbrella, since these theories retain the determinism characteristic of Newtonian mechanics. Quantum mechanics contains, in the conventional interpretation, a radical indeterminism and so is usually separated from deterministic theories. Here, however, I will follow the more popular usage in which classical physics signifies the period before the twentieth-century twin developments of relativity and quantum mechanics, which together are called *modern physics*.

3

THE
DEATH
OF
ABSOLUTE
NOW

*I see [in the special and general theories of relativity] not only a new
physics, but also, in certain respects, a new way of thinking.*

Henri Bergson (1922)

THE DESTRUCTION OF THE AETHER

As the twentieth century neared, it seemed to most physicists that we
live in a two-component universe. Matter was composed of atoms, but
these did not simply move about in an empty void but were embedded in a
continuous, elastic medium—the aether. The electrical force largely responsible for holding chemical atoms and molecules together reached out as field
lines in the aether between particles. The magnetic force, which also plays
an important role in matter, was distributed in similar fashion. Light was
transmitted as oscillating electromagnetic waves in this medium, as its
energy shifted back and forth between electric and magnetic components.
Some even speculated that "thought waves" might also be transmitted from
mind to mind as a kind of wireless telepathy analogous to the wireless teleg-
raphy made possible by electromagnetic waves (Stenger 1990a).

Perhaps the aether also constituted the medium for gravity, but most physicists still viewed gravity as an instantaneous, action-at-a-distance force. While the observation of electromagnetic transmissions had demonstrated that energy could be found in the space between bodies, comparable gravitational waves were not observed. In any case, the problem of light seemed to be solved. It remained to find direct evidence for this remarkably invisible and elastic substance that filled the universe from the tiny spaces between atoms to the great spaces between stars.

Around 1887, Albert Michelson and Edward Morley performed a series of experiments fully capable of finding direct evidence for the existence of the aether. Using an interferometer, a device invented by Michelson, they attempted to measure the difference in the speeds of light beams that traveled along two perpendicular paths within the instrument. A slight difference was expected to result from changes in the earth's velocity relative to the aether as it circled about the sun at 30 kilometers per second, one-ten-thousandth of the speed of light. The Michelson and Morley instrument was sufficiently precise to detect a tenth of the expected difference. They observed no effect. The speed of light did not seem to depend on the motion of the source or detector relative to the medium in which electromagnetic waves were supposedly propagating.

In 1892, George Fitzgerald tried to explain these results by postulating that the length of an object contracted in the direction of its motion through the aether, exactly compensating for the effect of its motion on the velocity of light. A few years later, Hendrik Lorentz expanded on the idea, which is now called the **Fitzgerald Lorentz contraction**.

Lorentz had made contributions to the theory of electromagnetism that rivaled those of Maxwell. While that theory was brilliantly successful, it seemed to depend on the point of view of the observer, the equations being different for an observer at rest and another observer moving a constant velocity with respect to the first. This appeared to violate a profound principle of physics that, more than any other, had characterized the break that the scientific revolution of the sixteenth and seventeenth centuries had made with previous thinking.

Copernicus suggested that the earth moved. Galileo explained why we do not notice this, establishing the **principle of relativity** which asserts that no observation can distinguish between being at rest and being in motion at constant velocity. This was a revolutionary idea, shattering Aristotle's concept of absolute space. Newton's laws of mechanics and all of the physics that emerged up until electrodynamics were consistent with Galileo's principle of relativity. Using an example familiar today, Newton's laws of motion are the same whether you sit in a chair in your living room or in the seat of a jet cruising at 350 miles per hour. Electrodynamics appeared to be the exception, suggesting that perhaps the principle of relativity is wrong and an absolute space, defined as the reference frame of the aether, may exist after all.

The formal way a theory is tested for consistency with the principle of Galilean relativity is to write it down in terms of the variables measured in one reference frame, and then use the **Galilean transformation** to translate to the corresponding variables that would be measured in a second reference frame. If the new variables still obey the original equations of the theory, then the theory is said to be **invariant**, obedient to the principle of relativity. The consequence is that no observation consistent with that theory can be used to determine a preferred reference frame.

Take Newton's second law. If you measure the force F on a body of mass m and acceleration a, you will find they obey the second law, $F = ma$. If I were moving at a constant velocity relative to you and made the same measurements, I would obtain the same values for the quantities F, m, and a. It needs to be noted, however, that this only applies when the relative velocity is constant (and, we will see, much less than the speed of light). We define an **inertial frame of reference** as one in which Newton's second law is observed, and a **noninertial frame of reference** as one where it is not. An accelerated reference frame, such as the rotating earth, is noninertial, although we often treat it as inertial in the first level of approximation.

By contrast to Newton's laws, Maxwell's equations of electromagnetism are *not* invariant under a Galilean transformation. The electric and magnetic fields obeying Maxwell's equations in one reference frame do not Galilean-transform to fields in another frame that still obey the same equations. A point charge viewed at rest has an electric field but no magnetic field. View that same charge in a reference frame moving at constant velocity and you will see a magnetic field but no electric field. If the Galilean transformation is correct, the laws of electromagnetism are different to observers moving with respect to one another.

Lorentz played around with the mathematics and came up with another way to translate between reference frames—the **Lorentz transformation**—for which the equations of electrodynamics are invariant. The Lorentz transformation implies the Fitzgerald-Lorentz contraction, consistent with the Michelson-Morley results. That is, the Lorentz transformation requires that the speed of light be the same in all reference frames.

This was all rather ad hoc at the time, and the physical meaning was unclear. In particular, the Lorentz transformation required that time intervals be different in the two references frames, and Lorentz was not ready to discard the powerful intuition we all have of an absolute, universal flow of time that is independent of the observer. Another smart guy had to come along to tell us that this was an illusion.

THE SPECIAL THEORY

In 1905, Einstein provided a physical foundation for the Lorentz transformation in his **special theory of relativity**. He examined the logical consequences that would follow if he hypothesized (1) that the principle of relativity is valid and (2) the speed of light is absolute. He was aware of the Michelson-Morley results and that Maxwell's equations were not Galilean invariant, but apparently had not heard of the Lorentz transformation (Pais 1982, 21). In any case, he independently derived it from his two hypotheses without making any assumptions about the nature of space and time other than their operational definition in terms of measurements made with meter sticks and clocks.

As we saw above, the Lorentz transformation required that time intervals change from one reference frame to the next. Einstein concluded what Lorentz was not quite prepared to contemplate, namely, that clock time is in fact relative. If time is what you measure on a clock, and if the Lorentz transformation is correct, then clocks moving with respect to one another will give different readings for the time intervals between two events. Since this effect is very small unless the relative clock speeds are a significant fraction of the speed of light, we do not notice this in common experience. Indeed, the Lorentz transformation becomes the Galilean transformation, as it should, at low relative speeds between reference frames. But an observer watching a clock in a frame moving at very high relative speed would, according to Einstein, see that clock running slower than another clock at rest with respect to the observer. This is called **time dilation**.

Note, however, that the clock seen to run slower is moving with respect to the observer. Let's call that observer Alf. Nothing funny happens to any clocks that are at rest with respect to Alf. They continue to read normally, what we call **proper time**. And, since another observer, Beth, can be riding along with a clock that Alf perceives to be moving, nothing funny happens to her clocks either. Otherwise the principle of relativity would be violated because one reference frame would then be distinguishable from another.

The equations of special relativity allow Alf to calculate Beth's proper time from space and time measurements he makes in his own reference frame.[1] It will agree with what Beth measures on her clock. And Beth can calculate Alf's proper time. Finally, a third observer, Cal, moving with respect to both Alf and Beth can calculate their proper times. They can calculate his, and they will all agree. While the clocks may give different values for the intervals between events (i.e., those times are relative), the proper time is an invariant.

Special relativity became the exemplar for the positivist school of philosophy that, as we saw in chapter 1, surfaced late in the nineteenth century. Time is what you measure on a clock; distance is what you measure with a meter stick; temperature is what you measure on a thermometer; and so on. To the positivist, physics (and, by extension, all science) concerns itself solely with measurements and their description. Einstein was probably never a strong positivist, and would later assert quite opposing views. However, philosophers of the logical positivist school that flourished at the same time in Vienna were so impressed by the results of relativity that they adopted strict empiricism as a working principle. (I will discuss logical positivism later in chapter 14.)

The Lorentz transformation implies that the measured distance between two points, such as the length of an object, will also depend on the reference frame. Space, like time, is relative. And, Einstein proved, energy and momentum are also relative. Furthermore, the mass m of an object at rest is equivalent to an energy $E_0 = mc^2$, where c is the speed of light. We call E_0 the **rest energy** of a particle. Since c is a constant, its presence in the equation relating mass and energy merely serves to change the units from the conventional ones of mass to those of energy. Basically, Einstein said, rest energy and mass are equivalent.

You will often read in the textbooks that a moving body has a "relativistic mass" that equates to its inertial and gravitational mass, as defined by Newtonian physics (see below). However, I will follow modern convention and generally use the term mass to refer to the **rest mass**, that is, the mass measured in the reference frame in which the body is at rest.

The equivalence of mass and rest energy was a radical new concept, as revolutionary as the relativity of time. However, it was also a simplifying, unifying concept. Prior to relativity, the vague notion existed that there were two kinds of stuff in the universe, the matter in concrete objects and something more tenuous and ethereal, the "pure energy" of electromagnetic radiation. Moving bodies, however, also contained kinetic energy. And bodies were also thought to have potential energy in the presence of a force field. These various energies could be converted to one another, consistent with energy conservation, as, for example, when a falling body has its potential energy converted into kinetic energy as it falls.

Einstein showed that the Lorentz transformation requires changes in our formulas for momentum and kinetic energy, although they properly reduce to the old ones at speeds low compared to light. The equations of special relativity were new in 1905, and while not immediately accepted, within a few years they had become standard physics. Today the Lorentz transformation and the equations of relativistic kinematics have stood the test of almost a century of application and experimentation. Time dilation and Fitzgerald-Lorentz contraction have been observed exactly as predicted, even at low speeds where the effects are tiny, as has mass-energy conversion and every other aspect of special relativity.

The dramatic success of Einstein's special theory of relativity, along with the parallel development of the quantum theory, put an end to speculations about the aether among most mainstream physicists. A different kind of aether is still speculated about today, as we will see in a later chapter. For now, suffice it to say that in the century since Michelson and Morley, no physically real, material aether has been observed in experiment and none has been needed in theory.

While quantum mechanics is required to fully understand the behavior of light as a stream of particles, we can still use classical ideas to understand some of the properties of photons—provided we use the equations of special relativity. Photons, by definition, travel at the speed of light. Relativistic kinematics demands that a particle moving at the speed of light have zero mass. Stating it more precisely, a particle of zero mass must move at the speed of light.[2]

Although photons have zero mass, they can and do carry energy and momentum. Again this follows from relativistic kinematics, where the energy of a massless particle is proportional to its momentum. This is exactly the same relation between the energy and momentum of an electromagnetic wave, as it should be if photons are to be associated with light. Thus, we have a natural explanation for the constancy of the speed of light. It can be seen to result from the fact that photons are massless.

As we will see later, *all* elementary particles may be fundamentally massless, picking up an effective mass as they move through space and scatter from background particles that are currently thought to fill all of space.

When a body is a source of light, it emits photons that travel through space until they encounter another body that absorbs or reflects them. In this way, energy and momentum are transferred through space from the first body to the second by purely contact action. No action at a distance or waving of an intervening medium is needed.

This provides a contact-only mechanism for the electromagnetic interaction between two bodies over distances both small and large. Suppose we have two electrons moving near each other. In the picture we will discuss in greater detail later, one electron emits a photon that travels across the intervening distance and is absorbed by the other electron. Energy and momentum are carried by the photon from one electron to the other and by this means they interact without the presence of any force field.

We will return to these ideas when we discuss quantum mechanics and develop the full picture of elementary particles. However, note that we have eliminated the notion of the aether as a real, physical reality and recognized light as a stream of photons, implying the nonreality of the electric and magnetic fields themselves. For if the aether does not exist, what medium is doing the vibrating to produce electromagnetic waves? And where are these electromagnetic waves anyway? With the observed discreteness of electromagnetic radiation, carried by photons, how can we still retain any notion of wavelike continuity? As we will see, the nuclear forces that were still unknown during the early period of modern physics are now also described in terms of "quanta" rather than continuous fields. Such fields remain an important part of their mathematical description, but do not exist as direct observables.

Einstein's special theory of relativity carries with it another implication that will be very important in our later discussions. According to special relativity, no particles or signals of any known type are **superluminal**, that is, they cannot be *accelerated* to move faster than the speed of light. Note the emphasis on *accelerated*. Relativity does not forbid superluminal particles, just the acceleration of subluminal ones to superluminal speeds. The theory does not rule out another class of objects, called **tachyons**, that *always* move faster than light. However, no such objects have yet been observed.

As we will discuss in detail later, Einstein introduced another assumption, now called **Einstein causality**, in order to forbid superluminal motion. He noticed that the time sequence of two events becomes reversed when you transform them to a reference frame moving at more than the speed of light with respect to the original frame. The additional hypothesis of Einstein causality, not part of the original axioms, asserts than cause must always *precede* effect, and so no object can move faster than the speed of light. This presumes the irreversibility of time.

Einstein used the term **local** to describe two events in space-time that can be connected by signals moving no faster than light. In this case, the events occur at the same place in the reference frame moving with the signal and the space-time separation is said to be **timelike** or "inside the light cone." When the separation is **spacelike**, no such reference frame can be found and the events are "outside the light cone." As we will see, quantum mechanics is widely thought to violate Einstein's speed limit, but the fact remains that no superluminal motion or signalling has ever been observed.

GRAVITY

During the age of classical physics, Newtonian gravity was presumed to act instantaneously over space. Newton's law of gravity was a simple equation that gave you the force between two bodies. From this single equation, planetary orbits and more complicated gravitational effects such as tidal forces could be calculated. The only other inputs needed to practice gravitational physics were the standard laws of motion and certain mathematical and numerical tools for carrying out more difficult calculations involving non-spherical objects and systems of more than two bodies.

When the notion of fields came into vogue long after Newton, mass became regarded as the source of a gravitational field that was imagined to radiate from a point mass in the way that the electric field radiates from a point charge. However, the gravitational field remained static while the dynamic (time-varying) effects of electromagnetism were seen and codified in Maxwell's equations. Electrodynamic theory led, as we have noted, to the prediction and observation of electromagnetic waves. In the case of gravity, no comparable dynamics evolved.

Gravitational waves were not predicted in Newtonian theory, and have not been observed to the present day. Gravity is forty orders of magnitude weaker than electromagnetism. Only by the rapid movement of huge masses on the astronomical scale, such as two neutron stars rotating around one another, can you expect a gravitational oscillation to be measurable with the most delicate of modern instruments. A huge experiment is currently underway to search for these gravitational waves.

The accurate representation of planetary motion was the great triumph of Newtonian gravity. Kepler's laws of planetary motion followed from the theory. Edmund Halley (d. 1742) showed that the comets seen in 1531, 1607, and 1682 were a single object in a highly elliptical orbit around the sun and predicted its reappearance in 1748. When Halley's comet made its spectacular appearance on schedule in 1749, there were few left who doubted the great power of science. The Age of Reason had reached its zenith.

More accurate observations of the planets indicated deviations from Kepler's elliptical orbits. Kepler's model assumed only a two body system: the sun and the planet in orbit. The deviations that were the result of the gravitational attraction of other planets could be calculated, with clever approximation and sufficient numerical effort. Independently making that effort, John Couch Adams (d. 1892) and Joseph Le Verrier (d. 1877) inferred that a planet beyond Uranus was perturbing the orbit of what was at the time the most distant known planet in the solar system. In 1846, astronomers verified this prediction with the discovery of the planet Neptune (Will 1986, 90).

In 1859, Le Verrier showed that problems with the orbit of Mercury, which he had first noticed in 1843, could not be explained by Newton's theory. The precession of the major axis of Mercury's orbit was measured to be 574 arcseconds per century (an arcsecond is $\frac{1}{3600}$ of a degree). Le Verrier estimated that 277 arcseconds were the result of perturbations from Venus, 90 arcseconds from Earth, 153 arcseconds from Jupiter, and 11 arcseconds from Mars and everything else. His calculation was thus 43 arcseconds short of accounting for Mercury's motion. (Here I am reporting a slightly more exact modern calculation; Le Verrier actually obtained a 38 arcsecond deficit [Will 1986, 91–92].) In short, after all this success, a problem with Newtonian gravity was found. It errs in predicting where Mercury will be seen in the sky a century in the future by an amount approximately the width of your small finger as viewed from a distance equal to the length of a football field.

GENERAL RELATIVITY

In 1915, Einstein calculated the orbital precession of Mercury in a new theory of gravity he was developing and compared it with the Newtonian result. His theory gave 43 arcseconds per century greater than the Newtonian prediction! He wrote that this discovery so excited him that he had heart palpitations (Will 1986, 93). Einstein's new theory was **general relativity**, published in 1916. By 1976, the anomalous precession of Mercury had been measured to be 43.11 ± 0.21 arcseconds per century, compared with the general relativity prediction of 42.98 (Will 1986, 95).

General relativity came about as a consequence of Einstein's attempt to generalize the results of special relativity to accelerated, or noninertial, reference frames. Recall that both Galilean and special relativity only apply to reference frames moving at constant relative velocity. These are, in practice, the inertial reference frames in which Newton's second law of motion, $F = ma$, is upheld. Einstein knew what all physicists know, that you can retrieve Newton's second law in a noninertial frame by introducing "fictitious" forces. For example, in a rotating—and thus accelerating—reference frame such as the earth, we can still make use of Newton's second law by adding to the external "real" forces, F, the fictitious centrifugal and coriolis forces. Einstein had an insight: perhaps gravity is also fictitious!

Consider a closed capsule far in outer space accelerating at 9.8 meters per second per second. An occupant of the capsule cannot, by any experiment performed inside without looking out, distinguish whether she is indeed out in space or sitting in a uniform gravitational field of "one g," where $g = 9.8$ meters per second per second is the acceleration due to gravity on the earth.[3]

The key to this observation is an assumption about mass. The measured force divided by the measured acceleration of a body a gives the **inertial mass** m_I. In principle, this can be different from the **gravitational mass** m_G that you get by measuring the force due to gravity, that is, the weight, and dividing by the measured acceleration due to gravity. Newton had assumed their equivalence, $m_I = m_G$ in his theory of gravity.

In the nineteenth century, Roland von Eötvös performed an experiment that accurately compared the gravitational and inertial masses. By the 1970s this equivalence had been shown to one part in a trillion (Will 1986, 29–31). In general relativity, Einstein hypothesized the **principle of equiv-**

alence of gravity and inertia. He then showed how the notion of a gravitational force can in fact be done away with and the phenomena we associate with gravity resulting from accelerations a body experiences as it follows its natural path in a curved space.

While we live in three-dimensional space, Einstein's idea can be more easily visualized in two dimensions by adopting the perspective of Thomas Nagel's "view from nowhere" (Nagel 1986). Imagine space as a rubber sheet stretched out in a flat plane. A small ball bearing is rolling along the plane. From Newton's laws of motion, it follows a straight path. We then place a billiard ball on the surface that sinks down into the sheet, leaving a depression in the surrounding rubber (the third dimension here is not space but potential energy). Neglecting any loss of kinetic energy to friction, three things can happen when the ball bearing approaches the depression caused by the billiard ball: it can roll into the depression and, if it has enough speed, pull back out again; it could roll around inside the depression, orbiting around the billiard ball and be captured by it; or the two balls can collide.

Each of these possible motions is observed in the solar system. In this picture, no force of gravity acts to capture bodies, like you and me, to the earth. Unless you are an astronaut reading this far from any star system, you are now sitting at the bottom of a depression in three-dimensional space that has been distorted ("warped," as they like to say in science fiction) from flatness by the presence of the massive earth. No force of gravity keeps the earth in its orbit around the sun. Instead it runs around the sides of the deep valley of the sun.

So general relativity does away with the need to introduce the gravitational field. However, in its place another field is introduced: the **metric field** of space-time. Einstein formulated general relativity in space-time, and allowed for it to be curved rather than flat. This nonflatness he described by a non-Euclidean metric field. Furthermore, this metric field was not the same form at every point in space-time, but varied from point-to-point. Masses, like the billiard ball on our rubber sheet, provide one means for introducing curvature into an otherwise flat space.

Thus the metric of space-time is a field, denoting the geometry at each point in space and time. But unlike the electric and magnetic vector fields, which are specified by three numbers at every point in space, the metric field requires ten numbers at every point in space-time and is represented mathematically by a **tensor**. Using the somewhat advanced mathematical methods of tensor algebra, which took him a few years to learn (now they are a part of every physicist's tool kit), Einstein applied a "principle of simplicity." He made the fewest number of assumptions possible to derive an equation for the metric that satisfied the requirements of the theory.

The main requirement was that the equation for the metric be invariant to transformations between all reference frames, inertial and noninertial.

The matter distribution was introduced by relating the metric to the *momentum-energy tensor* that in classical field theory describes the energy and momentum densities of a continuous elastic medium, like the stress and strain in rubber sheet used in our two-dimensional model. In this way, Newton's theory of gravity was recovered, meeting the requirement that new physics reverts to old physics when applied to the more limited domain. Although Einstein did not realize this at first, the form of his equation guaranteed energy and momentum conservation. That is, these followed from his assumptions of invariance and simplicity.

Einstein's equations allowed for another term to contribute to the curvature of space-time, one that is present even in the absence of matter or radiation: the so-called **cosmological constant**. He originally thought it might provide an effective repulsive force to balance the normal gravitational attraction between bodies, thus yielding the stable, biblical, "firmament" of stars that was believed to constitute the universe. When, within a year or two, Edwin Hubble discovered that the universe was in fact expanding, the cosmological constant was no longer needed. Einstein referred to this as his "biggest blunder," and until recent years the cosmological constant has been assumed to be zero.

Now it appears that Einstein did not blunder at all. The cosmological constant has reappeared on two fronts. It showed up in 1980 as the driving force for the exponential **inflation** that is believed to have occurred during the very first moments of the big bang, when the universe had no particles or radiation, just spatial curvature (see chapter 13). Even more recently, observations of distant supernova explosions by independent groups of observers have indicated that the expansion of the universe seems to be speeding up, ever so slightly. The universe may be falling up! The initial explanation for this was a cosmological constant in the gravitational equations, although, as we will see, other possibilities exist. The cosmological constant in general relativity is equivalent to a curvature of space that is inherent in that space and not the result of nearby massive bodies.

While inflation in the early universe and cosmic acceleration today are still questioned by skeptics, we need to recognize that the cosmological constant is not some arbitrary "fix" that was added to the theory to make it agree with data, as the story is usually portrayed in the press and often reported by physicists who should know better. The cosmological term was present in the original equations of general relativity and we have no reason, at this writing, to assume that it is necessarily zero. If future data should yield a zero value, then that would have to be explained in terms of a deeper theory. When Einstein originally included the constant, he was being true to his principle of assuming the simplest, most economical answer until he was required to do otherwise. Zero is not necessarily the simplest choice.

Let us return to the early century, when the first tests of general relativity were being performed. As we have seen, the correct precession of the orbit of Mercury was immediately obtained, even before Einstein had published his theory. In addition, Einstein made two other predictions: light in passing a massive object like the sun would be deflected from its path and experience a **red shift**, that is, decrease, in its frequency.

As we saw above, a photon carries momentum and energy. The usual explanation of the gravitational red shift, which I myself have used in previous writings, is that a photon loses energy as it climbs out of a gravitational field. However, recent analyses have indicated that the photon does not in fact change frequency, or energy in the presence of gravity. Rather, the clocks that are used to measure that frequency run slower, as observed from the outside, in the presence of higher gravity (Okun 1999).

The tests of general relativity have a long and fascinating history. Suffice it to say here that Einstein's particular version, the one based on the simplest assumptions, has survived with flying colors after eight decades of ever-more-accurate measurements. Several alternatives to Einstein's minimal form of the metric equation have also been tested in the interim and ruled out (Will 1986).

On the other hand, it must be noted that only a few tests of general relativity have actually been feasible with existing technology. This is to be compared with special relativity and quantum mechanics, which have been tested thousands of times in thousands of ways, directly and indirectly, over the same time period. Einstein tried unsuccessfully for many years to extend general relativity to electromagnetism. This did not work because electric charge does not participate in a principle of equivalence in a manner analogous to inertial mass. As time has passed, Einstein's approach to unification has largely been abandoned, although much effort continues to try to unite general relativity and quantum mechanics in a quantum theory of gravity. Non-Euclidean geometries still play a role in most approaches.

Similarly, the nuclear forces discovered after the publication of general relativity have not shown any signs of being describable as fictitious forces, the product of the geometry of space-time. On the other hand, as we will see, great success in understanding electromagnetism and nuclear forces has been obtained in a quantum particle theory approach, with only contact forces, Euclidean space is assumed, and continuous fields appear solely as abstractions in the mathematics. The Democritan picture of atoms and the void is strongly indicated by these results.

But, then, what shall we do with the metric field of space and time? Einstein seemed to believe that his non-Euclidean space-time continuum was "real." He rejected any application of positivism in favor of a Platonic view in which reality, whatever it is, may be more deeply manifested in the mathematical equations and concepts of theoretical physics than the objects, like particles, implied more directly from observations. Many theoretical physicists today lean in this direction and claim the authority of the greatest physicist of the age. However, Einstein's exact views are ambiguous and, besides, the argument from authority went out with the Medieval Church. Einstein might have been wrong on this one.

While positivism is no longer a fashionable philosophy, empirical-minded physicists still continue to insist that we can only give meaning to what can be directly tested by measurement. Nobelist Steven Weinberg, who has written definitive textbooks on both general relativity and quantum field theory, had this to say about the idea that gravity affects the geometry of space and time:

> At one time it was hoped that the rest of physics could be brought into a geometrical formulation, but this hope has met with disappointment, and the geometric interpretation of the theory of gravitation has dwindled to a mere analogy, which lingers in our language in terms like "metric" "affine connection," and "curvature," but is not otherwise very useful. The important thing is to be able to make predictions about images on the astronomers' photographic plates, frequencies of spectral lines, an so on, and it simply doesn't matter whether we ascribe these predictions to the physical effects of gravitational fields on the motion of planets and photons or to a curvature of space and time. (Weinberg 1972, 147)

Weinberg admits that these views are "heterodox," but they at least illustrate that disagreements about the interpretation of general relativity exist even among the top experts. We will see the same situation later with quantum mechanics. Most physicists view the two debates, about the ontological meaning of general relativity and quantum mechanics, much as Weinberg does here—as a waste of time and not really very important. And, we will see that Weinberg himself has since taken what appears to be a more Platonic view in which quantum fields are the primary reality.

Unlike the typical physicist (or, at least as the typical physicist purports), most of us want to do more than just solve problems. We want to "understand reality" as best as we can, to know where we fit into the scheme of things. The highly successful theories of twentieth-century physics, constructed on a solid base of the still very powerful and useful theories of classical physics, should provide us with as good a glimpse of reality as we are likely to find anywhere else. Of course, at bottom we must have logical consistency, and the working theories already demonstrate that. We have seen that invariance is also a prominent notion that offers a clue. Simplicity, beauty, and elegance, though less precisely specifiable and more arguable, are other clues. They were just about all Einstein needed for general relativity. As we progress to a better understanding of even our oldest ideas, we will find them increasingly simpler and more elegant. Often this simplicity is buried in some rather esoteric mathematics, as is the case with Einstein's equation for the metric of space-time, but mathematics is just a language we must learn to be able to articulate precisely these ultimately very simple concepts.

WHICH MODEL SHOULD WE CHOOSE?

Gravity remains an ontological enigma. The beauty and success of general relativity seems to imply a reality of a curved space-time framework to the universe. On the other hand, this curved framework is not at all required, and is indeed a hindrance, for the rest of physics. Furthermore, general relativity is not a quantum theory, and so it must break down at some level. Unfortunately, this level, called the **Planck scale**, is far beyond the reach of current or even conceivable experimentation. The distance involved, 10^{-35} meter, is twenty orders of magnitude smaller than the nucleus of the chemical atom. The typical energy at that scale, 10^{28} electron volts, is sixteen orders of magnitude greater than the energy produced by the largest particle accelerator on earth, and eight orders of magnitude higher than the highest energy cosmic rays ever observed. It will be a long wait before we can directly probe the Planck scale, where quantum effects on gravity become important. So, positivists and empiricists would say, why worry?

At the scale of the elementary particles of the current standard model, gravity is many orders of magnitude weaker than the electromagnetic and nuclear forces that dominate at that level. So, again, why worry?

We still worry. We have two very powerful theories that agree with all observations within their own domains. Yet they are so different that it seems unlikely that they both describe the same, intrinsic reality. If we are forced to make a choice and toss one out, general relativity would have to be the one. It probably does not represent a viable model for a good portion of ultimate reality. As Weinberg points out, the non-Euclidean geometrical picture can be replaced by a Euclidean one in which the gravitational force is described by a gravitational field, in the nineteenth-century sense. Ultimately, that field will likely experience the same fate as befell the electric and magnetic fields. It will be quantized and replaced by a quantum theory in which the gravitational force is mediated by the exchange of gravitational quanta.

Of course, physicists have been thinking about quantizing gravity since they have been in the business of quantization. The hypothetical gravitational quantum is called, unsurprisingly, the **graviton**. Sometime in the 1950s, I heard Richard Feynman lecture about a quantum theory of gravity, with graviton exchange. As I recall, he said that it gave identical predictions to general relativity, including the 43 arcsecond Mercury precession.

However, the theory was plagued by the types of infinities that Feynman had helped to remove in quantum electrodynamics, as we will learn about in a later chapter. Unfortunately, the same techniques did not work for quantum gravity. Furthermore, since the graviton theory made no testable new predictions, which we have seen may require measurements at the Planck scale, no practical reason could be given to prefer the graviton theory over general relativity. By that time, the tensor methods of Einstein's theory were widely used and worked just fine in gravitational calculations. While Feynman's famous enthusiasm was perhaps not justified in this case, for a logically consistent quantum theory of gravity has proven elusive to this day, the current frontier of theoretical physics is being driven forward by the need to encompass gravity within any future quantum framework.

MILNE'S KINEMATIC RELATIVITY

In the 1930s, Edward Milne was able to rederive the Lorentz transformation in a unique way that provides considerable insight into the fundamental character of space and time. He called his theory **kinematic relativity** (Milne 1935, 1948). Unfortunately, the applications of kinematic relativity, particularly to cosmology, never amounted to much, and were rather unjustly derided (for a good discussion, see Bondi 1960, 123–39).

Milne utilized the **cosmological principle** that all observers in the universe are equivalent, that is, they all see the same laws of nature. This is also called the **Copernican principle**, arising as it did out of Copernicus's recognition that the earth is not the center of the universe. Indeed, this idea is closely related to what we called above the principle of relativity, introduced by Galileo and used by one of Einstein's two basic postulates. Whatever the name, by now this is a standard assumption about the universe.

Milne argued that as observers of the world around us, all any of us do is record signals from outside ourselves and send out signals of our own that may or may not be reflected back. With a clock and a flashlight, we can send out light pulses at some repetition rate, or frequency, and measure the arrival times and frequencies of any signals that may have been reflected or transmitted from other observers. We are able to detect signals coming in from any direction, and so can use the radarlike echoes returned from other observers to set up a space-time framework in their own individual reference frames. We have no meter sticks or other devices to measure distances. All we measure are times. Nevertheless, within this framework this is sufficient to determine distances.

Consider our two observers Alf and Beth, as illustrated in figure 3.1. First they must synchronize their clocks. Alf sends out a series of evenly spaced pulses one nanosecond apart that Beth simply reflects back at the same frequency as they are received, as shown in (a). Beth then calibrates her clock

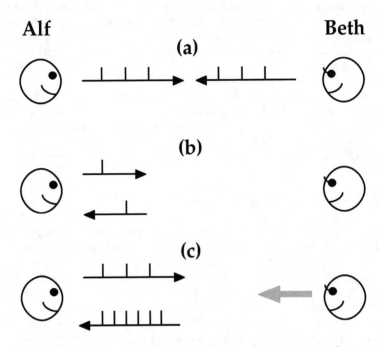

Fig. 3.1. In (a), Alf sends a series of pulses to Beth who reflects them back. Alf measures the same rate he transmitted, so they are defined to be at rest with respect to each other. Beth calibrates her clock to agree with Alf's. In (b), Alf sends a single pulse and measures the time for it to return. This time interval is defined as twice the "spatial interval" between the two. In (c), Alf measures a higher repetition rate for the return pulses and determines that Beth is approaching at some relative velocity.

so it reads one nanosecond as the time between pulses from Alf. The fact that the pulses each receive are returned at a constant repetition rate tells them (that is, defines) that they are at rest with respect to each other.

Alf defines his distance to Beth as half the time it takes on his clock to send a single pulse, as in (b), and receive an echo. He multiplies the time interval by some arbitrary constant c to change the units from seconds to meters or some other familiar unit of distance. Indeed, he can chose c = 1 and measure the distance in light-nanoseconds (about a foot). Beth does the same, so they agree on common time and distance scales. Note that the speed of light is *assumed* to be invariant—the same for both Alf and Beth. The distance from Alf to Beth is simply specified as the time it takes an electromagnetic pulse to go between the two, or half the time out and back. That spatial distance has no meaning independent of the time interval for a returned signal.

Now suppose that, on another occasion, Alf observes the pulses reflected by Beth are received at a higher frequency than those he transmitted, as in (c). He concludes that she is moving toward him along their common line-of-sight. This is the familiar *Doppler effect* and Alf can use it to measure their relative velocity along that direction. If the return frequency is lower, Beth is receding.

This is similar to how astronomers compute the velocity of galaxies, providing the prime evidence that the universe is expanding. They observe that the spectral lines from most galaxies are shifted toward lower frequencies. This **red shift** indicates the universe is expanding as farther galaxies exhibit greater red shifts. Although they do not use radar-ranging in this case, the principle is the same.

Milne showed that you get the Lorentz transformation when you relate the distances and times measured in two references frames moving with respect to one another. And once you have the Lorentz transformation, the rest of special relativity falls into place. This should not be too surprising, since he assumed the invariance of the light signals being sent back and forth.

Now you might ask, what if instead of electromagnetic pulses Alf had sent out low-speed particles like golf balls? You would run into trouble with the distance scale. For example, as shown in figure 3.2, Alf and Beth are moving apart and another observer, Cal, passes Beth moving in the opposite direction toward Alf. At the instant Beth and Cal pass one another they each send a golf ball to Alf. The golf ball moves at the same speed in their reference frames. However, Beth's golf ball is moving slower with respect to Alf than Cal's, since she is moving away and Cal is moving toward Alf. Alf then measures different distances to Beth and Cal.

If golf balls were used instead of light pulses for signalling between observers, spatial separations would then depend on the relative speeds of objects. Note that this is not the same as the Lorentz-Fitzgerald contraction. There we were talking about the relative distances measured in two reference frames moving with respect to one another. Milne's theory gives the correct result in that case. Here we have a single reference frame, Alf's, with two different distances being measured to what we would logically view as a single spatial point, the common position of Beth and Cal when they send off their golf balls.

I suppose the universe could have been that way, which would have made it very complicated indeed. Fortunately, we can define distance independent of relative velocity by using light pulses, that is, photons, since they move at the same speed for all observers. Put another way, if we had lived in a universe with no massless particles like photons, we would have had a devil of a time making sense out of it.

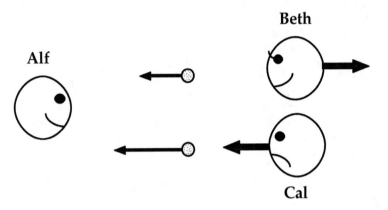

Fig. 3.2. Beth and Cal are at the same distance from Alf when they each send a golf ball to him. Since Beth is receding, her golf ball is moving slower than Cal's and so reaches Alf later. Alf concludes that Beth is farther away than Cal. This does not happen with photons since they always travel at the same speed.

One criticism of Milne's scheme was that we do not, in practice, measure the distances to stars and galaxies by bouncing light signals off them. But, as Hermann Bondi pointed out (1960, 127), we do not lay out rigid rulers to measure the distance either, and that was for many years the conventional definition of distance. Not any more. According to international agreement, the distance between two objects is now defined as the time it takes light to travel between them. The standard unit of time is the second, taken as 9,192,631,770 periods of oscillation of the electromagnetic energy emitted in a specific atomic transition of the isotope ^{133}Cs. The meter is then defined as the distance light travels in a vacuum in $^{1}/_{299,792,458}$ of a second. As for radar-ranging, the distance to the moon has been measured to within centimeters by bouncing laser beams off it. Milne's definition of space is now de facto and the constancy of the speed of light (or at last some universal speed c) is imbedded in that definition.

Parenthetically, note that the speed of light is now regarded as a constant by the very definition of distance. Occasionally one hears proposals in which the speed of light is a variable. Such proposals make no sense under the current convention and would require a new operational definition of distance.

If we insist on using meters for distance and seconds for time, the speed of light $c = 3 \times 10^8$ meters per second. But, as we have seen, we could just as well have decided to work in units of light-nanoseconds for distance, in which case $c = 1$. We must keep this in mind, and think in terms of a world where c = 1. This greatly simplifies our attempts to understand many relativistic ideas.

FOUR DIMENSIONS AND ARCHIMEDES POINT

In 1907, Hermann Minkowsky had shown how special relativity could be formulated in terms of a four-dimensional continuum in which the three dimensions of space, usually denoted by the cartesian axes, are joined by time as the fourth axis or "fourth dimension" (today we conventionally call it the "zeroth dimension"). An event occurring at (x,y,z) in space at a time t can be thought of as a point in **space-time**, (x,y,z,t). The motion of a body can be described as a path in space-time called a **worldline**. These are easiest to see in two dimensions, as illustrated in figure 3.3, where the body moves along the x-axis.

The spatial distance between two events depends on your frame of reference. For example, suppose you get into your car at home at 8 A.M. This is Event A. You arrive at your work place 15 miles straight down the road at 8:30 A.M. This is Event B. The distance between the events is 15 miles in a reference frame fixed to the earth, but *zero* in a reference frame fixed to your car. This is why we can call the separation between A and B "local." They are at the same place in at least one reference frame. Any two space-time points that can be connected by a signal moving at the speed of light or less is local.

In this example, since your speed relative to the earth was far less than the speed of light, clocks in the two reference frames will read an identical thirty minutes—unless they happen to be incredibly precise atomic clocks. In that case, they will read differently by about 2 picoseconds, that is, two-trillionths of a second.

The space-time picture in which time is a fourth coordinate implements the dramatic change in the status of time under relativity. Time moves from being an absolute number that is the same for all clocks in the universe to becoming another coordinate along an arbitrary axis. Just as the spatial coordinates of events depend on your choice of spatial axes, being different when those axes are moved from one origin to another or rotated so they point in some other direction, so time depends on your choice of time axes. Indeed, invariance under Lorentz transformations expresses nothing more than that fact that every "direction" in four-dimensional space-time is no different from any other.

Archimedes claimed that he could move the earth from a vantage point where it looks like a simple pebble, if he were provided with a suitable lever. We can imagine a modern version of Archimedes point in which an observer looks down on space-time (see figure 3.3). Archimedes point is not a place

Archimedes

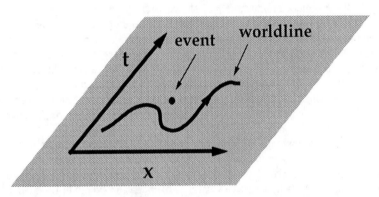

Fig. 3.3. Indicated are one of the three spatial axes, x, and the time axis, t. An event is a point (x,t) in space-time. The motion of a body is described by a worldline. In the modern Archimedean perspective, the observer watches from outside space-time, having the "view from nowhere" and the "view from nowhen."

in space but looks down on space from what philosopher Thomas Nagel (1986) called the **view from nowhere**. It is also not a point in time, but looks down on time as well from what I will call, following philosopher Huw Price (1996), the **view from nowhen**.

In modern relativistic kinematics, we also extend our notion of momentum to four dimensions. We take the total energy (kinetic plus rest plus potential) of a particle to be the fourth component of a "four-momentum space" in which the first three components are those of the particle's three-dimensional vector momentum.

In advanced classical mechanics, a deep connection is drawn between position and momentum. The state of a particle is defined by its position in an abstract **phase space** of six dimensions comprised of momentum and spatial coordinates. This gets extended in relativity to a phase space of eight dimensions: four coordinates of time and space, and four coordinates of energy and momentum. Each coordinate is paired with what is called its **conjugate momentum**. The energy is then interpreted as the momentum conjugate to the time coordinate.

We will get back to this when we discuss the connection between symmetry principles and conservation laws discovered in modern physics. Even in classical physics, however, when the description of a system does not require the use of a particular coordinate the corresponding momentum is conserved. By the same token, energy is conserved when time is an "ignorable" coordinate, in the language of classical physics.

Time, however, is not just another coordinate. Milne's discussion illustrates that time is in some ways more primitive than space. As mentioned, an international committee used a clock in defining the units by which we measure both time and space. Every observer can have a clock, even if that clock is simply one of her atoms. A single particle can constitute a clock; the unit of time could be defined as the period of rotation about the particle's axis. An elementary string can oscillate along its length, with the period of oscillation used to mark time. In short, clocks are elementary units from which we can build up the rest of physics.

As mentioned, the time that is measured on a clock in the reference frame in which that clock is at rest is called the proper time. The measured time interval for a clock perceived to be in motion will be longer than the proper time. This is time dilation. Since the clock does not move in its own references frame, its spatial coordinates are fixed and can be always taken to be zero; that is, the clock always sits at the origin of its own spatial coordinate system. The distinction between proper time and coordinate time, where the latter is what is measured in a reference frame that is moving relative to the first, is an important one. Proper time is an invariant, coordinate time is not.

Similarly, the mass of a body, or system of bodies, is an invariant. Energy and momentum, however, are coordinates that will be different in different reference frames.[4]

We can always use the Lorentz transformation to go between reference frames, so coordinate space, time, energy, and momentum are still meaningful concepts. They simply do not carry with them the same ontological weight as proper time and mass. That is, when we are seeking those concepts in science or elsewhere that might be more than simple human inventions but elements of reality itself, the best candidates would seem to be those that have a universality that transcends relativity and human subjectivity. This is why the invariants of physics serve as exemplary candidates for elements of reality. They are the same no matter who measures them, in whatever reference frame they happen to abide.

The relativity of coordinate time, now well established by almost a century of increasingly precise measurements, has deep philosophical implications. No pre-twentieth-century thinker, with the possible exception of Leibniz, even dreamed of this possibility. All cast their philosophies and theories within a universe governed in terms of the absolute time of common sense, and many still do. But, like the flat earth, this is another place where common sense has led people astray.

As Einstein discovered, the concept of a universal "now" is meaningless. Two events that are simultaneous in one reference frame are not so in all reference frames. What is "now" depends on the reference frame. Each reference frame has a unique present, and thus a unique past and future (under certain circumstances that we will discuss), but these are not universal. As one of many consequences that follow from the insight of Einstein, the notion of an entity "coming into being" is rendered meaningless within the framework of relativistic time (Grünbaum 1964, 1971). If you wish to observe the universe objectively, you can only do that with the view from nowhere/nowhen.

This is not to say that temporal concepts are subjective. All observers at rest with respect to one another (as defined in the Alf-Beth sense) will objectively measure the same proper time intervals between events. Those at the same place will agree on two events being simultaneous. Relativity is not synonymous with subjectivity. But any concept that is relative, like any concept that is subjective, cannot be universal. Only objective invariants are universal. Quantities such as the proper time and proper mass are objective invariants. While a clock ticks off the proper time in its rest frame, the proper time of a clock in other reference frame can be measured by correcting for time dilation. A similar correction can be made to determine a particle's proper mass from another frame with respect to which the particle is perceived to be moving. Other invariants can be found in physics. Since these invariants are both objective and universal, they are the best possibilities we can imagine for the stuff of objective reality.

NOTES

1. If Δx is the distance between two events measured in a given reference frame, and Δt is the time interval measured in the same reference frame, then the proper time interval, defined as the time interval measured in the reference frame in which the two events occur at the same place, is given by $\Delta t_p^2 = \Delta t^2 - (\Delta x/c)^2$.

2. Technically, what we call the speed of light in relativity is in principle a limiting speed. So, if photons were someday found to have a small mass, they would not move at this limiting speed and we will have to call it something other than the speed of light. However, massless photons are strongly predicted by fundamental theory and the current experimental limit is exceedingly small, less than 10^{-20} gram.

3. The "uniform" requirement is to avoid the fact that, in principle, you could tell if you were sitting on earth be observing how falling objects are not parallel but converge on a point, namely, the center of the earth.

4. The operational definition of the (proper, invariant, rest) mass of a system of one or more bodies is given by $m^2 = (E/c^2)^2 - (p/c)^2$, where E is the total measured energy of the systems and p is the magnitude of the total measured (vector) momentum.

4 THREE ARROWS OF TIME

The Moving Finger writes; and having writ,
Moves on: nor all your Piety nor Wit
Shall lure it back to cancel half a Line,
Nor all your Tears wash out a Word of it.
The Rubôayôat of Omar Khayyôam
Translated by Edward Fitzgerald (1953)

THE THERMODYNAMIC ARROW

In the 1870s, Boltzmann used statistical mechanics to prove his **H-Theorem** in which a quantity H, equivalent to negative entropy and to the modern quantity of **information** used in information theory, is shown to reach a minimum when a system of particles achieves thermal equilibrium. That is, the entropy naturally moves toward its maximum, toward greater disorder and less information.

The H-theorem looks very much like the second law. Thermodynamic systems initially out of equilibrium will, if left alone, eventually reach this condition which is characterized by the system having a uniform temperature throughout. Thus, two bodies of different initial temperatures will, after being placed in contact, gradually reach a common temperature someplace in between the two starting values. The reverse process in which two bodies at the same temperature acquire different temperatures does not happen in an isolated system.

Yet from a particle mechanics perspective, such a process is not forbidden. Energy can be transferred by molecular collisions in such a way as to increase a warmer body's temperature while lowering that of a colder one. That is, the system can just as well move away from equilibrium and not violate any of the principles of particle mechanics such as energy and momentum conservation. Indeed, on the microscopic scale a system that is macroscopically in some average state of equilibrium actually fluctuates about that equilibrium, although the average fluctuation is small when the number of particles is large. The H-theorem applies to systems initially far from equilibrium and deals only with average behavior.

Irreversibility seems to be associated with the many body systems of everyday experience. A simple illustration of this is air in a room. That air is composed of individual molecules of nitrogen, oxygen, carbon dioxide, and a few other substances that move around pretty much randomly. Their average kinetic energy is given by the temperature.[1]

Suppose you have a closed room full of people. Somebody decides to open the door. At that instant, all the air molecules just happen, by chance, to move out the door. As the air then rushes out of the room, everyone inside explodes and dies. Is that possible? Very definitely yes! No known principle of the mechanics of particle motion forbids it. Is it likely? Very definitely no. The probability that all the molecules are moving in the direction of the door when it opens is minuscule, and therefore not likely to happen even once on earth during the planet's entire existence.

But what about a room with just three molecules, as illustrated in figure 4.1? The probability that the molecules randomly fly out the door is not at all small. Suppose someone films three of the many molecules in the air outside an evacuated chamber moving inside when the chamber is opened. Without telling you, however, he runs the film backward through the projector so it looks like what we see in figure 4.1. Could you say for sure the film was playing in reverse? You would not bet your house on it.

On the other hand, suppose you are shown a film in which 10^{25} molecules in a chamber rush out the opening to the outside air, leaving a vacuum behind? In this case you could safely bet your house that you are really viewing a film of outside air rushing in to fill a vacuum that is being run in reverse through the projector. Irreversibility seems to hold true when there are many particles, while it is absent when there are only a few.

Clearly, it cannot be a fundamental law of physics that processes involving N particles or fewer are reversible, while those with greater than

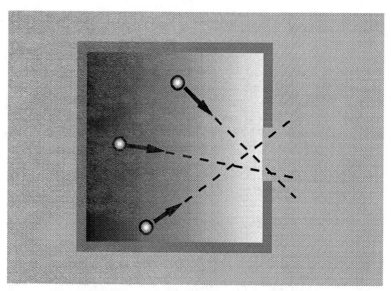

Fig. 4.1. It would not be surprising to see the air molecules in a room fly out the door if there are only three air molecules. This is highly unlikely since there is a much larger number of molecules in a normal room, but not impossible.

N particles are irreversible. What is N? 100? 10,000? 3,486,756,493? For the second law to be fundamental, it would have to apply for any value of N. As far as we can tell, it does not.

The second law is a statistical asymmetry and not a deterministic law of particle behavior. The air in a room is not forbidden from emptying out when a door is opened, killing everyone inside. A broken glass can reassemble and a dead man can spring to life if randomly moving molecules just happen to be moving in the right direction. However, these events are extremely unlikely, since our macroworld is composed of huge numbers of particles.

How, then, did Boltzmann derive irreversibility given the underlying reversibility of particle mechanics? Actually, he did not. As Price (1996, 26) shows, Boltzmann built irreversibility into the derivation of the H-theorem with his "assumption of molecular chaos." Particles in the system are assumed to be uncorrelated at the start; that is, they have random velocities *initially*. They are, however, not assumed to be randomly uncorrelated *finally*. This treats time asymmetrically to begin with, and so it is no surprise that asymmetric time comes out.

All the H-theorem really proves is that equilibrium is the most probable condition of an otherwise random system left to its own devices, and that this is a condition of maximum entropy. Regardless of time direction, a system well away from equilibrium will tend to move toward it. This was important, but it was not a proof of the second law. Price points out that Boltzmann's treatment was also useful in enabling us to shift the problem from why the second law is asymmetric in time to why we have molecular chaos in one time direction and not the other, or why the entropy is lower in one direction than the other. As we will see in the following sections, the same problem arises in the consideration of the cosmological, radiative, and quantum arrows of time. The issue is not why the universe, according to the second law, must move from a condition of greater to lesser orderliness, but why, when we consider the systems of many particles existing in our universe, one extreme of the time axis has a vastly different orderliness than the other.

In 1928, Arthur Eddington introduced the expression **arrow of time** to represent the one-way property of time that has no analogue in space (Eddington 1928, 68). He associated the arrow with an increasing random element in the state of the world, as expressed by the second law of thermodynamics. Eddington recognized that molecular motions were intrinsically reversible, but that they tend to lose their organization and become increasingly shuffled with time. However, he did not address the basic issue of why the molecules were organized in the first place. He regarded the second law as a fundamental principle of the universe, saying it occupied "the supreme position among the laws of Nature" (Eddington 1928, 74).

Today the conventional view of physicists is that the second law amounts to little more than a definition of the arrow of time as the direction of increasing entropy in a closed system. Open systems, that is, systems that exchange energy with their environment, can gain or lose entropy. Organization is viewed as a lowering of the entropy of a system. It requires outside energy to do the ordering and a corresponding increase in entropy of the outside system that is providing that energy. So, as the sun provides energy for ordering processes on earth, the entropy of the universe as a whole gains the entropy that the earth and sun lose.

But why is the universe's entropy so far below maximum that it can absorb some from the stars and their planets? Furthermore, another question remains: why is the arrow of time for all closed systems the same? If they are closed, they have no knowledge of each other and you would expect half to have the same arrow as the universe and half to be opposite. I will suggest answers below.

THE COSMOLOGICAL ARROW

We have now explored the sky by means of a wide variety of telescopes and other instruments that measure the full range of the electromagnetic spectrum, from radio waves to gamma rays. These observations uniformly agree on a picture of a hundred billion or so galaxies, each containing typically a hundred billion or so stars, flying away from one another as from an explosion that is dubbed the **big bang**. According to current best estimates, the big bang began about thirteen billion years ago.

The average mass density of matter in the universe appears to be insufficient for gravity to eventually pull the galaxies back together again into a "big crunch." As far as we can tell from current observations, the universe is likely to keep expanding forever. It would seem that this eternal expansion alone selects out one time direction for a **cosmological arrow of time**.

To illustrate the huge time asymmetry that exists on the cosmological scale, let us compare the entropy of the universe at three times: (1) the beginning of the big bang, (2) the present, and (3) the far future.

What was the entropy of the universe at the beginning of the big bang? If we push back to its very earliest periods, the first moment we can still describe even crudely in terms of known physics is the **Planck time**, $t_{PL} = 10^{-43}$ second after the moment $t = 0$ that we normally identify with the origin of the universe. At the Planck time, the universe, according to our best current cosmological theories, would have been confined to a sphere of radius equal to the **Planck length**, $R_{PL} = 10^{-35}$ meter. Note that the Planck length is just the distance travelled by light in the Planck time.[2]

The Planck length is the distance below which the domains of general relativity and quantum mechanics finally overlap and can no longer be treated separately. Since we do not yet have a viable theory of quantum gravity, the physics at times less than the Planck time is unknown and highly speculative. Perhaps there is no physics—for a good reason. The quantum uncertainty principle (see chapter 5) says that the uncertainty in the mass confined in that region is $M_{PL} = 10^{-5}$ grams, the **Planck mass**. General relativity says that an object whose mass is confined to this region of space will be a variety of black hole. And, as the saying goes, "black holes have no hair." That is, they have no discernable structure—no physics, chemistry, biology, psychology, or sociology.

Since we cannot see inside a black hole, we have no information about it, and so it has maximum disorder or maximum entropy. The entropy of a black hole of radius R is simply $S_{BH} = (R/R_{PL})^2$.[3]

In the case of a Planck-sized black hole, $R = R_{PL}$ and $S_{BH} = 1$. That is:

The entropy of the universe at the beginning of the big bang was S = 1.

What is the entropy of the universe today? Penrose (1989, 342) argues that this entropy should be dominated by black holes, which, as we just saw, have maximum entropy. He estimates that the entropy of the universe today is of the order $S = 10^{100}$ or 10^{101}, for reasonable assumptions on the number and size of black holes. The first figure is close enough. And so:

The entropy of the universe today is $S = 10^{100}$.

What will be the entropy of the universe far in the future? Penrose (1989, 343) considers the possibility of a big crunch and assumes that the final state of the universe in that case will be a single black hole. However, he does not take this to be another Planck-sized black hole like the one I have used to represent the universe at the Planck time. That is, the big crunch is not simply a time-reversed big bang. Rather, the black hole at the end of the big crunch is taken by Penrose to contain all the current matter in the universe, whose entropy he estimates to be 10^{123}.

As mentioned above, a big crunch now seems unlikely and we expect the universe to expand forever. Its entropy will continue to increase, ultimately approaching (if never exactly reaching) a maximum. But as we have seen, that maximum is equal to the entropy of a black hole containing the same amount of matter. So the ultimate entropy of an ever-expanding universe will be the same as Penrose estimated for the big crunch, provided we assume the same amount of matter. While we have good reason to question that the number of particles will remain the same order of magnitude in all that time, that number is not likely to reduce to one. Thus, for our purposes:

The entropy of the universe in the far future will be of the order $S = 10^{123}$.

Penrose argues that the huge entropy discrepancy at the two extremes of the time axis indicates the need for a new law of physics. He believes that quantum gravity is about the only place left to look for such a law. He calls his proposed law the **Weyl Curvature Hypothesis** (Penrose 1989, 345). Weyl curvature refers to the tidal portion of the curvature of space-time in general relativity, the part of that curvature that is present even in an empty universe. Penrose's hypothesis holds that this curvature is zero at the initial "singularity" that produced the universe, but not at any other singularities, where we can take "singularities" here to refer to black holes.

We have seen that the initial entropy of the universe was very low, as low as it can possibly be. The final entropy, if Penrose's calculation is correct, will be 123 orders of magnitude larger. But note: The initial entropy was *also* as large as it could have been, since it was also the entropy of a black hole. Thus, the universe has maximum entropy at the two extremes on the time axis. In each case, the universe is in equilibrium. At each time, the universe is in a state of total chaos. This is a point that has been missed by almost everybody, including Penrose. The universe will not only end in complete chaos, it also began that way!

Which extreme of the universe is more highly ordered? The one with unit entropy, or that with entropy 10^{123}? If you simply equate disorder with the absolute entropy, then you would conclude that the early universe carried far greater order than that of today or the future. But, both extremes are black holes, and black holes have maximum entropy with no room for order. Both are equally disordered.

Consider order formation on earth. The earth is an open system that receives energy from the sun and radiates energy back into space. In the process, both the sun and earth lose entropy, while the rest of the universe gains entropy. Let's put in some numbers here.

The average surface temperatures are about 6,000K for the sun and 300K for the earth. Thus the earth radiates 6,000/300 = 20 infrared (IR) photons for each visible photon it receives from the sun. Taking the entropy to be given by the number of particles, the universe gains 20 units of entropy lost by the earth for every photon the earth gets from the sun. Now, the earth absorbs 2.5×10^{36} photons from the sun each second, so in the four to five billion years of the earth's existence the entropy of the universe has increased by about 10^{54} as a result of ordering the earth.

Now, where did those 10^{54} units of entropy go? They were distributed to other matter in the universe. For example, the IR photons might have collided with the 3K microwave photons in the cosmic microwave background that is everywhere. After many collisions, equilibrium would be reached and the temperature of the background would have risen. However, since the background in the visible universe contains about 10^{88} photons, the temperature increase was negligible. That is, the microwave background would have had no trouble absorbing the entropy from the earth in its lifetime, nor that of all the 10^{23} or so other planets in the visible universe. So, we can rest comfortably that the current universe has plenty of room left for order to form, by at least ten orders of magnitude.

On the other hand, when we consider the two extreme times when the universe already has all the entropy it could hold, then no order can form. This was the case at the Planck time. Then the universe had only one unit of entropy and so was indeed as ordered as it could be. But it was also completely without order.

Penrose (1989, 343) argues that the early universe was considerably unlikely. He shows a drawing of "the Creator" pointing his finger to the tiny region of phase space, selecting one universe out of the $10^{10^{123}}$ universes that can be formed from the current matter of the universe.[4]

This has become one of the so-called anthropic coincidences that have been used by theists in recent years to argue that the universe shows evidence for intelligent design, with life and humanity as the purpose. Penrose (1989, 354) notes that the entire solar system and its inhabitants could have been created more "cheaply" by a selection from only $10^{10^{60}}$ universes, so an anthropocentric conclusion is hardly justified. But he still thinks something special happened, that the beginning was not just a random shot.

Did the hypothetical Creator really have $10^{10^{123}}$ choices in creating the universe? Not if the universe really was a Planck-sized black hole at the Planck time. As we have seen, the entropy in that case was unity and thus the phase space contained a single cell. The Creator in fact had no choice where to poke her finger! If there ever has been any external creative input to the universe, it must have happened after the Planck time.

I see nothing that prevents us from viewing the cosmological and thermodynamic arrows of time as being identical. In each case we have systems where entropy increases along one time direction and decreases along the other. We then arbitrarily choose the positive time axis to point in the direction of increased entropy in a closed system. Of course, we still have to understand the source of this entropy gradient. We will discuss that later. But first, a loose end needs to be tied up.

In the previous section I raised the issue of why all the closed systems in the universe, presumably out of contact with one another, agree with each other on their respective arrows of time. I think the reason is simply that they are in fact not out of contact. What we call closed systems are really, except perhaps for the universe as a whole, only approximately closed. A little heat will leak through any insulator. Even the "empty space" between stars is not empty but contains microwaves and other radiation. Furthermore, all these almost-closed systems are products of the same original big bang. They are in fact strongly correlated with one another and should be expected to share time's arrow with the universe as a whole.

We still have to explain how we can have an overall entropy gradient in a time-symmetric universe. Paul Davies (1993) has suggested that **inflation** in the early universe provides the way out. Price (1996, 85–86) argues that Davies is also applying the double standard of assuming a time direction to begin with and then having it appear as a result. However, I think Davies' basic idea can be made to work. The trick is to maintain underlying time symmetry while allowing "localized" violations.

Before we see how this can come about, I need to take a moment to describe the basic features of the inflationary universe.[5] More details are given in chapter 13. The idea of inflation is generally attributed to Alan Guth (1981). Demosthenes Kazanas (1980), published an earlier, less comprehensive, and largely unacknowledged version that had all the basic ingredients of inflation. Guth did go further in realizing the full implications of the idea. Andre Linde (1982) also seems to have had the idea independently and made important early contributions that are more widely recognized.

In the following, I will present the standard description of inflation in terms of general relativity. It should be noted, however, that general relativity is not a quantum theory and ultimately a quantum description will have to be presented. Later in the book I will indicate how the picture changes from a quantum mechanical viewpoint and in the light of very recent new data.

Einstein's equations of general relativity can be applied to a universe empty of matter and radiation. In that case the curvature of space is specified by the quantity known as the **cosmological constant**. When that constant is positive and nonzero, the universe undergoes a very rapid, exponential inflation during its first fraction of a second. Inflation can also result with zero cosmological constant if an energy field exists with negative pressure. These possibilities are hard to tell apart.

When first proposed, inflation provided a natural solution to a number of outstanding problems in cosmology. It explained why the geometry of the visible universe is so close to being Euclidean, the so-called flatness problem. With inflation, the universe within our horizon is like a tiny patch of rubber on the much larger surface of an expanding balloon. Note the implication that the visible universe, all 100 billion galaxies arrayed over 13 billion light years or so, is just a tiny portion of what emerged from the original explosion. Much, much more lies outside our horizon. What it is we will never know. But don't fret; it's probably more of the same since it is all from the same source.

Inflation also provided an explanation for another puzzle called the "horizon problem." When we look at the **cosmic microwave background**, we measure the same temperature (2.7 degrees Kelvin) and spectrum (pure black body) in all directions to four or five significant figures. Yet some of those regions, according to the old big bang theory, had to be out of causal contact in the early universe. That is, they never could have interacted with one another to achieve the thermal equilibrium implied by having the same temperature. Inflation puts them back into causal contact, and thermal equilibrium, in the early universe, along with all that other stuff beyond our horizon as well.

The level of anisotropy in the cosmic microwave background, where the temperature is slightly different in different directions, provided a critical test for the inflationary model. At some point the theory required a small anisotropy. Otherwise it could not be made to agree with the observed anisotropy of matter, clumped as it is into highly localized galaxies and stars. If not observed at some point, the inflationary model would be falsified.

Instead, inflation passed this test with flying colors. The expected anisotropy on the order of one part in 100,000 was confirmed by the COBE satellite and later observations. Increasingly precise data have continued to support inflation and rule out alternatives.

Still, inflation soon met with other problems not at all unusual in the early history of most theories that ultimately prove successful. New and better estimates of the average density of matter in the universe, including the still-unobserved **dark matter**, were too low to give a universe so extremely flat as inflation requires. Some opponents of inflation announced gleefully in the science media, which tends to overhype most cosmology stories, that inflation was dead. This was reminiscent of earlier claims, also hyped by the media, that the big bang itself was dead (Lerner 1991; see my review in Stenger 1992). But just as the big bang survived this earlier onslaught, so inflation may have been quickly rescued by the facts. Although the jury is still out, recent independent observations of distant supernovae indicate that the universe is accelerating, that is, "falling up"! This implies that the universe has a residual nonzero, positive cosmological constant or other form of **dark energy**, as will be discussed later. The best fit to all the data is still provided by the inflationary model supplemented by a cosmological constant, with no alternative coming close. Inflation is still alive and well.

As we will see in chapter 13, inflation provides us with a natural scenario for the creation of the universe "not by design" but by accident. The idea that the universe started as a random quantum fluctuation was suggested by Edward Tryon (1973). At that time, he had little theoretical basis for his momentous proposal. Then inflation came along to make the idea more plausible. Today we own sufficient knowledge to speculate rationally that the universe originated as an energy fluctuation, one of countless many, in a primordial background of empty space-time. This energy first appears not in particles or radiation, but in the curvature of space leading to inflation. That is, in the language of general relativity, the universe starts as a curvature fluctuation. All this is allowed by existing physics knowledge. (For a not-too-technical presentation of the basic physics, see Stenger 1990a). Since this is a chapter on time reversibility, let me try to describe what happens in a way that does not assume any particular direction of time, and in that way return to Davies' suggestion that inflation can account for the entropy asymmetry.

Suppose that the primordial fluctuation occurs at an arbitrary point on the time axis we label $t = 0$. The fluctuation is equivalent to giving the otherwise empty universe a cosmological constant, as allowed by the general theory of relativity discussed in the last chapter, or an equivalent field. Though empty of matter at this time, the universe exists in a state called the **false vacuum**, which contains energy (provided by the quantum fluctuation) that is stored in the curvature of space, sort of like the potential energy of a stretched bow (the de Sitter universe). The energy density is proportional to the cosmological constant. Einstein's equations then yield an exponential inflation e^{Ht} in which the exponent H is proportional to the square root of the energy density. In fact, H is just the **Hubble constant** for that epoch, the ratio of the velocity at which two bodies recede from one another as the result of the expansion of the universe divided by the distance between them.[6]

Now, the square root of a number can be negative just as well as positive. Thus, technically, the solution of Einstein's equations must contain a term with a negative Hubble constant as well. This will lead to exponential deflation instead of inflation on the positive side of the t-axis. The deflation can be simply neglected since it will be quickly overwhelmed by the inflationary term.

However, let us ask what happens on the negative side of the t-axis. There the positive Hubble constant term will deflate and become negligible, while the negative Hubble constant term will inflate. Thus, we get a completely time-symmetric inflation on both sides of the t-axis: two universes (really the same universe), one going forward in time and one going backward, as shown in figure 4.2.

Inflation ultimately leads to the entropy-producing processes that, as we will see later, in the early stages of the big bang could have eventually produced the universe as we know it. This entropy increase is associated with the formation of particles as the small energy that constituted the original quantum fluctuation expanded exponentially (without violating the first law of thermodynamics—see Stenger 1990a) and that greatly increased energy is converted into particles. In the time-symmetric picture suggested here, both sides of the time axis experience the entropy increase we characterize with time's arrow pointing in that direction. In this manner we obtain a cosmological time-asymmetry in an otherwise time-symmetric reality. However, since much of what happens during entropy generation is random, we would not expect the two universes to resemble one another for very long. The time symmetry, as we say, gets spontaneously broken.

In short, our current theories in cosmology are perfectly consistent with a reality having an underlying time symmetry and a cosmological arrow of time equivalent to the thermodynamic one, chosen to be the direction at which the global entropy increases. Current theories of the early universe explain entropy generation and particle production processes by which the matter and forces that populate our universe took form.

THE RADIATION ARROW

A third suggested arrow of time is provided by radiation. Like the other equations of physics, Maxwell's equations show no preference for the direction of time. They allow for electromagnetic waves that propagate backward in time as well as forward. The usual solutions are called **retarded**, arriving

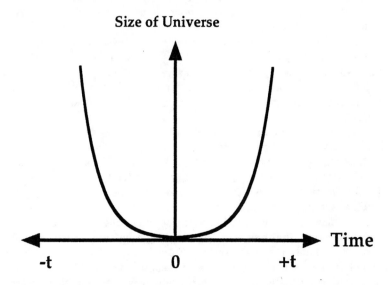

Fig. 4.2. The time-symmetric inflationary universe. Starting at t = 0, the universe under-goes a short period of exponential expansion on both the +t and −t side of the time axis, where time's arrow runs away from zero on both sides.

at the detector after they left the source. Solutions which arrive at the detector before they leave the source are called **advanced**. They are elimi-nated in practice by asserting, as a "boundary condition," the fact that they are not observed.

This apparent asymmetry is identified as the **arrow of radiation**. Like the other of time's arrows, this has been fully analyzed by Price (1996, chapter 2). Earlier analyses can be found in the books and articles by Davies (1974, 1977, 1983, 1995) and the book by Dieter Zeh (1989, 1992). These references can also be consulted for further discussions on the other arrows of time.

In 1956, philosopher Karl Popper wrote that the simple observation of water waves provides evidence for a temporal asymmetry other than the thermodynamic variety (Popper 1956). Toss a rock in a pond and you will see circular waves radiating outward. The time-reversed process of waves converging on a point is never observed.

Davies (1974) and Zeh (1992) both disagreed, arguing that the radia-tion arrow follows from the thermodynamic one, which is basically statistical. In principle, waves could be generated around the edge of the pool resulting in a converging wave front, but this would require coherence all the way around, which is statistically very unlikely. But, that is just what the thermodynamic arrow is all about—the low probability of certain phenomena to be seen running in reverse direction from which they are normally observed.

Price, however, criticizes these arguments as again applying a double standard by assuming that we only have diverging but no converging radiation in nature. Looking in reverse time we see converging waves and we have to explain why we see no diverging ones. The problem, in other words, is not with the convergence or divergence but with the highly special circumstances that exist in the center of the pool where the rock hit the water. As was the case for the thermodynamic and cosmological arrows already discussed, in order to explain the evident asymmetry we have to explain why, in our world, we have these special regions where entropy is exceptionally low.

Price traces the radiative arrow to the difference between sources and absorbers in the macroscopic world. Microscopically we see no difference. Speaking in classical terms (which sometimes can be applied to a microscopic domain—as long as it is not too microscopic), an oscillating charge is a source of a coherent electromagnetic wave that propagates through space and sets another charge at a different location oscillating with the same frequency. This is indistinguishable from the time-reversed process in which the second particle is the oscillating source and the first the receiver.

Macroscopically, coherent sources of radiation, whether water or electromagnetic waves, are far more prevalent than coherent absorbers. The rock dropped in the pool sets up a coherent wave in which many atoms in the water are set oscillating in unison. The atoms in the wall around the pool are rarely oscillating in unison so that they can emit a single coherent wave that converges back on the original source. Light detectors, such as the photomultiplier tube that I will discuss in detail in chapter 8, are never used as sources. Lasers act as sources of coherent light and are never used as detectors, although the original maser devices on which they were based were detectors. This, in fact, illustrates the major theme of this book—the time symmetry of quantum phenomena.

When we look at phenomena at the quantum level, the distinction between source and absorber disappears. A photon emitted by a quantum jump between energy levels in an atom can travel in a straight line and excite another atom in a nearby detector. This process can be readily reversed, with the second atom de-exciting and emitting a photon that goes back along the same path re-exciting the source, as illustrated in figure 4.3. At the quantum level, then, the processes of emission and absorption are perfectly reversible.

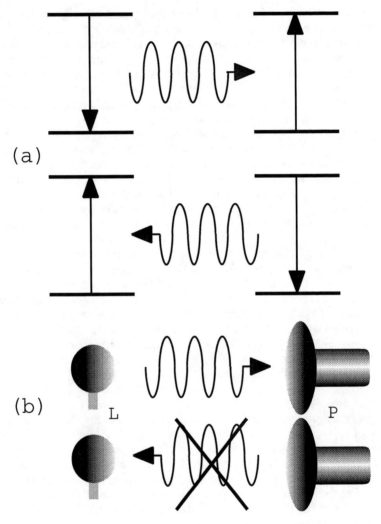

Fig. 4.3. (a) The basic quantum process of emission and absorption, a transition between energy levels in an atom is reversible. (b) Most macroscopic sources and detectors, like lamps L and photodetectors P, are irreversible.

As with the cosmological arrow, after careful analysis we find that the arrow of radiation is indistinguishable from the arrow of thermodynamics. Again it represents an arbitrary choice we make based on the fact that most macroscopic; that is, many-body processes exhibit an entropy gradient that arises from the large contribution that randomness makes to these phenomena. We can conceive of other similar arrows, such as the *arrow of evolution*, derived from an entropy gradient that results from the strongly random component within natural selection. And, for the reasons discussed previously, the evolutionary arrow can be expected to coincide with the thermodynamic and cosmological arrows.

In the following chapters we will consider in some detail the arrow of time that seems to be associated with quantum phenomena and show how that, too, is an artifact. Furthermore, we will see how time reversibility at the quantum scale makes it possible for us to understand some of the puzzling features of quantum mechanics. Indeed, quantum mechanics seems to be telling us that the universe has no intrinsic arrow of time.

NOTES

1. More precisely, absolute temperature T is defined by the average kinetic energy K of the N molecules in a body in thermal equilibrium: $K = 3NkT/2$, where k is Boltzmann's constant, which simply serves to change the units of absolute temperature (usually degrees Kelvin) to units of energy such as Joules or electron-volts.

2. Within a sphere of radius equal to the Planck length R_{PL}, the de Broglie-Compton wavelength of a particle equals the circumference of a black hole of the same mass: $2\pi R_{PL} = h/mc$. For a black hole, the rest energy equals the potential energy: $mc^2 = Gm^2/R_{PL}$, where G is Newton's constant. In natural units, $\hbar = c = 1$, $R_{PL} = t_{PL} \sqrt{G}$ and the $m = M_{PL} = 1/\sqrt{G}$ is the Planck mass.

3. The entropy of a black hole of radius R can be estimated using the approximation $S_{BH} = kN$, where N is the number of particles inside and k is Boltzmann's constant. That number will have a maximum value $N = Mc^2/E_{min}$, where $E_{min} = hc/\lambda_{max}$, where $\lambda_{max} = 2\pi R$. Since black holes have maximum entropy, we get, in natural units $\hbar = h/2\pi = c = k = 1$, $S_{BH} = MR$. Since $Mc^2 = GM^2/R$, and $R_{PL} = \sqrt{G}$, we find $S_{BH} = (R/R_{PL})^2$.

4. Recall that the number of states available to a system is related to the entropy by $n = e^S$. Penrose approximates this as $n = 10^S$.

5. For nonspecialist introductions to the subject, see Guth 1984, 1997, and Linde 1987, 1990, 1994. I have also written about inflation in Stenger 1988, 1990a.

6. Although the universe is empty, you can still think of placing two test particles in the expanding space and measure the ratio of their relative velocity and distance.

5

THE
QUANTUM
DOMAIN

If one wants to clarify what is meant by "position of an object," for example, of an electron, he has to describe an experiment by which the "position of an electron" can be measured; otherwise this term has no meaning at all.

Werner Heisenberg

MOSTLY VOID

The decades 1890–1910 saw the particulate nature of the familiar matter of rocks and trees established beyond any reasonable doubt. The chemical elements were associated with objects that are unimaginably small, typically one ten billionth of a meter in diameter. As inferred from the measurement of Avogadro's number, on the order of a trillion trillion of them were needed to make one gram of matter. Chemistry indicated that the elements were not further divisible, and so the name "atom" was used to designate these primary objects. However, they soon proved not to be the uncuttable atoms of Democritus.

The various forms of radiation observed to emanate from matter indicated its further divisibility. This radiation appeared to come from within atoms themselves. Cathode rays were found to be composed of electrons much smaller than the atom. The atom proved to be composed of electrons and other elementary constituents.

In 1895, Wilhelm Roentgen discovered **X-rays**, induced by accelerated electrons striking matter. These rays were identified as electromagnetic waves with frequencies even greater than those of ultraviolet light, beyond the high frequency end of the visible spectrum. The next year, Antoine Becquerel observed natural emanations from uranium, the heaviest element then known. Careful experiments by Marie and Pierre Curie established many of the properties of the remarkable phenomenon of radioactivity.

Three types of radioactivity were identified: Highly penetrating **alpha-rays** were found to be composed of doubly charged helium ions, helium atoms in which two electrons had been removed. **Beta-rays** were identified as electrons, again providing evidence for the electron being a fundamental constituent of matter. And **gamma-rays**, electromagnetic waves of even shorter wavelength than X-rays, were observed in some processes. The Curies discovered the previously unknown elements polonium and radium, produced as byproducts in the decay of uranium. Thus, they demonstrated the transmutation of elements, long sought after by alchemists; it was happening all the time in nature and soon would be achieved artificially in physics laboratories. Indeed, gold has been produced from "baser" materials, but not in sufficient quantities to be lucrative.

In 1910, alpha rays were seen to scatter at surprisingly large angles from thin gold foils. Ernest Rutherford inferred from this that the atom was composed of a positively charged nucleus that was thousands of times smaller than the atom itself, and yet contained most of the atom's mass. This demonstrated that matter was not only composed of atoms and the void, but was mostly void. In Rutherford's model, the atom is like a tiny solar system, the nuclear sun surrounded by planets of orbiting electrons. People imagined the universe repeating itself at different levels. However, the force holding the electrons in the atom was electrical, rather than gravitational, and the atom proved upon further study to only superficially resemble the planetary system.

THE OLD QUANTUM THEORY

Simultaneous with the confirmation of the corpuscular nature of ordinary matter came the recognition of the corpuscular nature of light. The photon theory, used by Einstein to explain the **photoelectric effect**, was equally successful in accounting for other observations. One example is the **Compton effect**, in which scattered X-rays experience a decrease in frequency. Classical matter waves do not change frequency when they bounce off objects, as can be corroborated by watching water waves striking a rock. According to the photon theory, however, photons must transfer energy to any object they bounce off. Since photon energy is proportional to the frequency of the associated waves, a decrease in X-ray frequency is to be expected.

Despite the success of the photon theory, the wave theory of light did not disappear. It was still invoked to explain the interference and diffraction of light. Localized bodies, in our normal experience, do not exhibit these phenomena. And so it seemed that light possesses the dual properties of particle and wave.

Classical particle mechanics could not explain all that was known about light. Newton's corpuscles did not diffract. But classical wave mechanics could not explain these new phenomena. Huygens's waves cannot account for the photoelectric and Compton effects. Neither classical particles nor classical waves can explain the narrow lines in the frequency spectrum for the emission and absorption of light by gases. In 1922, Niels Bohr was able to reproduce the observed spectral lines of hydrogen, the simplest chemical atom, by a combination of the Rutherford atomic model and quantum ideas that went beyond both classical particles and waves.

Bohr pictured the electron as moving in a circular orbit around the hydrogen nucleus. According to classical electrodynamics, such a system is unstable as the electron radiates away energy and spirals into the nucleus in a tiny fraction of a second. Bohr suggested that the electron cannot spiral all the way in because only discrete orbits are "allowed" and the lowest one is still well outside the nucleus. He hypothesized that the allowed orbits are those for which the angular momentum, or more generally the action, is an integral multiple of Planck's constant h divided by 2π. We call this the **quantum of action** and designate it by the symbol $\hbar$. Bohr's hypothesis was that angular momentum $L = n\hbar$, where n is an integer called the principle quantum number. Each Bohr orbit corresponded to a given energy. An electron in that orbit is said to be in a certain **energy level**. The lowest energy level, the ground state, has $n = 1$ and an isolated atom in this state is stable since there are no lower orbits in which it may move.

When low density hydrogen gas is exposed to a high voltage pulse, the electrons in its atoms are exited to higher energy levels. As they drop back down to lower levels, they emit photons whose energy is given exactly by the difference in the energy of the two levels. Thus, by Planck's relation $E = hf$, only certain specific frequencies f are emitted by these "quantum jumps." Bohr's quantization principle gave the exact frequencies that were observed.

In what is now termed the **old quantum theory**, Bohr's ideas were further refined to handle more complex atoms. Observations of **fine structure**, tiny splittings of spectral lines, led to the introduction of additional quantization rules analogous to $L = n\hbar$. This was interpreted to imply that the electron had an intrinsic angular momentum, or **spin**, that was quantized in half-units of $\hbar$.

Spin is the angular momentum of a rotating body, the sum of the angular momenta of all its constituent particles. If these constituents are electrically charged, then the resulting currents produce a magnetic field. Thus, an atom has a spin that results from the **orbital angular momentum** of the electrons rotating about the nucleus, plus any spins of the nucleus and electrons themselves.

Atoms are generally magnetic as a result of these circulating currents, as are their nuclei. Large scale magnetism, as observed in iron for example, is the consequence of the magnetism of the constituent atoms of the material. One would be inclined to think that a point particle, as the electron still appears to be today a century after its discovery, would have no spin or magnetic field. In fact, it has both and this observation has had profound consequences.

In 1921, Otto Stern and Walther Gerlach passed a beam of silver atoms through nonuniform magnetic field (see figure 5.1). Classical electromagnetic theory predicted a smooth spreading out of the beam as the external magnetic field acted on the magnetic fields of the atoms. Each atom was viewed as a little bar magnet, with a north and south pole that was expected to be randomly oriented with respect to the external field. But this is not what they saw. Instead, the beam of silver atoms was split by the magnet into two separate beams. Apparently the magnetic fields of the silver atoms in the beam had only two orientations, either "up" or "down" along the direction of the external field.

Further experiments showed that when bodies of spin s are sent through Stern-Gerlach magnets, 2s+1 beams result. Both odd and even numbers of beams are observed, odd for integer s and even for half-integer s. When electrons are passed through, two beams emerge confirming that they have s = $\frac{1}{2}$. Particles with half-integer spin are called **fermions**, while integer spin particles are called **bosons**. The photon is a boson of spin 1.

In 1925, Wolfgang Pauli proposed the **Pauli exclusion principle** which states that only one fermion can be found in any given quantum mechanical state. The Pauli principle accounted for the chemical periodic table of the elements, explaining why the lowest energy state of all atoms isn't simply the one where all the electrons are in the lowest Bohr orbit.

Without the Pauli principle we would not have the diverse structure in chemistry necessary for life. You and I would not exist. As more electrons are added to the region around a particular nucleus, they must move to higher energy levels, or "shells," as the lower levels get filled. Those elements, like helium and neon, with fully occupied shells are chemically inert, that is, tightly bound and nonreactive. Those, like sodium and chlorine, with one more or one less electron than needed to complete a shell, are highly reactive. Those atoms with half-filled shells, like carbon, silicon, and germanium, allow for the most diverse molecular structures that can assemble into complex objects such as living organisms and computer chips.

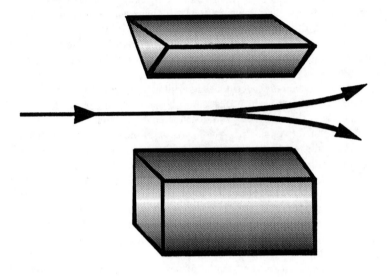

Fig. 5.1. The Stern-Gerlach experiment. A beam of silver atoms is split in two by an inhomogeneous magnetic field.

Chemistry is thus seen as the set of phenomena that is associated with the interactions between atoms, as they exchange or share electrons. Apart from providing most of the mass of matter, the nuclei of atoms have little to do with chemistry. Since biology and most of the other material phenomena are essentially chemistry and classical physics, little more is needed to understand the basic physical mechanisms behind the great bulk of human experience, including life and mind.

Beyond most of human experience lie the fundamental phenomena we associate with the quantum world. These phenomena seem strange and mysterious, but only because they are unfamiliar. I hope to convince you that the quantum world is within our understanding, as long as we do not insist on applying all our preconceptions based on common experience to that regime. We do not need to abandon every preconception. Many still apply. But we should not cling obstinately to those that lead to incorrect or absurd conclusions.

PARTICLES AND/OR WAVES

In 1923, Louis de Broglie suggested that perhaps objects like electrons, which we normally think of as particles, will also exhibit wave effects analogous to those for light. He proposed that the relationship between the wavelength λ and momentum p of a photon, $\lambda = h/p$, holds for *all* particles. This relation was soon confirmed with the observation of the diffraction of a beam of electrons, which was explained by treating the beam as an incoming wave with this wavelength. Since then, other particles and even atoms have been observed to produce diffraction effects, with wavelengths given by the de Broglie formula. As for everyday objects such as baseballs and people, they diffract, too, but the effect is far too small to be observed.

De Broglie's relation gave an intuitive explanation of Bohr's quantization rule for the angular momentum of the electron in an atom. Bohr's allowed orbits were just those in which the de Broglie waves resonated around the circumference, like standing waves on a guitar string.

De Broglie suggested that perhaps a particle was in reality a localized superposition of waves called a **wave packet**. This proved useful as a visual tool, but not as a complete theory. When, a few years later, Erwin Schrödinger introduced his wave mechanical version of quantum mechanics, his **wave function** could not be associated with a physical wave packet picture for more than one particle. When we have N particles, the wavefunction of the system is no longer a wave in three-dimensional space but rather in an abstract 3N-dimensional space. Furthermore, the dispersion (spreading out) of the wave packet with time was predicted to be so great in the case of electrons and like particles (but not photons) that they would not retain their localization for significant periods of time, in disagreement with the data.

In any case, neither de Broglie nor anyone else knew what these electron waves were. They seemed analogous to photon waves, that is, electromagnetic waves, but since the failure of the aether theory no one knew what photon waves were either. What is doing the waving, for photons, electrons, or anything else?

The source of most of the mystery of quantum physics lies in this clash between two classical concepts: the particle and the field. They are separate and distinct in classical physics. A particle is discrete and localized. A field is continuous and spread out over space. Everyday phenomena seem to fall into one class or the other. Rocks, people, and planets are like particles—discrete objects that are found in one place at one time. Water, air, and light are like fields—continuous media that appear simultaneously distributed over a large region.

Note that the meaning of a field is a bit more complicated in relativistic physics, where the notion of simultaneity is shown to be meaningless for describing events separated in space. The fields of classical electromagnetism are fully consistent with relativity, and indeed stimulated the discovery of the theory. However, instantaneous action-at-a-distance fields, like the Newtonian gravitational field, are inherently relativity-violating, that is, superluminal.[1] Indeed, as we will see, some of the so-called paradoxes of quantum mechanics arise from trying to maintain both Einstein's speed limit and the notion of a continuous action-at-a-distance field.

Particles and fields have become intertwined in the usual expressions of quantum physics. Photons and electrons are said to exhibit the properties of both particles and fields, appearing localized and distributed, discrete and continuous. This is usually called the **wave-particle duality**. You will often hear that something is a particle or a wave, depending on what you decide to measure. I have said this many times myself in the classroom, parroting what I heard in class and read in textbooks when I was a student. But in recent years I have recognized what a dangerously misleading and indeed incorrect statement this is. Indeed, it provides the main justification for "quantum mysticism," the belief that human consciousness controls the nature of reality, and similar bizarre ideas (Stenger 1995).

In fact, no experiment provides data that speak against the fundamental particulate nature of light. Whatever you decide to measure about light, if you do so with sufficiently sensitive photodetectors you will always detect something localized and discrete—a photon. You cannot detect light and at the same time decide not to measure the position of a photon. Every experiment you perform measures a photon position that lies within the bounds of the detector that registers the signal. The position accuracy is limited only by the pixel size of the detector, such as the grain size of photographic film.

To measure a wave property, such as the frequency of the associated electromagnetic wave, you can send the light through several small slits (one suffices, but the more the merrier) and have it fall on a detector matrix. You may have to wait a long time or turn up the intensity of light to gather a lot of hits. The distribution pattern of these hits can then be analyzed to obtain the frequency of the light.

But, even in this case, you measure photon particles, one-by-one, at localized positions. Often photographic film is used in the place of detectors, and we measure the integrated light intensity without counting individual photons. However, the photons are still there, as you can demonstrate with sufficiently sensitive equipment. They did not go away, magically changing into "waves" because we made a conscious decision to use crude photographic film rather than sensitive, high-tech photodetectors. Nor did the photon become a wave when some human experimenter made a conscious decision to measure frequency instead of an accurate position.

The phenomena we label "wave effects" are still there when we measure one photon at a time, determining its position each time. And the phenomena we label "particle effects" are still there when we decide to measure frequency. Furthermore, if we measure the energy of the individual photons and divide by Planck's constant, we will get a number $f = E/h$ that will agree with the wave frequency measured from the light intensity pattern.

More to the point is the relationship between particle and field, rather than particle and wave. Mathematically, a field is a quantity that has a value for each point in space. To make it relativistic, we have to add the time dimension and remember that what may be simultaneous in one reference fame will not be simultaneous in all reference frames. As we will see, quantum field theory contains mathematical fields that are directly related, one-to-one, to each elementary particle.

HEISENBERG'S QUANTUM MECHANICS

The mid-1920s witnessed the development of what is essentially the quantum theory still in use today. The old quantum theory was increasingly found to be unsatisfactory, in terms of its ad hoc nature, theoretical inconsistencies, and failure of many predictions. Two approaches to a new, more general quantum theory were independently and almost simultaneously proposed. Both proved highly successful in a wide range of applications. These two approaches, more-or-less, can be characterized by the alternative pictures of particle and wave.

A particle-based approach was taken in 1925 by the twenty-four-year-old Werner Heisenberg. Physicists in 1925 knew that something in classical mechanics had to give in order to explain quantum phenomena. But they were stymied by what that should be. Most assumed that a whole new set of dynamical principles would have to be uncovered, surely a daunting task.

Heisenberg had a different notion, at first only dimly perceived. He conjectured that the equations of classical particulate dynamics were still correct. They just had to be reexpressed in terms of a different set of algebraic rules. That is, it was not the physical ideas themselves that had to be changed, but the mathematics, or logic, by which these principles were applied.

In the traditional formulations of classical theories, the quantitative properties of objects, such as position, mass, or angular momentum, are represented by real numbers—the kinds of numbers we deal with in everyday life when we go to the market or look at a clock. These numbers are associated with the readings you would get on the dials of your instruments when you measure the corresponding quantities in the laboratory. Even when you do not measure a particular variable, the unwritten assumption in classical physics is that you *could* have, and so classical equations may still contain that unobserved quantity as a real number.

Heisenberg noted that many of the concepts that were represented in the old quantum mechanics, like the momentum of an electron in its orbit in an atom, were in fact not being directly observed in specific experiments. He realized that it was not necessary for these properties to be represented by real numbers in the theory, when the theory was not being asked to compare calculations of these quantities with measurements or dial readings. For example, the Bohr theory of hydrogen contains an expression for the radius of the electron's orbit. But you never actually measure the radius of an electron's orbit.

Heisenberg began with the problem of an oscillating electron in which you do not measure the electron's motion itself but merely detect the emitted electromagnetic radiation. This was basically the same problem originally studied by Planck, which led to his proposal of the quantization of the energy of radiation in units $E = hf$. The oscillator is a prototypical problem in physics, going back to Galileo's pendulum.

Since the electron's position is not measured, then according to Heisenberg's conjecture the mathematical symbol representing position in the theory need not obey all the usual algebraic rules of real numbers. No instrument dial reading could be identified with any strictly theoretical quantity. Heisenberg assumed, since he had no other viable choice, that the mathematical symbol for position still obeyed its classical equation of motion. He found he got answers that agreed with the data when he assumed that the product of two symbols did not always commute. That is, if Q and P represent two classical observables, the product QP was not always equal to the product PQ.

When he saw Heisenberg's paper, Max Born, who had studied more mathematics than the typical physicist of the day, recognized that the noncommutative symbol manipulation was "nothing but the matrix calculus, well known to me since my student days." Matrices are simply tables of numbers, and mathematicians had provided rules for their addition, subtraction, and multiplication. In particular, the products of two matrices P and Q do not in general commute: QP is not the same as PQ.

Born left his brilliant, shy assistant Pascual Jordan to work out the details. Jordan complied, Heisenberg jumped back into the act, and the result was the matrix formulation of quantum mechanics. In this formulation, the ad hoc quantization rules of the old quantum theory are replaced by *commutation relations*, that is, rules that give the value of the **commutator** $[Q, P] = QP - PQ$, for any two particle properties Q and P. For example, if x is a particle coordinate and p_x is the particle's momentum component along the same spatial axis, then $[x, p_x] = i\,\text{K}$, where $i = \sqrt{-1}$. The use of complex numbers (numbers that contain $\sqrt{-1}$) rather than exclusively real numbers, like the use of matrices, was also justified by the Heisenberg conjecture.

Jordan was able to demonstrate that the corpuscular properties of electromagnetic waves resulted from the noncommutivity of the associated dynamical variables. He also discovered the commutation rules for the components of angular momentum, and showed that these resulted in half-integer as well as integer quantum numbers. This gave a theoretical basis for the previous ad hoc introduction of half-integer electron spin in the old quantum theory.

Working independently, twenty-three-year-old Paul Dirac found that the quantum mechanical results could be obtained by taking the classical mechanical equations expressed in the particularly elegant notation of *Poisson brackets* and replacing the brackets with commutators. One important consequence of this was that a close connection was then established between the properties of a system whose symbols do not commute with each other. They are in fact just those pairs of properties in classical mechanics, like x and p_x, that are called **canonically conjugate**.

In classical mechanics, each coordinate of a system is associated with a canonically conjugate momentum that must also be specified to fully define the state of the system. The state of a classical system is then viewed as a point in **phase space**, with an axis for the coordinate and conjugate momentum of each degree of freedom of the system. For example, the phase space of a gas containing N point particles (with no internal degrees of freedom such as rotation or vibration) has 6N dimensions. The equations of motion are then used to predict a trajectory in phase space that fully specifies how the system will evolve with time. In the case of an N particle system, the future positions and momenta of each particle are determined by the initial point in phase space and the equations of motion.

Dirac's association of canonical variables with Heisenberg's noncommuting symbols emphasized that quantum mechanics was still closely connected to classical mechanics. As mentioned, the physics was the same; just the mathematics, or logic, was different.

SCHRÖDINGER'S QUANTUM MECHANICS

All this occurred in 1925. An alternative, wave-based approach was developed by Schrödinger in six papers published the very next year. Schrödinger took note of the fact that the motion of a particle in classical mechanics can be equally described in terms of the propagation of a wavefront whose phase was essentially the action of the particle, the same action quantized by Bohr. Starting from the classical Hamilton-Jacobi equation of motion where this wave interpretation is manifest, Schrödinger derived the famous equation now associated with his name.

Schrödinger considered the oscillator problem, which Heisenberg had used the preceding year in developing the ideas of his quantum theory, and obtained the same results. He also solved the hydrogen atom problem, deriving the Bohr energy levels. Pauli had already done the same with Heisenberg's scheme, but Schrödinger's solution was simpler and easier to understand. Most importantly, Schrödinger demonstrated the formal equivalence between his **wave mechanics** and the Heisenberg-Born-Jordan matrix mechanics.

The Schrödinger equation contains a quantity called the **wave function**, a complex number usually designated by ψ. He originally interpreted the square of the magnitude of the wave function at a particular position and time as the charge density of the particle, say the electron in an atom, at that time. Born showed that this could not be correct in general, and gave the interpretation of the wave function that has remained to the present day: $|\psi|^2$ is the *probability* for finding the particle in a unit volume centered at that point at that time. I will refer to this as the *Born axiom*.

Schrödinger's mathematical methods utilized partial differential equations that were far more familiar to most scientists of the day than the Heisenberg-Born Jordan matrix methods. Equations of this type are involved in all branches of classical physics. Maxwell's equations are partial differential equations, analogous to the equations of fluid mechanics. The formal solutions to Schrödinger's equation for many applications, such as the oscillator and hydrogen atom, had already been worked out for many classical problems and were in mathematical physics textbooks. Thus, Schrödinger's wave mechanics became the version of quantum mechanics that most people learned and put to use and the wave function took on a central role in quantum discussions that has lasted to the current day.

This happened, despite the fact that matrix mechanics, with no explicit wave function, was far more powerful and was soon further generalized and made axiomatic. But to most physicists at the time, the matrix theory with its commutation rules struck many, including such influential figures as Einstein and Pauli, as too abstract and nonintuitive. How does one visualize a commutation relation? Even if they did not know what was doing the waving, the wave picture was far more comfortable and familiar.

Heisenberg objected to wave mechanics, saying it could not be used to explain quantum jumps and radiation. However, he was shrugged off by his seniors. For example, William Wien, the director of the Munich Institute of Theoretical physics, remarked that Schrödinger had proved once and for all the absurdity of "quantum jumps" and thus put an end to a theory based on such notions (Jammer 1974, 56). And Schrödinger wrote how gratifying it would be to view the quantum transition as a change from one vibrational mode to another instead of in terms of quantum jumps. He thought this would preserve space-time continuity and classical causality (Holton 1972c).

And so, Schrödinger's wave mechanics became the quantum paradigm that was most widely learned, applied, and transmitted to the public. It remains today the quantum mechanics taught at the introductory level in universities and is usually the only quantum mechanics, if any, most students learn—even if they go on to Ph.D.s in chemistry or microbiology. Solving the hydrogen atom problem from Schrödinger's equation remains an exemplar of theoretical physics in action, providing results of immense utility. To the science-literate public, to many highly trained scientists in biology, chemistry, astronomy, and even to those active in some subfields of physics, the Schrödinger equation defines quantum mechanics.

However, two Schrödinger equations exist, which leads to some confusion. Schrödinger's original equation, the more familiar one, is precisely termed the **time-independent Schrödinger equation**. It is nonrelativistic and has limited applicability. It can be derived from a more generic relation called the **time-dependent Schrödinger equation**. This equation is the basic dynamical equation of quantum mechanics, replacing the classical laws of motion in telling us how a system evolves with time. The time-dependent Schrödinger equation is valid relativistically when written in the proper form, and remains part of the modern formalism of quantum mechanics (to be discussed below). It is more general and powerful than either of the original Schrödinger and Heisenberg formulations.

Despite the widespread use in many disciplines of Schrödinger's nonrelativistic, time-independent formula, wave mechanics and the wave function occupy only a small sector of the fully generalized quantum paradigm, as it already had developed by the early 1930s. The modern formulation of quantum mechanics still in use was developed mainly by Dirac and made axiomatic by John von Neumann, working from the original matrix mechanics. **Quantum electrodynamics (QED)**, the study of photons, relativistic electrons, and electromagnetic radiation that developed over the next twenty years, had barely any role for the wave function or the nonrelativistic Schrödinger's equation. Nor does relativistic quantum field theory and the current standard model of particle and fields. On the other hand, Heisenberg's notion that quantization arises from commutation rules was fundamental to the development of QED and theories that emerged from it.

UNCERTAINTY AND COMPLEMENTARITY

Shortly after Heisenberg completed his paper on the new quantum mechanics, he joined Bohr in Copenhagen on a temporary fellowship. There, they and a host of prominent visitors, including Schrödinger, debated what all the new quantum mechanics meant. They achieved no convergence of views. That debate continues to the present day, long after the deaths of the original participants. If anything, perspectives on the "meaning" of quantum mechanics have diverged, rather than converged, in all this time.

In February 1927, while Bohr was off skiing, Heisenberg developed his famous **uncertainty principle**. He did so without direct recourse to the wave picture. However, the wave analogy remains the simplest way to understand the idea. So, with some reluctance, I will present that approach. Please remember that the conclusions in no way depend uniquely on the wave picture.

To measure a particle's position to some accuracy, Δx, you must use light, or some equivalent probe of wavelength $\lambda < \Delta x$. Imagine a *gamma-ray microscope* in which very short wavelengths are used to determine the position of an electron in an atom. Gamma-ray wavelengths are very much smaller than an atom and so can be used to probe its details. From the de Broglie relation, $p = h/\lambda$, the momentum of the probing photon must then be greater than $h/\Delta x$. Since we cannot control how much momentum is transferred in the collision between the photon and the struck electron, the electron will have an uncertainty in momentum, Δp, of about this amount. Thus, we have, approximately, $\Delta x \Delta p > h$. More exactly, it can be shown that $\Delta x \Delta p \geq \hbar/2$, where $\hbar = h/2\pi$ is Bohr's quantum of action and the "uncertainties" Δx and Δp are precisely equal to the usual standard errors on x and p given in statistical theory, the so-called σ values.

We see that the more accurately we try to determine a particle's position, the less accurately we know its momentum, and vice versa. In the case of a gamma-ray photon, the momentum will be so high as to knock the electron out of the atom. Using a lower momentum photon, say an X-ray, might not destroy the atom but provide a much poorer position measurement. Heisenberg realized that it was not possible to speak meaningfully of the exact position of an electron in an atom since that position cannot be accurately measured. Does this mean it has no precise position? Physicists and philosophers will probably still be arguing about this in the year 2027.

Heisenberg considered the example of the path of an electron in a cloud chamber, which unlike the orbit of an electron in an atom is obviously measurable. He noted that what was being measured in the cloud chamber was not the precise path of the electron, but a discrete sequence of imprecise positions indicated by water droplets. The effect of the uncertainty principle is small on the scale of the droplets. So we can see fairly definite tracks of particles in a cloud chamber, even though their momenta and positions are not exact. On the other hand, at the atomic scale we cannot build a device analogous to a cloud chamber that allows us to "see" the paths of particles.

Heisenberg immediately recognized the dramatic philosophical implications of his uncertainty principle. The determinacy of classical mechanics, that is, the in-principle predictivity of physical events, is rendered inoperative by the uncertainty principle. In order to apply the equations of motion of classical mechanics to predict the motion of a particle, you have to know both the particle's initial position and momentum. Now, Heisenberg said, this was impossible. All you can do is make approximate measurements and thus approximate predictions—although the statistical behavior of a large number of similar measurements is highly predictable.

As a consequence, the whole notion of causality, so fundamental to almost all arenas of human thinking, from psychology to theology, was called into question. The deterministic *Newtonian Clockwork Universe* was replaced by one in which only the average motion of a body is calculated by the paradigms of physics. Randomness and limited predictability are thus incorporated as unavoidable ingredients of nature.

Arguments continue to this day over the interpretation of the uncertainty principle. Heisenberg seems to have simply viewed it as a practical matter. You cannot observe a system without disturbing it to some extent. This is rather obvious when you think about it, and people as diverse as Plato and Marx had made similar sounding but nonquantitative statements. Here the quantities are important. At the macroscopic scale, the effect is usually negligible. On the other hand, at the atomic level, gamma-ray photons are like machine gun bullets that wreak havoc with anything they "look" at.

You will often hear about applications of the uncertainty principle to the social systems studied in economics or anthropology, where the interaction between observer and observed is very strong. I have no opinion on whether this is a useful extrapolation of physics or not. I do know, however, that everyday physical objects, like tables and chairs, do not recoil appreciably when we shine light on them, while atoms do.

Heisenberg seems to have held the view that particles still have a definite position and momentum at a given time. We simply cannot measure the two of them simultaneously with unlimited precision. Here he appeared to disagree with Bohr, who tended to adopt the more extreme positivist doctrine that an unmeasured property is too meaningless, too metaphysical, to even talk about. This interpretation of quantum mechanics has produced unfortunate consequences, as mentioned, leading to the claim that quantum mechanics supports the mystical notion that the properties of bodies do not even exist until they are observed, that reality is a construct of human consciousness.

Bohr was not completely happy with what Heisenberg had come up with in his absence, and they argued vehemently after Bohr returned from skiing. Heisenberg had made a few minor mistakes that were corrected, and he burst into tears at one point under the assault of the famous senior professor. Heisenberg was, after all, still a twenty-five-year-old post-doc looking for a permanent position. By fall, however, he had landed a job in Leipzig and Bohr had not only reconciled the uncertainty principle but become its primary champion. Previously, he had been contemplating a vague notion called **complementarity**, and he now merged this with the uncertainty principle into a new philosophical doctrine.

The uncertainty principle implies that you cannot describe a system simultaneously in terms of both its coordinates and momenta. In the particle description of the system, you use coordinates. In the wave description, you use wavelengths. But, since $p = h/\lambda$, the wave description is equivalent to a description in terms of momenta. You might argue that this, too, is a particle description, though waves also carry momentum. In any case, according to Bohr's principle of complementarity either choice provides an equally valid, *fully complete* description. You can describe a quantum system in terms of either coordinates or momenta—never both. Neither particles nor waves have any special claim on the truth, according to Bohr's doctrine.

THE MODERN QUANTUM THEORY

We have seen how once Schrödinger's wave mechanics surfaced it gained wide acceptance as the simpler and more intuitive way to do quantum mechanics. However, the more abstract and unfamiliar matrix mechanics was not discarded by the few who understood it and appreciated its greater power. Most significantly, a connection between the uncertainty principle and the commutation relations of matrix mechanics was soon established. It was proven that for any two observables A and B, the product of their uncertainties is given by $\Delta A \Delta B \geq |[A,B]|/2$, where $|[A,B]|$ is the magnitude of commutator of A and B, where we recall $[A,B] = AB - BA$. For example, recall that $[x, p] = i\hbar$. From this it follows that $\Delta x \Delta p \geq \hbar/2$, as stated in the uncertainty principle.

So, at least mathematically, the uncertainty principle was seen as a direct consequence of the commutation relations among conjugate properties. These formulas contained the essence of quantum mechanics. As mentioned, the commutation relations for the observables carried over from classical physics are obtained directly from their classical Poisson brackets. Classical *physics* was retained in matrix mechanics, with only the *mathematics* changed, a point generally misunderstood because of the incessant hype about the quantum "revolution" overthrowing classical mechanics.

Perhaps some feeling for the connection of noncommutivity to the uncertainty principle can be obtained by imagining a sequence of measurements. Because one measurement can upset the result of another, you may get different values if you measure x and then p than if you measure p and then x. If you repeat such an experiment many times, you will get a well-defined momentum with a wide distribution of positions in the first case, and well-defined position with a wide distribution of momenta in the second, with the widths of these distributions given by the uncertainty principle. Another way to say this is that a measurement of the quantity x•p, in the sequence given, is not the same as a measurement of the quantity p•x, in the opposite sequence, so we cannot equate the two. Thus, we cannot represent x and p by numbers in the theory, since numbers commute.

Still, quantum mechanics was much more than classical mechanics done in a funny way. As the new quantum theory evolved, other properties were discovered that had no classical analogue. For example, Dirac introduced **annihilation** and **creation operators** for photons, which were ultimately extended to include electrons and other elementary particles. These operators have no classical analogue. The Dirac formalism enabled a proper treatment of electromagnetic radiation and eventually evolved into quantum electrodynamics and quantum field theory, which will be discussed in chapter 7. The key notion, again, was noncommutivity. To "quantize" something means to write down its commutation relations. They are the starting point of any quantum theory and lead to the dynamical formulas and uncertainty relations for those properties.

Dirac's version of quantum mechanics, which he called *transformation theory*, is (like so much of the Master's work) a thing of aesthetic beauty. Once a physics graduate student has read Dirac's *Principles of Quantum Mechanics* (first edition 1930), and learned its elegant notation of "bras" and "kets," that student will want to do quantum mechanics no other way. For example, the **ket** |"label"> represents a quantum state, where "label" is whatever descriptors we need to specify the state.

It is significant for the theme of this book to note that the term "wave function" is used only once in Dirac's book, in the following early footnote: "The reason for this name [wave function] is that in the early days of quantum mechanics all the examples of these functions were in the form of waves. The name is not a descriptive one from the point of view of the modern general theory" (Dirac 1989, 80).

Considerable mathematical abstraction remains in what has become the conventional formalism of quantum mechanics. But the basic ideas are not too difficult to grasp if you have just a little familiarity with vectors and coordinate systems. One of the axioms is the **principle of superposition**. This postulates that the state of a quantum system can be represented by a **state vector** in an abstract **Hilbert space** that contains one coordinate axis, and thus one dimension, for each possible outcome of a measurement of all the observables of the system. The set of observables is defined by all those variables whose symbols commute with one another, such as the coordinates of a particle x, y, and z, where $xy = yx$, and so on. This set can contain only one member of any pair of variables that fail to commute with one another, for example, x or its conjugate momentum component p_x. Momenta are part of an alternative observable set, in compliance with Bohr complementarity.

The number of dimensions of the Hilbert space can be as low as two, as when the only variable is the spin component of a single electron. In that case the two axes correspond to "spin up" or "spin down." At the other extreme, the dimensionality can be infinite, as when a variable like the position of a particle is assumed to have a continuous range of values.

Of course, an infinite dimensional space is difficult to comprehend. It may help to recognize that, in principle, all operationally defined physical variables, including space and time, are ultimately discrete. The discreteness of space and time just happens to occur at such a small scale, what is called the **Planck scale**, that we currently neglect it even in subnuclear physics. It represents the smallest distance that can be operationally defined, that is, defined in terms of some measuring procedure. Assuming discreteness, at whatever scale, just think of the possible positions of a particle as x_1, x_2, . . . , x_n. The Hilbert space in this case is an n-dimensional space with axes 1, 2, . . . , n in which the state vector is aligned along the "1" axis when the particle position is x_1, along the "2" axis when the particle position is x_2, and so forth. Note that for each coordinate dimension in normal three dimensional space, there are n dimensions in Hilbert space. How big is n? Don't even think about it. Just visualize simple cases like n = 2 or 3.

As is the case for the more familiar x, y, z coordinates of three-dimensional space, the coordinate axes in Hilbert space can be changed from one set to another without necessarily changing the vector's magnitude or direction. When the magnitude remains unchanged, the process is called **unitary**. Recall that the time evolution of a quantum state is governed by the time-dependent Schrödinger equation. This is a unitary process. Following Penrose (1979, 250), I will designate this process as **U**.

Those familiar with vectors may remember that the rotation of a vector with respect to some coordinate system is mathematically equivalent to the vector remaining fixed and the coordinate system rotating. The same applies to Hilbert space. The first, in which the state vector evolves in some absolute Hilbert space, is called the **Schrödinger picture**. The equivalent alternative in which the state vector remains fixed while the coordinate system evolves is called the **Heisenberg picture**.

Each coordinate set corresponds to a different complete set of observables. Thus the state of the system can be described in any number of ways, depending on the coordinate system you prefer. Making measurements on a complete set of commuting observables of the system determines the state of the system at the time the measurements are performed.

In the conventional view, the act of measurement of an observable results in the state vector "collapsing" so that it now points along the axis that corresponds to the quantity and value measured. This collapse is into a new Hilbert space that is a subspace of the original one. This is not as formidable as it sounds. Just imagine the state vector as a pencil originally held up in an angle from a table, in three-dimensional space, falling over to lie in the two-dimensional subspace of the surface of a table.

However, the pencil analogy is not perfect. When the state vector falls over in the act of measurement, its length decreases. It becomes as long as its original projection on the axis. That is, the measurement process is "non-unitary." Following Penrose again, I will designate this process as **R**, for **reduction**, a term equivalent to "collapse." This nonunitary process is the source of much of the debate about quantum mechanics. It is not predicted by the Schrödinger equation, indeed not even allowed by it since the Schrödinger equation only permits unitary evolution. The **R** process is inserted in the formalism of standard quantum mechanics as an added assumption. In the next chapter I will discuss the alternatives, in particular, the many worlds interpretation in which the state vector does not collapse. For now, let me continue with the conventional view.

The square of the projection of the original state vector on a particular axis gives the probability of finding the value of the observable specified by that axis. If the state vector points along the axis, that probability is 100 percent.

Consider the example of electron spin, which is the easiest to picture in Hilbert space, despite involving the unfamiliar notion of spin. Suppose the state vector points at 45 degrees in two-dimensional Hilbert space (figure 5.2). This implies that the spin has a 50 percent chance of being measured to be spin up or spin down along the corresponding axis in normal three-dimensional space. Once a measurement of the spin component along that axis is made, the state vector in Hilbert space collapses or reduces to point in the direction that corresponds to the measurement result, "up," or "down," whichever the outcome of the experiment may be. It moves into a subspace of one dimension. Again, quantum mechanics does not predict the outcome of that measurement, only its probability based on our knowledge of the original state vector, 50 percent in the example given.

In this example, note that a rotation of 180 degrees ("up" to "down") in ordinary space corresponds to a 90 degree rotation in Hilbert space. This means that it takes two complete rotations of the spin vector in ordinary space to return to the original state vector in Hilbert space. It is important not to confuse the two spaces, one familiar from everyday experience, the other a mathematical abstraction but still based on a generalization of that experience.

Recall that we could view the unitary evolution of the state vector, the **U** process, in Hilbert space in either the Schrödinger picture, in which the state vector rotates, or the Heisenberg picture, in which the coordinate system rotates. This is true for the **R** process as well. The two perspectives are not distinguishable by experiment, but the Heisenberg picture offers, I think, a more satisfying ontological view. The act of measurement on a quantum system can be likened to photographs we might take of an object from different perspectives. The object is out there, "in reality," fixed and immutable, and we simply move around it to capture different angles. Although the object looks different in these photos, we still associate them with the same object.

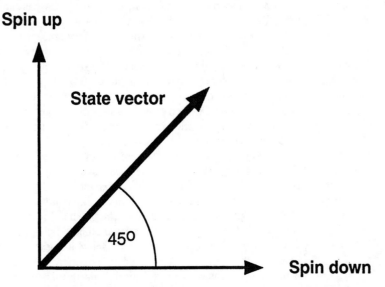

Fig. 5.2. The Hilbert space state vector for an electron with equal probability of being spin up and spin down. When the spin is measured, one or the other result occurs and the state vector rotates to the appropriate axis. Note that a 180° rotation in normal space corresponds to a 90° rotation in Hilbert space, in this case. In general, each axis in Hilbert space corresponds to the possible result of an observable of the system. Thus, the dimension of Hilbert space is infinite for continuous observables such as the single dimension position coordinate x of a particle.

Where does the wave function fit into this scheme? In the state vector formalism, the wave function is simply one of many different ways the state of the system can be represented.[2] As mentioned, Dirac wrote the definitive textbook on quantum mechanics with "wave function" mentioned only once, in a dismissive footnote. The de Broglie-Schrödinger notion that the wave function is a real, physical field in three-dimensional space was already questionable by the late 1920s.

THE COPENHAGEN INTERPRETATION

We can now see why only a complete set of noncommuting properties can be used to exactly specify the state of a quantum system. These are the most that can be simultaneously measured with precisions limited only by the accuracy of the measuring equipment. This differs from classical mechanics, where nothing but the apparatus limits the accuracy with which you can measure any and all properties.

Now here's the rub. Recall that those pairs of properties that do not mutually commute are just those that we call canonically conjugate pairs. Such pairs are needed in classical mechanics for the full specification of the state of the system. You must know the values of both halves of a pair to calculate their future values. For example, you need to know both position and momentum to predict future position and momentum with the equations of motion. As mentioned above, Heisenberg had recognized how his uncertainty principle implied that the universe was not fully deterministic, the "clockwork universe" implied by Newtonian mechanics.

Quantum mechanics does not seem to supply sufficient information to specify all that is in principle measurable about a system. Here we have the source of a major disagreement that has never been reconciled over the years. By 1927, two camps had emerged that can be identified by the names of two great physicists: Bohr and Einstein. The Bohr camp insisted that quantum mechanics is complete, that it says all we can say about a system. The Einstein camp said quantum mechanics is incomplete because there is much more physics should be able to say. Of the other great fathers of quantum mechanics, Heisenberg and Dirac agreed with Bohr, while Schrödinger and de Broglie initially aligned themselves with Einstein. De Broglie, however, got cold feet. The discoverer of the wave-particle "duality" eventually went along with the crowd, who mostly supported Bohr. Many years latter, after some of his original ideas were resurrected by David Bohm, de Broglie regretted that he had not stuck to his guns and tried to reinstall his old theory (de Broglie 1964).

Einstein was never happy with the direction quantum mechanics had taken, with its probabilities ("God does not play dice"), and the fact that at a given time we can only specify half or less of the classical properties of a system. Until his dying day, he took the position that quantum mechanics was incomplete, that it was correct as far as it went, as a statistical theory, but eventually a deeper theory will be found that restores the classical concepts of a well-defined reality that behaves according to natural law and simultaneously accounts for all the variables of a system.

In what became the *Copenhagen interpretation of quantum mechanics*, accepted as standard by the bulk of physicists over the years, Bohr insisted that quantum mechanics was complete. This did not mean quantum mechanics was "final" in the sense of answering all possible questions about nature. Completeness here refers to the assertion that the measurement of a full set of commuting variables of a system is sufficient to completely specify its quantum state and provide all the information that is possible to check by experiment. The complete arrangement of the experimental apparatus must be used in specifying the system. As we will see in the next chapter, the future as well as the past arrangement must be included in this specification.

The Copenhagen interpretation, as conventionally applied, is positivist and instrumentalist. All we can know about the world is what we measure. We describe these measurements with theories containing abstract quantities that are not themselves measurable and thus not to be taken as elements of reality. The state vector (or wave function) represents our knowledge of the system. It evolves deterministically according to the time-dependent Schrödinger equation, but only allows for statistical predictions about the outcome of measurements. According to Bohr, complementary representations of the state vector exist that are equally valid, suggesting a dual reality in which matter is both particlelike and wavelike but, since these properties are incompatible, we can only "see" one or the other, depending on what we decide to look for.

But perhaps the best way to present the Bohr view is to use his own words, which need to be read carefully because he never wrote very succinctly:

> On the lines of objective description, it is indeed more appropriate to use the word phenomenon to refer only to observations obtained under circumstances whose description includes an account of the whole experimental arrangement. In such terminology, the observational problem in quantum physics is deprived of any special intricacy and we are, moreover, directly reminded that every atomic phenomenon is closed in the sense that its observation is based on registrations obtained by means of suitable amplification devices with irreversible functioning such as, for example, permanent marks on a photographic plate, caused by the penetration of electrons into the emulsion. In this connection, it is important to realize that the quantum mechanical formalism permits well-defined applications referring only to such closed phenomena. (Bohr in Schilpp 1949)

And, let me add some clearer words from Heisenberg:

> If we want to describe what happens in an atomic event, we have to realize that the word "happens" can apply only to the observation, not to the state of affairs between the two observations. It applies to the physical, not the psychical act of observation, and we may say that the transition from the "possible" to the "actual" takes place as soon as the interaction of the object with the measuring device, and therefore with the rest of the world, has come into play; it is not connected with the act of registration of the result in the mind of the observer. The discontinuous change in the probability function, however, occurs with the act of registration, because it is the discontinuous change in our knowledge in the instant of recognition that has its image in the discontinuous change in the probability function. (Heisenberg 1958)

NOTES

1. Gravity waves in general relativity move at the speed of light.

2. Let |x> be the state vector that corresponds to the measurement of a particle's position being x. Let |Ψ> be the state vector in the space of all possible |x>. Then the wave function Ψ(x) = <x|Ψ>, the scalar product of |x> and |Ψ>, that is, the projection of |Ψ> along the |x> axis.

6

PARADOXES
AND
INTERPRETATIONS

I still believe in the possibility of a model of reality, that is to say a theory, which shall represent the events in themselves and not merely the probability of their occurrence.

Albert Einstein

INTERFERENCE

Nowhere is the strange behavior of quantum phenomena more celebrated than in the familiar Young's double slit interference experiment, which I described in chapter 2. First performed in 1801, this experiment seemed to firmly establish the classical wave theory of light that had been proposed by Huygens in the seventeenth century. The observed bright and dark bands on the detecting screen could now be explained in terms of the interference of coherent, monochromatic light waves from each of the slits. Indeed, the analogous experiment can be done with a pan of water in a simple classroom demonstration that visually demonstrates how waves interfere.

This would have been the end of it if light were in fact the product of ethereal vibrations, as was believed by nineteenth-century physicists. However, as we have seen, the aether does not seem to exist and light appears to be composed of material particles called photons. If we place an array of very sensitive photon detectors along a screen where the beams from the two slits are brought together, then individual, localized photons are registered. The statistical distribution of detector hits follows the interference pattern expected for waves. This seems to be impossible to explain in terms of familiar localized objects such as baseballs, which travel along specific paths from pitcher to catcher, or pitcher to batter to the bleachers. Photons would be expected to go through one slit or the other, not both. But if a photon goes through one slit, how can the interference pattern arise? How can the photon carry with it information about the location of the other slit?

In the conventional quantum mechanical explanation, the uncertainty principle prevents us from localizing the photon to the position of a particular slit. And so, the path of the photon is indeterminate. Most physicists leave it at that, saying that this is the best we can do with our theories—describe observations (see, for example, Fuchs 2000). Thus the photon cannot be thought of as passing through either slit. You are not even permitted to ask the question, since it cannot be answered experimentally, at least with certainty. But if you are the typical reader, you are still likely to puzzle: "Where did the photon go?"

Feel free to ask the question. While any attempt to observe which path a particle takes will interact with it and thus make the interference pattern less sharp, we can still make approximate observations of the path taken without wiping out the pattern completely (Wootters and Zurek 1979, Neumaier 1999). Such a measurement is consistent with the uncertainty principle, since it gives up some of the experimenter's knowledge of the particle's momentum, or equivalently, its de Broglie wavelength, in order to gain some information about its position. In that manner, both the particle and wave aspects of light are made manifest in the same experiment. This invalidates the oft-repeated, solipsistic claim that an object is either a particle or a wave depending on what you decide to measure.

Experiments with electrons, neutrons, and other particles also exhibit interference effects. These are conventionally "explained" in terms of the wave-particle dualism, where a particle has a de Broglie wavelength that supposedly accounts for this wavelike behavior. But, as I have emphasized, no waves are ever directly observed, like the water waves in the pan. Particles are detected locally. Individual particle counters still show the interference pattern.

The Mach-Zehnder interferometer, shown in figure 6.1, is a common example used to illustrate this puzzle. A beam of monochromatic light strikes a half-silvered mirror M1 that splits it into two beams. These beams are then reflected by M2 or M3 so that they are brought together and passed

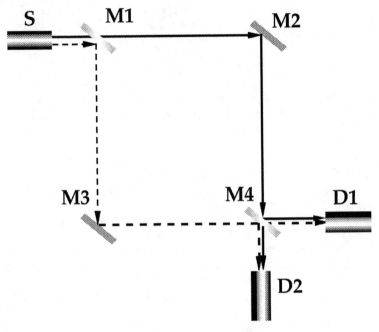

Fig. 6.1. The Mach-Zehnder interferometer. Light from source S is split into two beams by half-silvered mirror M1. Mirrors M2, M3, and M4, where M4 is half-silvered, bring the beams together. The relative path length is such that interference leads a signal in detector D1 and none in detector D2. The effect is observed even for single photons.

though another half-silvered mirror, M4. The path lengths and thicknesses of the mirrors can be adjusted so that the beams constructively interfere in one direction and destructively interfere in another. Thus, a positive signal is registered in photodetector D1, while none is found in photodetector D2.

This can be easily understood in terms of the wave theory of light. The beams are coherent and can interfere like water waves. However, the source beam can be so attenuated that only a single photon passes through the apparatus in a given time interval. That photon will always be detected by D1 and never at D2. This contradicts the commonsense expectation in which D1 and D2 will each, on average, count half the photons. However, this is not what is seen. Somehow a single photon is interfering with itself.

As I have said, the conventional wisdom of Copenhagen quantum mechanics regards it as meaningless to even talk about the photon following either path, since no observation is made along those paths. Reading onto-logical significance into this opinion, we are forced to conclude that, as far as Copenhagen is concerned, "in reality" the photon follows neither path. Alternatively, we might somehow think of the photons as following both paths. In either case, we have difficulty trying to form a commonsense picture of what is happening.

From the quantum theoretical standpoint, no problem exists. The wave function that enters the final mirror M4 is a superposition of the photon states along each path, which constructively interfere at D1 and destructively interfere at D2.

Many types of interference experiments have been done over the years, with photons and other particles. Recently, so-called quantum erasers have demonstrated further remarkable effects of quantum interference in double slit type experiments (see Seager 1996 for a discussion). In particular, these demonstrate that interference can be turned on and off by changes in the arrangement of the apparatus made after the photons have already passed through the slits or separated into different beams by mirrors or other means. Quantum teleportation is another extraordinary phenomenon that has actually been performed in the laboratory, not with Captain Kirk and his crew, but with photons (Zeilinger 2000).

Still, the effects that strike many people as bizarre are exactly as predicted by conventional quantum theory. Although every new experiment exhibiting quantum phenomena is breathlessly reported by the science media, in no instance has it been suggested that quantum mechanics, as a mathematical theory, requires revision.

Wheeler (in Elvee 1982, 1) pointed out that an interferometer experiment can be carried out over cosmic distances by using the observed gravitational lensing effects of supermassive black holes. Lensing is now well established, with many examples of multiple images of galaxies now published in the astronomical literature. These images can be registered in separated photodetectors.

Two beams from the same galactic source will be coherent, and mirrors can be used to bring these beams together in the laboratory. As in the Mach-Zehnder interferometer described above, these mirrors can be so arranged as to produce constructive interference in one detector and destructive interference in the other.

When the mirrors are removed from the apparatus, a photon will be detected in one photodetector or the other, unambiguously specifying whether it came from one side of the intervening black hole or the other. When the mirrors are present, the observed interference would seem to imply that each photon in the pattern somehow came by both routes. But the decision whether to include or exclude the mirrors is made tens or hundreds of millions of years after the photon actually passes the black hole. The implication is that human actions today can affect what happened in distant space during the time of the dinosaurs.

Wheeler interprets his proposed experiment in the customary way: the photons do not travel by any definite route until they are detected. As he famously put it (in Elvee 1982, 17): "No elementary quantum phenomenon is a phenomenon until it is a registered phenomenon. . . . In some strange sense this is a participatory universe." While this has been interpreted by some to mean that human consciousness controls the reality of photon paths, this was decidedly not Wheeler's view (for more discussion of this point, see Stenger 1995b, 96–97).

Even without a "participatory universe," many laboratory observations over the past decades have seemed to imply that quantum objects somehow "know" (speaking metaphorically, of course) what will be the arrangement of the apparatus by which they someday will be observed. As I will argue, the data cry out for an interpretation in which causal processes move in both time directions at the quantum level. However, this is not say that the future determines the present with certainty, any more than the past determines the present with certainty.

EPR

In *The Unconscious Quantum*, I told the story of the profound intellectual debate on the meaning of quantum mechanics that was carried on between Bohr and Einstein from the time of the watershed Solvay meeting in 1927 until their deaths (for the full history, see Jammer 1974). Indeed, this debate continues in their spirits to this very day. I need not go through all this again here. Rather, I will move right to the class of experiments that have remained at the center of the controversy since they were first proposed in 1935.

In that year, Einstein and two younger colleagues, Nathan Rosen and Boris Podolsky, wrote a paper claiming that the conventional interpretation of quantum mechanics was either incomplete or required superluminal signaling (Einstein 1935). To illustrate this, the authors, now commonly referred to as "EPR," proposed a simple experiment, shown in figure 6.2. There we see two particles emerging from a common source in opposite directions. For simplicity, they have the same mass. EPR pointed out that either the position or momentum of one particle can be predicted with certainty from measurements performed on the other. For example, if the source is at rest at $x = 0$ and you measure the position x for one particle, the position of the other will be $-x$. And, from momentum conservation, if you measure the momentum component p along the x axis for one particle, the other will have a component $-p$.

Einstein and his coauthors argued that since both position and momentum of a particle are in principle predictable in this experiment, they

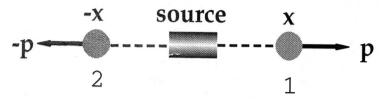

Fig. 6.2. The original EPR experiment. Two particles go off in opposite directions from the same source. An observer measuring the position x of particle 1 can predict that the position of particle 2 is −x. An observer measuring the momentum p of particle 1 can predict that the momentum of particle 2 is −p. Thus, the unmeasured position and momentum of particle 2, though incompatible by the uncertainty principle, must each have "an element of physical reality."

must each simultaneously possess "an element of physical reality," even though neither has been measured and the uncertainty principle says they cannot be measured simultaneously. In their exact words: "If without in any way disturbing a system we can predict with certainty (i.e., with a probability equal to unity) the value of a physical quantity, then there exists an element of physical reality corresponding to this physical quantity" (Einstein 1935).

The "physical realities" of the unmeasured x and p are inconsistent with Bohr's philosophy, which asserts that certain pairs of quantities like position and momentum are not simultaneously real because they are not simultaneously knowable. In his complementarity view, we cannot describe a system in terms of both positions and momenta, only one or the other. Each alternative view is equivalent and provides a *complete* description, that is, tells you all you can know about that system.

What is predicted in the EPR experiment depends upon what is measured, either position or momentum. That choice can be made at the last instant so that insufficient time exists for a signal to be sent from one particle to another without exceeding the speed of light. EPR concluded that either quantum mechanics is *incomplete* in not containing full information about all the physically real properties of a system, or it is *superluminal*, requiring signaling faster than the speed of light. This is referred to as the **EPR paradox**.

Over the years this paradox has been more of a challenge to philosophy than to physics, since no actual measurements have ever proved to be inconsistent with quantum mechanics or out of its realm of application, and no superluminal motion or signalling has been demonstrated.

Even before the 1935 EPR paper, Einstein had objected to what seemed to him to be the superluminal implications of quantum mechanics. The collapse of the wave function struck him as a "spooky action-at-a-distance," since a measurement performed at one place in space instantaneously changes the wave function throughout the universe. But, this is only a problem if you regard the wave function as a real, physical field. The mathematical abstractions we invent as part of a scientific theory can be as superluminal as we want them to be. They are no more physically real than any other figment of the imagination, just generally more useful. As we saw in the last chapter, the Copenhagen interpretation of quantum mechanics, as conventionally understood, does not regard the wave function as real.

Bohr countered EPR immediately. He argued that the EPR definition of reality "contains an essential ambiguity" when applied to quantum phenomena. He said the authors were unjustified in assigning the term "physical reality" to properties without considering their measurement. While the majority of physicists sided with Bohr, this hardly ended the dispute.

For the typical physicist, all a scientific theory is required to do is predict what is measured in the laboratory or observed in nature. And quantum mechanics does that for the EPR experiment, or at least for the equivalent versions that have been actually been carried out. As a result, most physicists have generally ignored the philosophical discussions on the meaning of quantum mechanics that have been carried on over the decades. These issues were never mentioned in any of the courses I took on my way to a Ph.D. in physics, or in any of the research into the basic structure of matter that I was paid to do for forty years. Only when, on my own time, I became sensitized to philosophical disputes, did I notice that a few modern graduate textbooks on quantum mechanics mention EPR in the last chapter, usually skipped or rushed through at the end of the term.

In 1951, David Bohm devised a way to perform the EPR experiment using electron spins which proved more practical than positions and momenta (Bohm 1951, 611–23). His proposal is illustrated in figure 6.3. A source S produces pairs of electrons that go off in opposite directions. The electrons are prepared in such a way that they form a **singlet** state, the state in which the total spin of the pair is zero. Observers with electron detectors at the ends of the two beam lines A and B are able to measure the spins of the electrons. They can do this with Stern Gerlach magnets (see figure 5.1). Bohm thought this could provide for an empirical comparison between the points of view of Bohr and Einstein, but it was not until 1964 that John Bell proved explicitly how this could be done.

Bell showed how the Bohm EPR experiment can be used to distinguish between conventional quantum mechanics and any theory that assumes definite values for quantum mechanically incompatible properties of the system (Bell 1964). The incompatible properties in this case are the spin components of the electron. Quantum mechanics requires that spin components along two different axes, for example x and y, cannot be measured simulta-

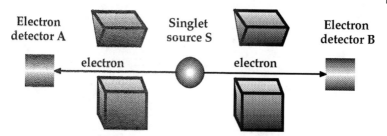

Fig. 6.3. The Bohm EPR experiment. A source S emits two electrons in the singlet spin state (total spin zero) in opposite directions. Observers at A and B can measure the spins of the electrons with respect to any axis of their choosing by rotating the magnets to the desired angle. According to Bell's theorem, quantum mechanics predicts a greater correlation of these measurements than expected from conventional notions of local reality.

neously with unlimited precision in the same manner that the position and momentum along a given axis cannot be measured simultaneously. Bell showed that if you assume definite values of two incompatible spin components, then a certain experimental quantity, a *correlation function*, cannot exceed a particular value. This is known as *Bell's theorem*.

Bell's theorem has been analyzed, revised, and extended by a host of authors, notably Kochen and Specker (1967; see also Clauser 1974, 1978, and Redhead 1987, 1995; I give a simple derivation in Stenger 1995b, 113–18). The theorem shows that it is impossible to formulate a local theory consistent with the usual, statistical predictions of quantum mechanics that at the same time assumes all potentially measurable quantities, including any "hidden variables," have definite values before they are measured. As EPR had claimed, either quantum mechanics is incomplete or superluminal connections exist.

The *hidden variables* referred to here are quantities, such as the position and momentum of the unobserved particles in the original EPR experiment. Recall EPR argued that these must be physically real. Similarly, the three spin components of an electron in the Bohm experiment, which cannot be measured simultaneously, would be examples of hidden variables.

As EPR had asserted for their proposed experiment, and Bell had proven mathematically for Bohm's version of that experiment, quantum mechanics implies that observations made at the end of one beam line depend on the arrangement of the apparatus at the end of the other beam line. This violates the commonsense notion that once two bodies become separated in space and are no longer in communication with one another, anything you do to one cannot possible affect the other. Quantum theory, however, anticipated that the two electrons in the Bohm experiment would

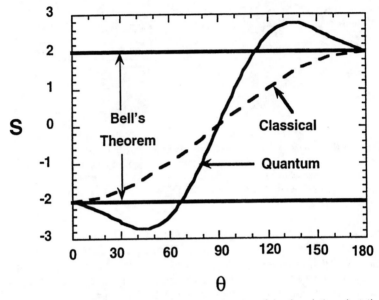

Fig. 6.4. Bell's correlation function S plotted as a function of the the relative orientation angles of the polarizer in the Aspect et al. experiment. Shown are the classical and quantum predictions. Also shown is the range allowed by Bell's theorem for any theory of local, incompatible hidden variables. The data follow the quantum prediction with errors smaller than the width of the line.

show a stronger correlation between measurements performed on each than was possible in any local theory assuming definite values for any incompatible hidden variables. No ambiguity existed in Bell's correlation function, which had a maximum absolute value of two for such hidden variables theories while quantum mechanics predicted a greater value under some circumstances.

The Bohm-EPR experiments has now been performed with great precision, using photons as well as electrons, with results in perfect agreement with the quantum calculation and strongly ruling out any theory of local, realistic hidden variables. The definitive work was done by Alain Aspect and his collaborators in an experiment with photons (Aspect 1982). The results are shown in see figure 6.4. The quantity S shown is Bell's correlation function, and the angle θ is the relative angle of the two polarizers at the ends of the beam lines in the experiment. That experiment was performed with photons rather than electrons, a detail that need not concern us. Basically the results show complete agreement with quantum mechanics and a rejection of both classical mechanics and any quantum theory that assumes both locality and the EPR reality of incompatible quantities.

This result has led many people to the conclusion that quantum mechanics is necessarily superluminal, or what is generally termed "nonlocal." However, note that in the Copenhagen interpretation, incompatible quantities are not simultaneously "real." Thus, Copenhagen remains perfectly consistent with locality, in the sense of not requiring superluminal motion or signalling. However, it raises deep metaphysical issues about what constitutes reality. The positivist Bohr was not disturbed by this, nor are the overwhelming majority of physicists since who have followed his lead. Again, quantum mechanics gives exactly what is observed and Einstein's speed limit remains enforced. So what's the big deal?

BOHMIAN QUANTUM MECHANICS

In 1952, the year after he proposed his version of an EPR experiment, Bohm resurrected the original idea of de Broglie that had been abandoned in the rising tide of Copenhagen. De Broglie had conjectured that the classical picture of a particle following a definite path in space, with definite position and momentum at each point on its path, can be retained if the particle also carries with it a wave field, a *pilot wave*, that can reach out instantaneously in space (see De Broglie 1964 for his later exposition of this idea and earlier references). In Bohm's later variation, the field carried along with the particle is called the **quantum potential** and provides for quantum effects (Bohm 1952). He did not explain the source of this field. But the implication was that it corresponded to some yet-to-be discovered subquantum force.

The de Broglie and Bohm theories gave no predictions distinguishable from the conventional theory (however, see Neumaier 2000, where the claim is made that Bohmiam mechanics is, in principle, empirically distinguishable from conventional quantum mechanics). But they did have one virtue in demonstrating that alternatives existed to the conventional Copenhagen interpretation. At the time of Bohm's work it was thought that hidden variables alternatives were impossible; von Neumann had proved a "no-hidden variables theorem." Bohm's model provided a counter example and Bell later showed that some of the assumptions that went into von Neumann's proof were unjustified (Bell 1966).

When the so-called local, realistic hidden variables theories were ruled out by observations in EPR experiments, Bohm and his followers continued to pursue superluminal hidden variable theories in what they eventually dubbed the "ontological interpretation of quantum theory" (Bohm 1993). This designation was presumably chosen to contrast with conventional quantum mechanics, which they refer to as the "epistemological interpretation of quantum theory." That is, Bohmians claim to deal directly with true reality, while the conventional, Copenhagen interpretation deals only with what we can learn from experiment. As with the many other alternative interpretations of quantum mechanics that have appeared over the past half-century, a consensus has not yet adopted the Bohm view and none seems in sight.

In Bohmian (and de Broglie) quantum mechanics, particles follow definite paths in space. In the double slit experiment, the particle passes though only one slit, but an associated quantum potential or pilot wave passes superluminally through both (Dewdney and Hiley 1982, Bohm and Hiley 1993). This may sound simple, but it requires us to discard the notion that nothing can travel faster than light. Bohmian mechanics is admittedly not Lorentz invariant, since it requires an absolute reference frame in which everything is simultaneous (Bohm 1993, 271).

The results of EPR experiments and Bell's theorem demand that the deterministic quantum potential in the Bohm theory is necessarily superluminal.[1] Somehow this field reaches out instantaneously throughout space. Its configuration depends on everything else that happened in the universe, at every place and every time. For all practical purposes, everything else summed over leads to what appears to be random results and the statistical behavior of quantum systems is thus "explained." In the Bohmian scheme, particles are simultaneously scattering off everything in the universe.

Before his death in 1993, Bohm became deeply involved in mysticism, writing about a holistic universe in which everything acts together in one great cosmic dance called the "holomovement" (Bohm 1980). Gary Zukav called this the "Dance of the Wu Li Masters" in his book on the new physics and Eastern mysticism (Zukav 1979), a literary genre that began with physicist Fritjof Capra's popular *Tao of Physics* (Capra 1975). I will not add further here to what I said about all this in *The Unconscious Quantum* (Stenger 1995b). But you can see how these ideas grew out of Bohm's quantum mechanics.

The Bohm quantum theory is complete in the EPR sense, allowing for a specification of all possible measurables. Bohm's theory solves the EPR paradox by being superluminal. It is so far consistent with all experimental results. Bohmian quantum mechanics requires the existence of some kind of superluminal force operating at the subquantum level. No evidence for such a force has yet been found, and no signals moving faster than the speed of light have ever been observed. According to the Bohm theory, "true reality" lies beyond current observation in a holistic quantum potential field that appears to tell us nothing more than we already know.

According to the Bohmian perspective, standard quantum mechanics is incomplete, which is sufficient to satisfy Bell's inequality and EPR's objections. But Bohm's theory is an uneconomical double-kill solution to the EPR problem. Incompleteness *and* superluminality are not needed together to solve the EPR paradox; one or the other will suffice, and probably neither is required.

INSEPARABILITY AND CONTEXTUALITY

The conventional wisdom holds that quantum mechanics is complete but superluminal in some way. However, superluminal signals have been proven to be impossible in any theory consistent with the axioms of relativistic quantum field theory (Ghirardi 1980, Bussey 1982, Jordan 1983, Shimony 1985, Redhead 1987, Eberhard 1989, Sherer 1993; but see Kennedy 1995). Thus, superluminal motion or signaling is inconsistent with the most modern application of relativistic quantum mechanics.

Assume, for the sake of argument, that quantum mechanics is complete. How can it still satisfy the EPR paradox without superluminality? First, we need to clarify our terms. Discussions of the EPR paradox often use a term **nonlocality** that one would reasonably expect to mean the opposite of **locality** and directly imply superluminal motion. Instead, nonlocality is often used in these discussions to mean something else that is better designated by another term: **inseparability**.

The best way to appreciate the distinction to again consider the Bohm-EPR experiment (figure 6.3). The results of observations made at the end of one beam line depend on the arrangement of the apparatus at the end of the other beam line. This does not mean that the actions taken at one end *determine* the results of observations at the other end. That would imply superluminal signal transmission. In fact, the EPR experiment cannot be used to send signals from one beam end to the other; no information is transferred, consistent with the theorem mentioned above (see also Mermin 1985 and Stenger 1995b, 135–39). Nevertheless, the two regions around the beam ends cannot be simply separated and treated independently. Two particles from the same source somehow "know" what is happening to each other, even after they have separated by a large distance.

Another way to say this is that quantum mechanics is *contextual*. Its results depend on the full context in which a measurement is made—the complete experimental setup. In some way that we still must come to grips with, information from the detector reaches backward in time to the source.

The conventional wisdom says: that's the way it is. Our quantum mechanical calculations give observations that agree with experiment, so why worry about it? Some of us nevertheless keep worrying about it and seek ontological propositions that Copenhagen dismisses as meaningless. These propositions are *metaphysics*, which is still a term of derision in physics. But if hard-nosed quantum physicists refuse to provide realistic ontological models that account for inseparability and contextuality, then the occult ontologies of soft-nosed quantum mystics will win by default.

THE CAT IN THE BOX

Erwin Schrödinger, still a young man in 1935, was undoubtedly gratified by both the professional and popular attention given to his famous equation and the wave formulation of quantum mechanics that he had pioneered ten years earlier. But, like Einstein, he was none too happy with the developing consensus in favor of the Bohr's positivist interpretation of what quantum mechanics implied about the nature of reality. So, when EPR came up with their "paradox" in 1935, Schrödinger was delighted and quick to follow with a paradox of his own: *Schrödinger's cat*.

In Schrödinger's thought experiment, a cat is placed in a closed steel chamber with some radioactive material that has a fifty percent chance of decaying in one hour, activating a circuit that electrocutes the cat (see figure 6.5). After an hour has elapsed, the box is opened and the cat is observed to be dead or alive. Schrödinger was concerned with the quantum description of this event, which seemed to imply that the cat is neither dead or alive until the chamber is opened and someone can look in and thereby collapse the cat's wave function. Prior to this act of observation, if you are to take the conventional quantum description seriously, the cat exists in some kind of "limbo," half dead and half alive.

Certainly Schrödinger did not intend his experiment to be taken literally but rather used as a colorful metaphor for what he viewed as a serious problem with the then developing standard formalism of quantum mechanics. Although other metaphors can be invented that would be more tasteful in today's more sensitive climate, Schrödinger's cat has become such a well-known puzzle that I think I will stick with it. No animals have been harmed in writing this book.

The puzzle Schrödinger highlighted, as Einstein had before him, has come to be called the "measurement problem." It is being debated to this very day. Schrödinger himself had originally introduced the wave function to represent the state of a quantum system. As we have seen, that state can be specified more elegantly by a vector in abstract Hilbert space in which, by the principle of superposition, an axiom of quantum mechanics, we have one axis for each possible result of a measurement.

The state vector representation is more suitable for the discussion of Schrödinger's cat. We imagine a two-dimensional abstract space with axes labeled L and R. The state vector of the cat points along L when it is alive and R when it is dead (see figure 6.6). My reason for use of "L" and "R" will become clear later.

Before the cat is placed in the chamber, it is in state L. When the chamber is later opened, the cat is found to be either L or R with equal likelihood. Since we do not know whether the cat is dead or alive until we open

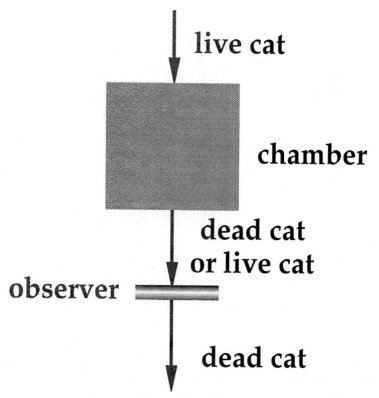

live cat

chamber

dead cat
or live cat

observer

dead cat

Fig. 6.5. The Schrödinger cat experiment. A live cat is sent into in a chamber where it has a 50 percent probability of being killed in one hour. When the box is opened, the cat is observed, in this case, to be dead. The paradox: quantum mechanics describes the state of the cat prior to its being observed as a limbo in which the cat is half dead and half alive. The act of observation puts the cat in the dead state.

the box and look, the quantum state of the cat in the box is neither L nor R, but the state of limbo I will label H, again for reasons we will see. The state vector H can be viewed as a superposition of L and R, as in figure 6.6.

According to the Copenhagen interpretation, not until the cat is observed does its state vector collapse to L or R, rotating instantaneously to either axis in the figure. Schrödinger thought that this description of quantum states was *the* characteristic trait of quantum mechanics that "enforces its entire departure from classical lines of thought." He regarded it to be of "sinister importance" (Jammer 1973, 212). Many have interpreted this role of observation in

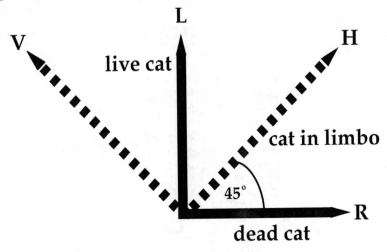

Fig. 6.6. The state vector of a cat points along L when the cat is alive, R when it is dead. Inside the chamber the cat is in "limbo" with its state vector either H or V, which are different superpositions of L and R in equal amounts. The same picture describes the relations between the state vectors of circularly and linearly polarized photons.

changing the state of a system to mean that reality is controlled by human consciousness (Wigner 1961; Capra 1975; Jahn 1986, 1987; Chopra 1989, 1993; Kafatos 1990; Squires 1990; Goswami 1993).

While Schrödinger raised legitimate issues concerning the conventional description of quantum states that are still unsettled today, the Schrödinger's cat experiment itself is not so troublesome as much popular literature would lead you to believe. Most physicists in Schrödinger's day, as today, agreed that the cat dies when common sense says it dies—when the electric shock is received. To believe otherwise is to believe an absurdity, such as a tree not falling in the forest when no one is looking. Quantum mechanics may be strange, but it is not preposterous.

A real cat in a box would be a macroscopic object, no different generically from the human eye that looks in when the box is opened, or a Geiger counter inside the box that triggers when a particle is emitted from the radioactive source. Such objects are not pure quantum states but incoherent mixtures of many quantum states. Thus, the cat experiment must be understood to be a metaphor for the representation of pure quantum objects, such as the photons in a coherent, polarized beam of light.

Let us look at the case of polarized light in more detail, because its strange behavior is probably closer to what Schrödinger had in mind. Suppose we have a beam of left circularly polarized light, which can be generated by simply sending light through a polarizing filter. Classically, this is described as an electromagnetic wave in which the electric and magnetic field vectors rotate counterclockwise, as seen when you are looking toward an oncoming beam. Either field vector can be represented as a linear combination of horizontal and vertical field vectors corresponding to waves that are linearly polarized in the horizontal and vertical directions respectively.

Quantum mechanically, an electromagnetic wave is described as a beam of photons. We can picture a left circularly polarized photon as a particle spinning along its axis like an American football thrown by a right-handed quarterback, turning in the direction at which a right-handed screw turns as it is screwed into a piece of wood. The confusion of left and right here results from an unfortunate definition of terms that were not originally known to be related. In any case, the state vector of the right-handed single photon in a left circularly polarized beam can be written as a superposition of horizontally and vertically polarized single photon states.

A linearly polarized photon can be represented as a superposition of right and left circularly polarized photon states, as the vectors H, for horizontal, and V, for vertical, in figure 6.6. Thus, a horizontally or vertically polarized photon is, in some sense, simultaneously circularly polarized in both the left and right directions. In the quantum mechanical formalism, the states H and V represent single photons, and indeed an experiment will measure individual photons passing through linear polarizers. Yet we cannot picture them as particles with spins aligned along some axis. If we try to measure the spin component of an H or V photon along its direction of motion, we will obtain a value +1 (L) half the time and −1 (R) the other half. Furthermore, we will convert the photon to circularly polarization in the process of measurement. (Massless unit spin particles like photons can only spin along or opposite their directions of motion.)

Thus, we have an indisputably quantum analogue for Schrödinger's cat. A live or dead cat corresponds to a left or right circularly polarized photon, while the cat in limbo is likened to a linearly polarized photon. (There are actually two states of limbo, H and V).

Let us look at the photon analogue of the Schrödinger cat experiment, as depicted in figure 6.7. A left circularly polarized photon in a state L enters a linear polarizer, for example, the simple lens in a pair of sunglasses. The state vector L is a superposition of linearly polarized states H and V. Say the polarizer selects out the horizontal component H, which is a superposition of L and R. The light is then passed through a circular polarizer that selects out the R component and passes that on to the observer.

In both the photon example and the original cat experiment, we are describing an experiment on a single object—although the experiment could be repeated many times to measure probabilities. The puzzle again is: How

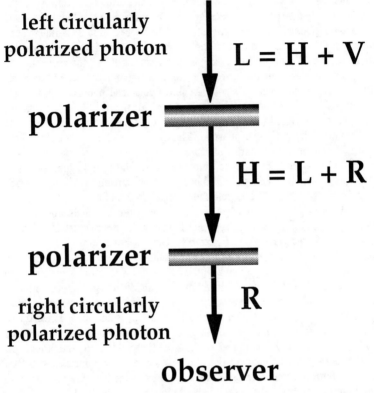

Fig. 6.7. The Schrödinger cat experiment viewed in terms of polarized photons. A left circularly polarized photon, state L, is sent into in a polarizer where its horizontally polarized component, state H, is selected out. It is then passed through another polarizer that selects out the right circularly polarized component R. The state L corresponds to a live cat, R to a dead one, and H to the state of limbo. If only one photon is involved, how can it be half L and half R when it is in the state H? Note: the notation L = H + V is not meant to refer to the mathematical sum of state vectors but just the fact that L is an superposition of H and V.

can the intermediate state H of this object be both L and R and the same time? Recall that an L photon can be viewed as a particle spinning along its direction of motion, like a right-handed screw. An R photon spins like a left-handed screw. Somehow H and V photons are half of the time left handed and half of the time right handed.

One answer to this seeming paradox is to regard the state vector itself as the "true reality." Recall, however, that the state vector does not reside in familiar physical space like a position or momentum vector, but rather in abstract Hilbert space. And so we have a Platonic ontology in which the true Form of the photon, and the true Form of the cat, exist in another realm, a space not at all like the space in which we locate our observations. In this view, the Platonic realm is the true reality, while the world of our observations is a distorted or veiled image of reality. I will talk more about Platonic ontologies in chapter 10.

GHOSTS OF PARALLEL WORLDS

Over forty years ago, Hugh Everett III reformulated quantum mechanics in a way that eliminated state vector collapse from the mathematics, and the collapse postulate from the axioms (Everett 1957). Originally called the "relative state formulation," the new methodology discarded the arbitrary distinction between a quantum system and a classical measuring apparatus assumed in the orthodox picture. Everett included the measuring apparatus as part of the quantum system, where we expect it should be in any complete theory. Furthermore, since the conventional formalism treats the observer as existing outside the system being observed, it cannot be applied to quantum cosmology where the whole universe is the quantum system and no outside observer exists. Everett's method thus provided a way to address quantum cosmology that was previously unavailable.

In the Everett scheme, the entire universe is represented by a single state vector that evolves according to the time-dependent Schrödinger equation, just as the state vectors or wave functions of ordinary quantum mechanics. In chapter 5, I followed Penrose and defined this evolution of the state vector as the **U** process. Recall that the state vector resides in abstract Hilbert space whose axes represent all the possible results of measurements. These notions are retained in the Everett formalism, along with the principle of superposition in which a state vectors can be expressed as a linear combination of the state vectors that point along the axes. And, just as a familiar three-dimensional vectors can be equivalently expressed in terms of different coordinate axes, so, too, do we have great flexibility in deciding what axes in Hilbert space to use.

Where Everett departed from convention was in eliminating the **R** process, which Penrose defined as the collapse (reduction) of state vector upon the act of measurement that occurs in the conventional quantum formalism. Instead, Everett's universal state vector continues to evolve according to **U** even after measurement. Recall that state vector collapse is not predicted by the Schrödinger equation, and indeed is inconsistent with it. This inconsistency results from the fact that the length of the state vector in Hilbert space remains fixed in the **U** process, which is what is meant by the term "unitary." By contrast, a collapsed state vector is usually "reduced" in length in the **R** process to reflect the less than certain probability of the outcome.

The various components of Everett's universal state vector are interpreted as representing other universes (see, for example, Deutsch 1997). However, this is a confusing terminology that I would prefer to avoid, since we still have a single universe being described by a single state vector. Instead I will used the original designation of parallel "worlds." Each outcome of a measurement is represented by a different world within the same universe. For example, in the Schrödinger cat experiment, one world exists in which the cat lives and another when the cat dies. The Everett view has come to be called the **many worlds interpretation of quantum mechanics (MWI)** (DeWitt 1973).

In MWI, the parallel worlds exist simultaneously, as "ghosts" that can interfere with one another and produce coherent quantum effects. They are decidedly not separate universes that have no effect on one another. For example, in the double slit experiment the photon passes through one slit in one world and the other slit in another world. These photons in different worlds interact with one another to produce the observed interference pattern in our particular universe (Deutsch 1997, 43).

Deutsch (1997, 205) uses the many worlds idea to explain the Mach-Zehnder interferometer discussed above (see figure 6.1). Two worlds exist with a photon in each. The photon in one world follows one path, the photon in the other world follows the second path. The two photons can interfere with one another and each world sees a single photon hitting detector D1, with none hitting detector D2.

The Everett formalism can be use to calculate exactly the effect observed. A single state vector for both worlds exists. It has a component corresponding to each world that is operated on by the state vectors of the detectors and mirrors, which shift the phase of one of the components by 180 degrees. The two components are combined in phase when computing the probability for detections at D1 and out of phase for detection at D2.

Many worlds quantum mechanics should not be confused with many universes cosmology, where our universe is part of a vast multiuniverse, each one with different physical laws. This cosmology will be discussed in chapter 13. As I have emphasized, the many worlds we are examining here are part of a single universe, with all worlds having the same physical laws. Ultimately, perhaps, the two ideas will be combined with a single "multiverse" state vector.

According to MWI, we observers have the impression that we live in a single world since each of our observations represents a particular branching along a tree in which worlds split off with every observational act. An ensemble of observers then experience what seems to be a statistical distribution of outcomes.

For example, imagine the Schrödinger cat experiment done with ten cats, each in its own box. Each time a box is opened the universe splits in two worlds with an observer in one world seeing a live cat and an observer in another seeing a dead one. This is repeated nine more times for a total of $2^{10} = 1,024$ branches and thus 1,024 different worlds. Now an observer in one world will find ten dead cats. Another will find ten live ones. But more often the result will be a mixed bag of dead and live cats, for example: DDLDLLDLDD. The statistics when all 1,024 worlds are put together will be exactly what would be expected from probability theory for random events.

In this way, everything that can happen happens. All possible worlds exist and the one any of us live in is simply the one that follows a particular thread through what Deutsch (1997) calls the "Fabric of Reality" in the title of his book. Of course, many of the threads in the fabric will contain worlds without humans or any form of life, sentient of otherwise. Obviously we live along a thread that does, by definition. In this way, the so called "fine tuning" of the laws of nature that some claim shows evidence for design in the universe is simply explained. The notion of fine tuning will be discussed in detail in chapter 13.

If we were only concerned with formal theories, then this would be the end of it. Thanks to Everett's brilliant work, we now have at our disposal a mathematical formalism without state vector collapse that works, and if that is all you want, you need go no further. However, if we are seeking ontological significance, trying to understand what it all means, then we must push on and begin exploring some fantastic new territory.

If MWI is providing us with an insight into the way reality is actually structured, then we are led to the remarkable view that the world we each experience is one of an infinity of parallel worlds. These other worlds sit like ghosts in our background, containing all possible variations of everything that can happen, all described by a single supercosmic state vector. I will leave it to other authors to speculate about what this all means to us as humans. Religions are surely confronted with a whole new range of moral issues. It has been suggested, presumably in jest, that an individual should just commit suicide when things do not go well, since in some other world he must be living the happiest of all possible lives. I certainly feel no compulsion to try this.

Let me again focus on the consideration of the more boring but simpler system of linearly polarized photons. In the many worlds interpretation, an H or V photon is really an L photon and its R ghost in the parallel world. But then, L and R photon states can be written as superpositions of H and V. So are the two parallel-world circular photons each an H and a V from another pair of parallel worlds? And are these composed of more pairs? Like Stephen Hawking's turtles (Hawking 1988, 1), it's parallel worlds "all the way down."

Actually, a way to break this infinite regression, and at the same time provide a simple picture of a photon, is to take the L and R photons to be the "real" ones. These correspond to the situations where the photon's spin is aligned either along or against the direction of motion, something more readily visualized than linearly polarized photons.[2] If we adopt this view, then when we have an H or V photon we need two parallel worlds—one for L and one for R. In chapter 8 we will see how, with time reversibility, we can put them in a single world.

The many worlds interpretation has a cadre of ardent supporters who, justifiably impressed by the Everett formalism and its elimination of the undesirable collapse postulate, claim that the infinity of parallel worlds contained within the theory actually exist. Deutsch is particularly adamant about this, insisting that all other pictures are untenable. He uses the example of the double slit interference experiment in which a single photon passing through one slit seems to interfere with itself passing through another. He says this is just the interference of a photon with its ghost in the parallel universe. As Deutsch puts it, "we do not need deep theories to tell us that parallel universes exist—single particle interference phenomena tell us that" (Deutsch 1997, 51).

One criticism of many worlds is that it does not tell us why our observers never "see" linear combinations, like a cat that is half dead and half alive. So while Everett's relative state formalism provides us with a viable alternative to the conventual theory, it may not, by itself, solve all the interpretational and ontological problems of quantum mechanics. Something else may still be needed.

No doubt Everett's contribution was a substantial one, demonstrating that you can do quantum mechanics without making one of its most controversial assumptions and, as an added bonus, explain its apparent statistical nature. While Everett's methods gave all the results of the conventional theory, like Bohm's theory it so far has made no unique, risky predictions that can be used to test it against the alternatives. Some proponents claim this may be possible. In any case, the success of a theoretical formalism does not imply the correctness of the ontological interpretation people may put to it. Maxwell's equations gave correct results, but this did not demonstrate the existence of the aether. Still, it is important to know that there are many ways to skin Schrödinger's cat.

HISTORIES

Alternate formulations of quantum mechanics have been developed that follow Everett's lead in including the detector as part of the quantum system, but try to keep it all in one world. These include Robert Griffiths's *consistent histories* (Griffiths 1984) and the related but not identical *alternate histories* of Murray Gell-Mann and James Hartle (1990). This approach, when combined with the ideas of **decoherence** that will be covered in the next section, has been termed "The Interpretation of Quantum Mechanics" by Roland Omnès in his book by that name (Omnès 1994). In Omnès view, these latest developments are essentially elaborations on the Copenhagen scheme, placing it on a firmer logical foundation but still following Bohr's intuitions. However, the way in which the new formulations incorporate the detector as part of the quantum system seems to mark a fundamental deviation from Copenhagen, and a substantial improvement.

Griffiths and Omnès (1999) have provided a short, semitechnical description of the consistent histories approach which I will try to summarize minitechnically. In consistent histories, the state vector does not evolve deterministically with time, as in the conventional U process described above (already a major deviation from Copenhagen). Instead, the Schrödinger equation is used to assign probabilities to so-called quantum histories, where a history is a sequence of events at a succession of times. Only those histories that can lead to possible measurements are considered "consistent." The consistency conditions are provided in the mathematical and logical formalism, but basically they require that a meaningful probability can be calculated for a history to be consistent. Histories that imply the measurements of incompatible variables, like a position coordinate and its conjugate momentum, are inconsistent. As Griffiths and Omnès (1999) put it, with positivistic overtones: "What is meaningless does not exist, and what does not exist cannot be measured."

Let us compare the consistent histories interpretation of the Mach-Zehnder experiment given by Griffiths and Omnès (1999) with the Copenhagen and MWI interpretations of the same experiment that were described above. In contrast to the Copenhagen interpretation, consistent histories allows the photon to follow one path or the other. We just do not know which path it is. In contrast to MWI, consistent histories does this in a single world. Indeed, a photon following both paths, as it does in MWI, fails the Griffiths consistency test.

In the Griffiths scheme, the history in which a photon passes along one particular path and directly through the mirror M4 to D1 in figure 6.1 without further modification of its state is inconsistent. However, the history in which a photon passes along one particular path and then emerges from

M4 in either a constructive or destructive superposition of detector states is consistent. Similarly for a photon following the other path. This should be become clearer in chapter 8, where I discuss the analogous experiment in terms of polarized photons.

Consistent histories is thus able to maintain the notion of particle following a definite path and still produce interference effects. The interference results from the particle state being placed in a superposition of the possible detection states by an element of the apparatus, in this case M4. When we detect a particle, say a photon in detector D1 in figure 6.1, we know that the superposition for the observed photon was constructive. The photon with this superposition followed one of the two paths, but we do not know which.

This situation is similar to the explanation I gave above of the polarized photon version of Schrödinger's cat. There the polarizers act as the elements of the apparatus that place the photon in particular superposed states. This is consistent with the principle that the act of measurement in quantum mechanics affects changes in the state of the system.

Like Everett's MWI, Griffiths' consistent histories formulation has a well-developed logical and mathematical structure that properly includes the detector in the quantum system—the great failing of the conventional von Neumann formalism. And, like MWI, the only argument, except perhaps for technical details, can be over what it all means in terms of the reality behind the phenomena. Unlike the dramatic depiction of multiple worlds given Everett's scheme, proponents of consistent or alternate histories have not attempted to provide a model that allows us to form a picture of the underlying reality free of the formalities of mathematics and logic. Even if the photon follows a definite path in space-time, it is still described in terms of some strange linear combination of quantum states specified by the detection apparatus. The arrangement of the apparatus could be modified before the photon gets to it. How can this photon "know" what state it must be in when the arrangement of the apparatus is still to be determined in the future?

Consider the Schrödinger cat experiment from the viewpoint of consistent histories. The intermediate state H is a superposition of the L and R states that correspond to the two possible detection states. While Griffiths insists his scheme is logically rigorous, it still leaves us with that troublesome cat in limbo that is half dead and half alive.

DECOHERENCE

As I discussed in detail in *The Unconscious Quantum*, the positivist language of Bohr and the Copenhagen school has suggested to some authors that human consciousness acts as the agent of state vector collapse. Perhaps the most reputable of physicists to adopt this view was Nobelist Eugene Wigner, who developed the general idea of a quantum-consciousness connection over thirty years ago. In an oft-quoted line, Wigner said: "The laws of quantum mechanics itself cannot be formulated. . . . Without recourse to the concept of consciousness (Wigner 1961). Wigner's notion was that the state vector of nonconscious (or perhaps simply inanimate) matter evolves according to the equations of normal quantum mechanics. Then it encounters something conscious (or perhaps simply alive) that *physically* collapses the state. As I understand Wigner's idea, Schrödinger's cat can collapse a state vector, and so can a tree falling in the forest. But until a human, or perhaps a cockroach, interacts with a polarizer, a photon remains half horizontally and half vertically polarized.

It should be emphasized that Wigner was being very speculative here. He had no empirical reason for suggesting such a profound role for life or consciousness. Furthermore, he did not propose something supernatural, or spiritual. The collapse mechanism was still a physical, natural one, although many authors since have attempted to make a spiritual connection (to list just a few, Capra 1975, Goswami 1993, Kafatos 1990, Zohar 1990, Zukav 1979).

In recent years, alternative notions have been developed in which state vector collapse is accounted for not by an act of human or supernatural consciousness but by natural processes. A generic model for physical state vector collapse was analyzed by Giancarlo Ghirardi, Alberto Rimini, and Tulio Weber in 1986 (Ghirardi 1986; for a good discussion, see Penrose 1994, 331). This is called the *GRW model*. It assumes that some unspecified random fluctuation initiates collapse. Putting in some numbers, the authors estimated that the fluctuation needs to occur only about once every 100 billion years to account for observations. Thus, free individual particles maintain their quantum states for a long time. However, a macroscopic object like a cat has so many particles that a particle state vector will collapse every 10^{-16} second and the state vector of the cat will disentangle in less than a billionth of a second.

The GRW model did not specify the exact physical process responsible for the fluctuation that supposedly induces the collapse of the state vector. It can just as well be human consciousness, or anything else, which makes it all rather ad hoc. Penrose has suggested that state vector collapse may be accounted for whenever a theory of quantum gravity is eventually developed. While he has written several books which argue massively that some new physics is required to understand consciousness, and that this new physics may lie in quantum gravity, he says he is not thinking here in terms of "conscious free will" taking an active role in state vector collapse. As he explains: "It would be a very strange picture of a 'real' physical universe in which physical objects evolve in totally different ways depending on whether or not they are within sight or sound or touch of one of its conscious inhabitants" (Penrose 1994, 330).

Quantum gravity, which most physicists believe is only important at the Planck scale, may not be necessary for state vector collapse. Wojciech Zurek, among others, has argued that the process of state vector collapse is simply the interaction of the particle and its environment or the measuring apparatus as part of the environment (Zurek 1991, 1993).

When a quantum state is viewed as a linear combination of other states, those states are said to be "coherent," a familiar term from wave optics. Collapse can thus be viewed as a process of decoherence. The process of particle decoherence corresponds to the quantum-to-classical transition. Quantum particles are coherent, classical particles are not. Classical waves, of course, exhibit coherent effects, but I am considering the ontology in which these do not exist "in reality."

In orthodox quantum mechanics, pure states and their evolution with time are usually assumed to happen within a closed system, that is, a system in which no energy, momentum, or other quantities are exchanged with the environment. This approximates many carefully designed experiments, but represents few systems in the world outside the laboratory where particles interact and exchange energy with the media in which they are moving. Even the photons in the cosmic microwave background in space are sufficient to decohere the photons from the sun that bounce off the moon. As a result, the moon does not appear two places at once.

Decoherence may be all that is necessary to solve the interpretational problems of quantum mechanics, although this is uncertain at this writing. When decoherence is put together with recent developments in quantum logic, an argument can be made that this essentially brings up-to-date the standard Copenhagen interpretation, stripping it of its questionable elements but retaining its overview. However, decoherence seems to imply a greater reality for the state function than is usually recognized in the orthodox view, so it seems to go beyond Copenhagen.

Max Tegmark (1998) sees decoherence as a way to explain why "we don't perceive weird superpositions" in the many worlds interpretation, which I noted above is an unsolved problem with MWI. He claims that decoherence during the act of observation provides the mechanism by which the coherent effects are washed out for many worlds observers. But if this is the case, why do we need many worlds? Decoherence may work equally well in rescuing the Copenhagen interpretation from its major flaw of not providing a mechanism for wave function collapse. Perhaps a modified Copenhagen, without the collapse postulate but supplemented by decoherence, may be adequate. Indeed this seems to be the view of some who promote the consistent or alternate histories views (Omnès 1994).

The decoherence explanation of the measurement problem, like the other interpretations mentioned and the many others that I have neglected to mention, has not yet developed a supporting scientific consensus. Each has its own fervent supporters who think theirs is the answer. Being rather unromantic, decoherence has not received the same play in the press and pop literature as its more fantabulous competitors of ghost worlds, holistic Bohmian holomovements, and conscious quanta. Nevertheless, it carries considerable appeal to those who believe nature is inherently simple.

NOTES

1. An indeterministic version of Bohmian mechanics exists, but this is hardly distinguishable from conventional quantum mechanics.

2. This is keeping with my desire to see how far I can push a localized particle ontology. Someone preferring a wave or field ontology would argue the opposite, since linearly polarized waves are more easily visualized than circularly polarized waves.

7

TAMING
INFINITY

Quantum electrodynamics gives us a complete description of what an electron does; therefore in a certain sense it gives us an understanding of what an electron is.

Freeman Dyson (1953)

THE QUANTIZATION OF FIELDS

In the classical theory of fields, the equations of motion for field quantities such as the pressure or density of a fluid are used to predict the behavior of the fields in much the same way that the particle equations of motion predict the motion of particles. Indeed, these field equations are derived from the particle equations, where infinitesimal elements of the fluid are treated as particles and assumed to evolve with time according to Newton's equations. However, an infinite number of such elements exists for a continuous field. Each is free to move in three-dimensional space, and has other degrees of freedom, such as rotation about its center of mass. The classical field, then, is a system with an infinite number of degrees of freedom. And, as we will see, infinities are the plague of all field theories, classical and quantum.

Consider the case of a vibrating string. We can think of it as a string of beads, each bead bobbing up an down in simple harmonic motion like a mass on a spring. No problem there, as long as the number of beads is finite. But now consider the motion of a violin string. We treat it as a infinite string of infinitesimal beads, even though we know that the string is in fact composed of a huge but still finite number of atoms. Please keep note of the fact that the assumption of continuity is what leads to the infinity. If we were simply to view a vibrating string as composed of the finite number of beads *it really is*, in the atomic picture, we would never have to introduce infinities or infinitesimals into the description.

So why do physicists make themselves all this trouble? They do so, not because the string *really* is continuous, but because this allows us to use calculus—a calculational method without which we could not solve many problems with even the speediest modern computer. Solving problems with a finite number of elements when that number is very large, like the number of atoms in a violin string, is in practice usually impossible using finite techniques. But thanks to the mathematical geniuses of Newton and Leibniz, we have the marvelous tool of calculus that enables us to get answers for many problems that would otherwise be unanswerable. In calculus, we replace noncalculable sums over large numbers of discrete elements with integrals over an infinite number of infinitesimal, continuous elements. And in a significant number of cases, those integrals are solvable. That is, we can sum the infinite series of terms that appear in our equations. And that is what the student of calculus is trained to do. But remember, a large finite number is not infinity. Infinity is an approximation.

Calculus an approximation? That's not what I learned in school! I was always taught that the calculus expression was the exact one and any discrete representation the approximation. In classical physics, where continuity is assumed to form part of the underlying reality, the calculus procedure is regarded as giving us the exact answer. Derivatives, integrals, and differential equations are the "true reality" in this assumed ontology. But, as we have seen, the universe appears to be composed of discrete elements, not continuous ones. In a discrete universe, the calculus derivatives, like velocity, are *approximations* to the reality of the division of one small number, like a distance interval, by another, like a time interval. Similarly, calculus integrals, like the volume of a sphere, are *approximations* to the reality of the finite sum of many small numbers, like the total number of volume elements inside a sphere.

In classical physics, the field description can always be replaced with a particle one that is often actually more convenient. All you have to do is perform a *Fourier transformation* on the fields, which is the mathematical equivalent of describing a vibrating string as a set of oscillating beads. You take the field, which is originally a function of the assumed continuous spatial coordinates and time, and write it as a sum of sine waves of different wavelengths. When you do this, a remarkable thing occurs: the amplitude of each of the sine waves of different wavelengths is found to obey the standard *particle* equations of motion. The continuous field, then, is equivalent to an infinite number of particles.

Most problems are easier to solve in the particle representation. The complication is that, when we assume continuity, the number of coordinates must be regarded as infinite. As we will see, this nineteenth-century hangup on continuity resulted in endless problems for those in the twentieth century who were attempting to develop the quantum theory of photons and electrons. Experiments kept telling physicists that the universe was discrete, and they kept refusing to believe it. Many physicists still deny the message from nature that it contains no continuous fields, just as they keep denying the absence of an intrinsic arrow of time.

The particle description of fields provided a natural place for quantum mechanics to enter in the description of what in the nineteenth century was viewed as field phenomena. The very term "quantum" means discrete. In 1927, Dirac had generalized the quantization procedures of quantum mechanics, which take you from the classical to quantum description of phenomena, and applied them to the abstract coordinates and momenta representing the classical electromagnetic field (Dirac 1927). He thereby obtained a quantum mechanical description of that field. In the process, he discovered that the mathematical objects corresponding to these coordinates and momenta could be described as **annihilation** and **creation operators** that removed or added one degree of freedom to the field. Dirac interpreted these particlelike degrees of freedom as the Planck quanta of the field. In the case of the electromagnetic field, these quanta were identified with photons. That is, the electromagnetic field was envisaged as being composed of an infinite number of photons, each behaving like a little harmonic oscillator.

Dirac called his procedure "second quantization." More accurately, it might have been termed "first quantization," since the Dirac procedure leads directly to the radiation law that Planck had used to describe the spectrum for black body radiation. Planck had derived the black body spectrum by *assuming* that the field is composed of oscillators whose energy was quantized in units $E = hf$, where f is the frequency of the radiation and h is Planck's constant. Dirac *derived* this connection from the rules of quantum mechanics. He also rederived, in a more fundamental way, the equations for the transition probabilities describing the emission and absorption of radiation, first obtained by Einstein, that could be used to predict the intensities of spectral lines. Thus, Dirac provided a precise theoretical foundation for the transition from classical field to quantum particle implied by the Planck notion that electromagnetic waves occur in discrete bundles.

Note the very important fact that the number of photons in Dirac's quantized electromagnetic field is infinite when that field is regarded as continuous. Dirac defined the vacuum state as the one in which the total energy is zero, but this he represented as an infinite number of photons of zero energy that are not observed. It takes energy to excite any detector, whether it be the eye or a far more sensitive photomultiplier tube. These photons could be imagined sitting out there everyplace in space, unseen because of their zero energy.

What has been more accurately termed "second quantization" was a similar procedure that Jordan applied a bit later to the matter field, that is, the wave function that Schrödinger had introduced in 1925 to represent electrons. Electrons are interpreted as the quanta of the matter field that is represented by the wave function. Since electrons already existed as classical objects, they were regarded as "first quantized" already by the Schrödinger equation. So Jordan's procedure was called "second quantization."[1]

And so, a common theoretical picture evolved in which all quantum fields are represented equivalently in terms of particulate quanta. This is the wave-particle duality, and Bohr complementarity, in another form. It raises the question as to what is the "true reality," particle or fields. If they are complementary, then it would seem that reality is not a combination of both, as was thought in the nineteenth century.

To many theoretical physicists today, fields represent the true reality while particles are just an abstraction. Robert Mills expresses this typical viewpoint: "The word 'photon' is seen as a linguistic device to describe the particlelike character of the excitations of the EM field" (Mills 1993, 60).

However, this cannot be demonstrated by proof or data. At this stage in our knowledge, it is equally plausible, and perhaps simpler and more intuitive, to regard particulate quanta as real and fields as abstractions that appear only in the mathematical theory. That is, the term *field* is the "linguistic device" or mathematical tool. It need only refer to a region of space containing many particles, like a field of pebbles on a beach or the beads on a string. Fields may not even be retained in some sectors of the mathematical theory, as they are replaced by particle coordinates and momenta.

Dirac's second quantization was not well-received at first, but when a few years later positive electrons were found, the formalism to handle the transitions between states of photons, electrons, and positrons was already in place.

THE DIRAC EQUATION AND ANTIMATTER

When Schrödinger first began to develop his wave mechanical approach to quantum mechanics, he treated the electron relativistically. That is, he did not originally limit himself to electron speeds much lower than the speed of light. However, he obtained wrong answers for the observed fine structure of the spectral lines of hydrogen with his relativistic theory. So Schrödinger abandoned this approach and instead proceeded to develop the nonrelativistic equation bearing his name. We always move one step at a time. The Schrödinger equation reproduced the energy levels for hydrogen, earlier derived by Niels Bohr in ad hoc fashion, and solved a few other problems, which was no minor accomplishment.

As it turned out, Schrödinger's original relativistic theory was not far off-base; it simply did not incorporate electron spin. Nor did his nonrelativistic equation. Neither equation was capable of giving the correct spectral fine structure, which very much depends on spin, the intrinsic angular momentum of a particle. The concept of spin was still being elucidated at the time by Pauli and others. Schrödinger's original relativistic equation, now called the Klein-Gordon equation, was eventually seen to apply to spinless particles and thus not relevant to the electron, which has spin $\frac{1}{2}$.

The solution of the nonrelativistic Schrödinger equation for the hydrogen atom nevertheless provided a sound basis for the quantum numbers that had been introduced in older quantum theory of Bohr and others. Schrödinger showed that for each Bohr level n, an "orbital" quantum number L occurs that takes on the values $0, 1, 2, \ldots$, terminating at n-1. The value of L gives the orbital angular momentum of the electron. Additionally, for each L a "magnetic" quantum number m_L ranging from $-L$ to $+L$ in unit steps gives the component of the orbital angular momentum projected on any particular axis.

The observed spectral lines were found to imply another quantum number m_s, corresponding to the component of spin along a given axis, with values $-\frac{1}{2}$ and $+\frac{1}{2}$. When this was incorporated, the fine structure of spectral lines, as well as other tiny but measurable effects that result when external electric and magnetic fields are applied to atoms, could be understood. Furthermore, when the Pauli exclusion principle was employed, the basis for the chemical periodic table was established.

The relativistic equation that accurately described the spin $\frac{1}{2}$ electron was discovered by Dirac in 1928. Dirac's equation did even better than Schrödinger's in describing the hydrogen spectrum. But that was the least of Dirac's triumphs. The spin of the electron fell naturally out of the Dirac equation, along with the strength of the electron's magnetic field, which is twice that expected for a rotating spherical charge. These had to be put in by hand in the Schrödinger theory. Furthermore, Dirac's equation implied the existence of a whole new form of matter—*antimatter*.

In retrospect, antimatter is required in any relativistic theory. But that was not obvious at the time and the Dirac equation was where the concept first appeared. Put simply, relativistic kinematics allows for a particle at rest of have an energy $E = -mc^2$ as well as $E = +mc^2$. Dirac necessarily found that his equations gave negative as well as positive energy solutions.

However, negative energy electrons had never been observed. The Dirac equation implied that the negative energy states should be as common as positive ones. Dirac attempted to explain this uncomfortable fact by proposing that all the negative energy electron states were filled. That is, the universe is bathed in a sea of negative energy electrons that do not interact with familiar particles since these particles lack the energy to lift the electrons out of their deeply bound states. Recall that Dirac had already suggested that the universe was bathed in zero energy photons that were also unobserved, so the suggestion was not as unnatural as it is often presented in the literature.

The negative electrons, in this picture, are like coins down a deep wishing well that cannot be spent without climbing down and lifting them out of the well. But unlike a wishing well, which welcomes every new coin that is tossed in, the Pauli exclusion principle forbids placing any more coins in the well when all the allowed coin-places are occupied. In the case of electrons, no negative energy levels are normally free for positive energy electrons to drop into, with the telltale emission of a million electron-volt photon.

Dirac realized that if a photon of a little more than a million electron-volts came along, it could kick an electron out of a negative energy state producing a "hole" in the electron sea. This hole, like the holes in modern semiconductors, would appear as a particle of positive electric charge, opposite to the ordinary electron. The hole would be observed as an "antielectron" or **positron**.

Dirac originally thought these holes might be protons, the nuclei of hydrogen atoms. But he soon realized that they must have the same mass as electrons. As it happened, cosmic ray physicists were seeing tracks of particles in their cloud chamber detectors that appeared to be positive electrons. After carefully ruling out alternatives, they soon confirmed the existence of the positron.

Despite the observation of positrons, Dirac's hole theory was questioned by Pauli and others. The notion of a universe bathed in an infinite sea of negative electrons was not very appealing. Physicists had no trouble with Dirac's picture of the vacuum as forming an infinite sea of photons. Photons carry no charge. But where were the charges and electric field of all those electrons?

A more palatable model for antimatter did not come along until 1949, when Richard Feynman published his theory of positrons. Instead of holes in a negative energy sea, Feynman proposed that positrons are electrons moving backward in time. An electron moving back in time is indistinguishable from a positron moving forward in time, negative charge going back corresponding to positive charge going forward.

Most physicists have continued to follow the familiar convention of a single time direction, determined by human experience, treating positrons as particles distinct from electrons. However, the Dirac theory, and indeed any relativistic theory, requires either negative energies or reverse time. Any model of reality consistent with relativity must contain either negative energies or motion in both time directions. And, as we have already argued, time reversibility at the fundamental level is deeply implied by everything we have learned about nature to the current day.

PREWAR QED

Feynman's clarification of the meaning of negative energy states would not happen for two decades. Returning to the historical sequence of events, the existence of positrons had a profound effect on the thinking of theoretical physicists in the 1930s. The process in which two photons with a little over a million electron-volts of energy could produce an electron-positron pair implied that the number of electrons was not conserved in high energy reactions.

When, in the nineteenth century, the atomic picture was developed and the electron recognized as a key constituent of matter, chemists had a ready explanation for many of the regularities observed in chemical reactions. While electrons might be exchanged among reacting atoms and molecules, the total number remained fixed. For example, when two hydrogen atoms and one oxygen atom united to form H_2O, the ten electrons in the original atoms remained present in the final molecule.

Now, however, we have a process that contradicts this chemical view. In pair production, an extra electron and positron are produced in the reaction. In the reverse process, pair annihilation, they disappear. Matter no longer seems to be conserved.

In fact, photons are also particles of matter, a point not usually conceded in many discussions of chemical phenomena. From the early days of the quantum, we have known that the number of photons is not conserved in physical processes. Photons are absorbed in the photoelectric effect. In atomic transitions, photons are emitted or absorbed by atoms. However, even today people tend to think of electrons and other subatomic particles as objects of a different nature from photons, to be classified as conserved "matter" while photons are classified as nonconserved "radiation." In fact, this distinction is an artificial one. While certain properties of matter are indeed conserved, in the case of both electrons and photons, particle number is not one of these properties—at least not in the normal time-directed way physicists describe photon and electron interactions at high energy.

In any case, the apparent nonconservation of particles in fundamental processes led to a dramatic alteration in the theoretical structure of physics. Only the two-body problem was completely solvable in classical mechanics.

To calculate the motion of three or more bodies, even when you knew exactly how they interacted, required approximation techniques. Now physicists found themselves in the situation where the number of particles was not only more than two, that number was indefinite and could very well be infinite! But all was not lost. Dirac had already described the electromagnetic field in terms of photons that are created and destroyed in electrodynamic processes. Jordan had done the same for electrons, and so the mathematical structure was in place for the next step: the interaction of photons and electrons.

Both the Schrödinger and Dirac equations had been used to describe an electron in the presence of static electric and magnetic fields, Dirac more accurately than Schrödinger. However, in those applications the fields were represented by their classical potentials. Thus, the calculations were semiclassical, that is, half quantum and half classical. Any complete *quantum electrodynamics*, or *QED*, would have to be fully quantum, describing, as one of its fundamental processes, the interaction of a quantized electron and a photon from a nearby quantized electromagnetic field.

In 1932, Hans Bethe and Enrico Fermi provided the picture for the interaction between particles that would become the standard conceptualization of fundamental physics. Instead of particles interacting with continuous fields, or fields with each other, as in classical physics, they visualized two electrons as interacting by the exchange of photons. As seen in figure 7.1, an electron emits or absorbs a photon that is, in turn, absorbed or emitted by the other electron. In this manner, the electrons exchange energy and momentum, which is just what constitutes an interaction. The force on a particle is the momentum it receives in a unit time interval. As we will see, particle exchange is exactly the scheme that was developed by Feynman with his **Feynman diagrams** and the precise rules for their calculation. All fundamental interactions, as far as we can tell, can be reduced to such particle exchange processes.

In 1933, Fermi introduced his theory of **beta decay**, a process by which atomic nuclei emit electrons. This eventually grew into the theory of weak interactions and today is explained in terms of the exchange of heavy particles, the W and Z bosons. As we will find when we discuss today's **standard model** of particles and forces, these particles were observed at exactly the masses predicted by a theory developed in the 1970s. In this theory, quantum electrodynamics is (more or less) unified with the weak and strong nuclear interactions.

In 1935, Hideki Yukawa suggested that the strong nuclear force that binds protons and neutrons in nuclei resulted from the exchange of a particle, called the **meson**, whose mass was intermediate between that of the electron and proton. That particle was eventually found in 1948, although the full understanding of the strong nuclear force would also have to await the discovery of quarks and the development of the standard model. As it worked out, Yukawa's meson was not the quantum of the strong nuclear force.

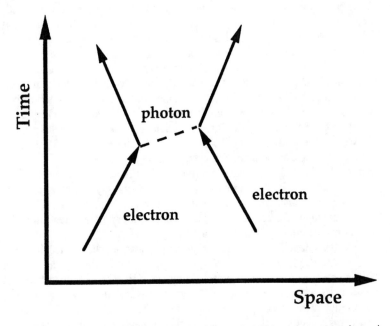

Fig. 7.1. The basic interaction between two electrons in which a photon is exchanged, resulting in a transfer of energy and momentum from one particle to another. Although pictures of this type are now called Feynman diagrams, the idea of particle exchange was first proposed by Bethe and Fermi in 1932.

Let us go back to the basic photon-electron interaction. As noted, it became very clear in the 1930s that a simple two-body photon-electron system cannot be isolated. The tiniest bit of kinetic energy carried by either particle can be transferred to one or more of the many zero energy photons Dirac envisaged filling the vacuum, making them nonignorable, and indeed measurable, energetic photons. Furthermore, at sufficiency high energy (a bit over one million electron-volts) a photon can lift an electron out of Dirac's negative energy electron sea, creating a positron-electron pair. Few applications were found where QED could be treated as a simple two-body process.

In 1928, Oscar Klein and Yoshio Nishina used the Dirac equation to calculate the rate for the scattering of photons by free electrons. This agreed very well with the data for low energy X-ray photons scattered by the electrons in light atoms such as carbon, but failed to correctly predict the observations of the scattering of higher energy gamma rays from the electrons in heavy atoms such as lead. Only after the discovery of the positron was it recognized that this discrepancy resulted from the gamma-ray photons creating positron-electron pairs, the positron annihilating with another electron, and other multibody effects.

To make matters even worse, the Heisenberg uncertainty principle allowed energy to be violated for sufficiently small periods of time. Thus, the vacuum had to be regarded as teeming with randomly occurring, short-lived electron-positron pairs, and just about everything else, even Mack and anti-Mack trucks, as some wag has put it. All of these vacuum particles had some probability for interacting with the original particles, and so any calculation of quantities such as transition rates had to take these possibilities into account. Fortunately the Mack/anti-Mack truck contribution can be neglected.

Quantum theorists in the 1930s worked hard on this problem, with limited success. Because of the infinite number of degrees of freedom that were assumed present in quantum fields, infinities kept showing up in their calculations. Heisenberg despaired and started working on other problems. Dirac was also ready to give up. To make matters worse, global war was on the horizon. It would take a new generation of remarkable men in the period immediately following World War II to show us how to handle these infinities.

CLASSICAL INFINITIES

The infinities that plagued QED were not to be laid exclusively at the door of quantum mechanics. In fact, quantization, that is, discreteness, ultimately turned out to provide the way out of infinities that were already present in classical electrodynamics and shoved under the rug. As Lorentz had shown, classical electrodynamics predicts that an accelerated charged particle will emit radiation and consequently lose energy. In order to conserve energy, work must be done on the particle to account for that energy loss. The force that does this work is called the radiation resistance, or the radiation damping force that opposes any external forces that are acting to accelerate the particle. But if the particle is all by itself, with no external forces present other than the force producing the acceleration, then what is doing this resistive work? The standard answer was that the particle does work on itself.

Radiation resistance was interpreted by Lorentz and other nineteenth-century physicists as a *self-action* in which the radiation emitted by the particle acts back on the particle itself. Actually, the notion is not that mysterious. Electromagnetic radiation carries momentum, so if radiation is being emitted forward by a particle, that particle must necessarily recoil backward to conserve momentum.

Lorentz had studied the radiation resistance problem in great detail, but we can understand its physical source fairly simply. Consider two particles within a charged body. Suppose the body is being accelerated along the line between these particles by some external force, such as a static electric field. The two particles interact with one another. In normal Newtonian mechanics, the internal forces between the constituents of composite bodies are canceled by Newton's third law: for every action there is an equal and opposite reaction. The internal forces of a body all cancel out and the motion of the body as a whole can be treated as if it were a single mass acted on by whatever external forces are present.

However, Lorentz recognized that this would not be the case when the finite propagation speed of electromagnetic waves was taken into account. Imagine our two particles within the body initially at rest. Particle A emits a pulse of electromagnetic radiation in the direction of B, which we will take to be the direction of the acceleration of the system. Simultaneously, B emits a pulse in the direction of A. Each pulse moves at the speed of light and carries a certain momentum that is to be transferred to the struck particle.

Although the particles start at rest, they begin moving as the result of their acceleration. In the time it takes the pulse to reach B, B will have moved relatively farther from A's position at the time A emitted the pulse. On the other hand, A will have moved relatively closer to the point where B omitted its pulse. If a series of pulses is emitted at a fixed rate by each particle, the rate of momentum transfer to the other particle will be the average force on that particle. That rate will be lower for the pulses received by particle B than for particle A, resulting in a net force opposite to the direction of acceleration. This is the radiation resistance. It results from the finite speed of light and seems to indicate that Newton's third law is wrong, or only an approximation.

Indeed, you can turn the picture around and argue that the accelerated charge would not emit any radiation if it where not for this difference between the internal forces that results from the finite speed of light. Since the internal work done by those forces does not cancel, some energy must be emitted.

The total force of radiation resistance can be written as a sum of terms, one proportional to the acceleration, another proportional to the time rate of change of acceleration, another proportional to the time rate of change of that, and so on. The term proportional to acceleration can be taken over to the other side of $F = ma$, the ma side, resulting in an effective increase in the mass of the body. We call this added mass the **self mass**. This quantity, which reflects the increased inertia that a charged particle exhibits when accelerated, was first calculated by J. J. Thomson in 1881 (Dresden 1993, 40). Unsurprisingly from a relativistic viewpoint, the added mass is just the electrical potential energy of the body, the repulsive energy that is ready to blow the particle apart, divided by the speed of light squared. Unhappily, the electromagnetic self mass goes to infinity when the diameter if the body goes to zero, that is, when the body becomes a point particle.

This is how classical electrodynamics was left at the turn of the century, when attention turned toward the quantum. Charged particles, such as electrons, had infinite electromagnetic self mass. Since this was contrary to observation, the conclusion was that electrons must have finite size that was otherwise not indicated by any observations. Fortunately, the size needed to make everything come out all right was very small, about the size of an atomic nucleus. Classical electrodynamics was known to breakdown anyway at such small distances, and at larger distances where it could properly be applied, the electron could be treated as a point particle.

SUBTRACTING INFINITY

As we have seen, the problem of infinities in QED came about as the consequence of the attempt to sum up, to "integrate," all the contributions of the large number of number of photons and electron pairs that live in the space surrounding a lone electron. When that number was infinity, the sum gave infinite transition rates. We can easily see why this is so. When it takes no energy to make something happen, then that something is going to happen at a very high rate—unless forbidden by some other basic principle. Furthermore, just taking the electron by itself and figuring in the contributions from interactions with the surrounding particles generated by the electron's own field resulted in an infinite self mass for the electron, just as occurs in classical electrodynamics.

In quantum theory, as you move closer to an electron and the energy density grows larger, you begin producing particle pairs. The masses of these particles add to the mass of the electron, which becomes infinite as you approach a point. A reasonable suggestion is that perhaps QED also fails at some small distance. Indeed, many physicists in the 1930s were of the opinion that QED fails at nuclear dimensions. But, unlike the classical case, there was no good reason to assume this a priori. Theories need to be pushed as far as they can go before deciding they no longer apply in a given domain. As it turned out, QED can be pushed to distances far smaller than a nucleus, indeed to the smallest distances that have been probed to date. But it took considerable effort from many people, and not a little genius from a few, to figure out how to do it.

The idea gradually developed that the infinities could be removed and the mass of the electron recalculated by a procedure called **renormalization**. In this process, infinite terms are absorbed into the calculation of the mass of the assumed point electron, which are themselves infinite for the reasons we saw above. A similar procedure was applied to the charge, which also had infinities problems. Again, it was classical physics that pointed the way.

In 1938, Hendrick Kramers, who up to this time had been buried in quantum physics, decided to take another look at the classical situation. He realized that the electromagnetic self mass was being added to the particle's "intrinsic mass," a quantity that is never actually measured. As we saw above, this added mass results from taking that part of the radiation reaction that is proportional to acceleration and moving it over to the other side of F = ma, that is, including it in the mass m. What exactly is the m in F = ma? It is not some abstract, metaphysical "bare" mass of a particle. Rather, m is the inertial mass that we measure in an experiment when we apply a measured force F to a particle and measure the resulting acceleration a. That is m = F/a, an operational quantity.

Kramers suggested that the self mass of a charged particle is simply absorbed in its measured inertial mass. Thus, when you do physics, such as when you put a charged particle in an electric field, the self mass is already included in the measured mass. The fact that someone calculates infinity for the self mass in his theory is an artifact (that is, a stupidity) of the theory, something some silly person put in there that does not exist in reality. If that person put it there, he can take it out.

Since Kramers made his suggestion within the framework of classical physics, its significance was not immediately recognized. However, he presented his proposal ten years later at the 1948 Shelter Island conference that marked the starting point of the explosive developments in postwar QED for which renormalization was the key (Schweber 1994, 189). The renormalization program that was then instituted for QED followed the lines that Kramers had outlined. The infinities that people kept calculating were recognized as of their own doing. They did not appear in observations, and so were clearly wrong. They had to be moved over to the other side of the equation and subsumed in what we actually observe in nature.

THE FEYNMAN-WHEELER INTERACTION THEORY

When Richard Feynman was still an undergraduate at MIT In the late 1930s, he fell "deeply in love" with an idea on how to solve the electron self-mass problem. As he described it in his 1966 Nobel Prize lecture:

> Well it seemed to me quite evident that the idea that a particle acts on itself, that the electrical force acts on the same particle that generates it, is not a necessary one—it is a sort of silly one, as a matter of fact. And so I suggested to myself, that electrons cannot act on themselves, they can only act on other electrons. That means there is no field at all. . . . There was a direct interaction between charges, albeit with a delay. Shake this one, that one shakes later. The sun atom shakes; my eye shakes eight min-

utes later, because of a direct interaction across. (Feynman 1965a)

Feynman reasoned that a field does not have any degrees of freedom independent of those of the particle source. You do not see light or electromagnetic waves. You see the particles that are the source of the light and the particles that react to it.

In the fall of 1938, twenty-year-old Feynman entered graduate school at Princeton where he was assigned to assistant professor John Archibald Wheeler, who was just four years older. Wheeler would independently make a great name for himself in physics, but his mentorship of Feynman alone merits lasting admiration.

As Feynman related in his Nobel lecture in Stockholm, he realized he may have been wrong about the electron not acting on itself when he learned about the radiation resistance that a charged particle experiences when it accelerates, which was described above. As we saw, the problem of infinite electron self mass was never really satisfactorily solved in classical physics, and Feynman still felt that he might be able to do so by eliminating self action. He conjectured that perhaps the radiation resistance of an electron might be due to the interaction with another electron. However, Wheeler showed that this would not work since the force between electrons falls off as the inverse square of the distance between them. Furthermore, there is a time delay between the action of the source and the second charge acting back on the source.

Wheeler then had the insight that perhaps "advanced waves" may be the explanation. In ordinary classical electrodynamics, the electric and magnetic fields detected at a given position and time are calculated from the distribution of charges and currents at the source at an earlier time called the **retarded time**. This allows for the finite time of propagation of the signal from the source to the detector. However, as we saw in chapter 2, Maxwell's equations are time-symmetric; they do not single out a particular direction of time. Thus, the same results can be obtained by computing the distribution of charges and currents at the source at a later time, called the **advanced time**. That is, you can equivalently treat the problem as if the signal arrives at the detector before it leaves the source. This solution is conventionally ignored by making the assumption, not usually recognized as an assumption, that cause must always precede effect.

Wheeler recognized that causal precedence is indeed an assumption, and Feynman was the last person in the world to shy away from discarding any sacred notion—even something so sacred as the unidirectional flow of causality. Thus was born the Wheeler-Feynman *interaction theory* of electromagnetic radiation.

Feynman and Wheeler hypothesized that charged particles do not act on themselves, only on each other. However, the motion of a charge is not only determined by the *previous* motion of other charges, but by their *later* motion as well. Furthermore, all the particles in the universe must be considered in understanding the motion of one particle.

Let us go back to the situation described earlier in which the self force of a two particle system was seen to result from the differing time delays between the emission and absorption of the electromagnetic pulses sent from one particle to the other. The different delays in the two directions resulted in a net force opposite to the direction of acceleration that was interpreted as the radiation resistance.

Now imagine what happens if we include effects from the *future* as well as the past, as we are logically required to do by the time-symmetry of electrodynamics. The pulses from the advanced time, that is, the future, will exactly cancel those from the retarded time, the past, leaving a zero net self force. Instead of being infinite, the electromagnetic correction to the electron's inertial mass is zero!

But if the electromagnetic self mass and energy are zero, Feynman and Wheeler still had to account for radiation. The fact is that accelerated charged particles radiate and experience radiation resistance. The young physicists reasoned that a lone charge, alone by itself in the universe, will not radiate. This will happen only when other charges are present. Radiation requires both a sender and a receiver. Furthermore, the sender is also a receiver while the receiver is simultaneously a sender. The jiggling of charge A "causes" a "later" jiggling of charge B; but this jiggling of charge B "causes" the "earlier" jiggling of charge A. Radiation is thus an *interaction* between charges, a kind of handshake, and not a property of a single charge.

Wheeler realized that the advanced and retarded effects of all the charges in the universe must be considered even in the interaction between two charges, at least in an average way. If all the charged particles in the universe contribute to the motion of a single particle, then the inverse-square falloff is compensated exactly by the fact that the number of charged particles in a spherical shell, assuming uniformity, increases as the square of the radius of the shell.

In the Feynman-Wheeler interaction theory, a charged particle simultaneously emits two electromagnetic waves, one forward in time and one backward. This happens continually. Any given emission results in no change in the particle's energy or momentum, since these cancel for the two waves. To see this in our more familiar way of looking at things, in one time direction, the wave emitted backward in time is equivalent to one moving forward in time that is absorbed. Thus, one wave is absorbed the instant another is emitted, the momentum and energy of the emitted wave coming from the absorbed one with no net change in these quantities.

Now imagine two charged particles relatively close to one another and far from all the other particles in the universe. The rest of the universe can be viewed as an absorbing spherical wall surrounding the particles some distance away. Each particle is continuously emitting electromagnetic waves in all directions, in forward and backward time pairs. Most of these waves get absorbed by the wall, both in the far future and in the far past. The wall is also sending out radiation, to the future and the past.

In the near future and past, one wave from each of the particles is not absorbed by the rest of the universe, but by the other particle. This upsets the balance and a net force between the two results. Basically, the two particles receive impulses at a higher rate than they would if each was by itself, far from other particles and the wall. Wheeler showed, rather simply, that if you assume the force on a particle is one half the sum of retarded and advanced forces, then the net force was exactly equal to the retarded force that we get conventional electrodynamics, *including the radiation resistance,* when interpreted as a self force. So the result of the interaction of two particles, in the Feynman-Wheeler interaction theory, was exactly as given in conventional electrodynamics, including radiation resistance but without self action or infinite self mass.

Imagine the scene at Princeton in the fall of 1940. The twenty-two-year-old Feynman stands in front of an audience including Einstein, Pauli, and Wigner, among other notables, to present the Feynman-Wheeler theory. Pauli objected, but no one remembers exactly what his objections were. Feynman recalled Einstein commenting that the theory was inconsistent with general relativity, but since electrodynamics was better established, perhaps a better way to do gravity would be developed that made the two consistent.

The theory worked. It gave the correct formula for radiation resistance without self action. This meant that classical electrodynamics does in fact provide a consistent picture of nonself-acting charged particles, provided you include the impulses propagated backward in time as well as forward—exactly as the equations tell you that you must do! Even today, this fact is not appreciated. The great outstanding, unsolved problem of classical electrodynamics, which had been left hanging at the turn of the century, was in fact solved by Feynman and Wheeler by their allowing for electromagnetic effects to propagate backward as well as forward in time. Or, to put it another way, they allowed the equations to speak for themselves and did not impose on these equations the psychological prejudice of a single direction of time. Even in the nineteenth century, before quantum mechanics, the equations of physics were pointing to a time-reversible universe. In their refusal to believe their own equations, physicists were running into problems and paradoxes that would require extraordinary means to solve within the framework of directed time. It took fifty years to find two physicists with the courage to propose time reversibility—and even they demurred from drawing its full implications.

FEYNMAN'S SPACE-TIME PHYSICS

The natural next step for Wheeler and Feynman was to try to make their classical interaction theory into a quantum one. Wheeler tried to quantize the theory in the conventional way, with no success. Feynman realized that the standard procedure of Dirac and Jordan, which I described above, though relativistic, was still based on an intrinsically nonrelativistic perspective in which quantities that describe systems at different places in space are always defined at the same time. Interaction theory, with its signals from the future as well as the past, just did not lend itself to quantization within this framework.

All formulations of quantum mechanics up to this point had relied heavily on analogies with the way of doing classical physics developed in the nineteenth century by Hamilton (see chapter 2). As a result, they were bound to traditional notions about the nature of time, in particular, the notion that time is absolute (that is, invariant—the same in all reference frames). Hamilton's equations of motion are used to predict the time evolution of the generalized coordinates and momenta of a classical system in terms of a quantity called the Hamiltonian that is usually, but not always, the energy of the system written as a function of the coordinates and momenta. Time is the independent parameter in these calculations, and we assume that the Hamiltonian can be specified at all positions in space at a given time. For example, in the case of a charged particle in an electric field, one term in the Hamiltonian will be the electrical potential energy at each point in space at a given, absolute time.

Both Schrödinger's and Heisenberg's versions of quantum mechanics, and the Dirac-Jordan methods of second quantization, were based on the Hamiltonian approach in which time is a parameter. By singling out time as something special, Hamiltonian formulations forced themselves to be intrinsically nonrelativistic, at least in their conceptual framework. They can be patched up to be relativistic, but only with great effort and in an ad hoc fashion. This was probably the major reason for the limited success of 1930s QED. Only a theory in which relativity is built in from the ground up could be expected to readily give correct answers when dealing with particles travelling near, or as in the case of the photon, at the speed of light. As we will see, QED ultimately flowered when it was made manifestly relativistic. This happened almost simultaneously with three independent but equivalent approaches. One of these, chronologically the last but computationally the easiest, would be Feynman's space-time approach.

Feynman sought a method of calculation that inherently recognized the relativistic insight that time and each coordinate of space must be placed on equal footing. Einstein had shown that the concept of "now" is meaningless except for events occurring "here." As Feynman explained in his Nobel speech: "I was becoming used to a physical point of view different from the more customary point of view. In the customary view, things are discussed as a function of time in very great detail. For example, if you have the field at this moment, a differential equation gives you the field at the next moment and so on." He called this the "Hamilton method, the time differential method."

Feynman explained that with the Hamilton method "you need a lot of bookkeeping variables to keep track of what a particle did in the past. These are called field variables." He broke with the Hamiltonian tradition by going back to what was a deeper principle of classical mechanics, the **principle of least action**. He explained: "From the overall space-time view of the least action principle, the field disappears as nothing but bookkeeping variables insisted on by the Hamilton method" (Feynman 1965b, 162–63). In its place we have a principle that describes the paths of particles throughout all of space and time. And it was intrinsically relativistic.

Author James Gleick, in his Feynman biography *Genius* (1993), tells how Feynman first learned about least action in high school in Far Rockaway, New York. Least action cannot be found in the usual high school or even lower-level undergraduate college physics textbooks. But Feynman's astute high school physics teacher, Abram Bader, took the bored student aside one day and had him calculate the kinetic and potential energies for the path of a ball thrown up to a friend in a second-floor window. Under Bader's guidance, Feynman showed that the sum of the two, the total energy, was constant all along the path, as he expected it to be from the textbook.

Then Bader had him also calculate the *difference* between the two, what Gleick calls the action but is technically the Lagrangian, which I introduced in chapter 2. The action is the Lagrangian averaged over the path multiplied by the time interval for the trip. In any case, the difference in kinetic and potential energies varied continually along the path. Bader then pointed out "what seemed to Feynman a miracle." The path the ball followed was the one for which the action was the least. For all other possible paths between the two points, he found a greater value (Gleick 1993, 60–61).

At Princeton, Feynman looked for a way to get the principle of least action into quantum mechanics and avoid the use of Hamiltonians. While enjoying a beer party there in 1941, he learned from a foreign visitor, Herbert Jehle, that Dirac had written a paper on this. The next day the two looked up the paper. Dirac (1933) had suggested that the classical action was the "analogue" of the phase of the quantum mechanical wave function. On the blackboard in front of the astonished Jehle, Feynman proceeded to derive the Schrödinger equation assuming the phase of the wave function was *proportional* to the action. When Feynman approached Dirac at a Princeton meeting later, in 1946, he asked him if by "analogous" he had meant "proportional." Dirac asked, "Are they?" Feynman replied, "Yes." The laconic Dirac responded, "Oh, that's interesting" and walked away (Schweber 1994, 390 and references therein).

The principle of least action says that the path a particle takes from a point A to another point B is the one for which the action is "stationary," that is, a maximum or minimum. Feynman elaborated this unto a quantum mechanical principle by postulating that the **probability amplitude** for the particle to go from A to B is the sum of the amplitudes for all possible paths between A and B (Feynman 1948, 1965a). The probability amplitude is a complex number whose phase is equal to the action in units of the quantum of action, $\hbar$. The absolute value of the amplitude is squared to get the probability. Phase differences between various terms in the amplitude lead to the interference effects that characterize quantum measurements.

This represented no less than another way to do quantum mechanics: the **Feynman path integral** method (Feynman 1942, 1948, 1949b, 1965a; Beard 1963). Like the classical least action principle that inspired it, Feynman's formulation is in many ways more deeply revealing of the nature of the principles involved than the more traditional, Hamiltonian formulation. Furthermore, it exhibits an intrinsic time symmetry and provides a means to understand the quantum-to-classical transition.

While suggestive of wave functions, Feynman amplitudes are not quite the same mathematical object. The wave function is very much tied to the Hamiltonian picture and evolves with time according to the time-dependent Schrödinger equation in conventional quantum mechanics. In the usual visualization, we imagine a quantum particle as a wave packet propagating from A to B. While the wave function only gives the probability for finding a particle at a given place, it still propagates deterministically between the two points.

Indeed, as we have seen, some alternative interpretations of quantum mechanics, notably the ones due to de Broglie and Bohm, assign ontological meaning to the wave functions and restore determinism at the expense of requiring superluminal connections. Bohm showed, as Erwin Mandelung had before him, that the phase of the Schrödinger wave function, which we have seen is proportional to the action, obeys the classical Hamilton-Jacobi equation of motion with quantum effects incorporated in a "quantum potential" (Mandelung 1927, Bohm 1952).

The Feynman amplitude, by contrast, does not specify a unique history for a wave or particle. In fact, it has no waves or fields of any kind. It simply considers all the paths of a particle between two points and computes the likelihood of each. The proportionality constant that relates the phase to the action is simply $\hbar$, Bohr's "quantum of action." That is, the phase is simply the action in units of $\hbar$. The classical principle of least action then follows in the limit where the action is many action quanta, which is the normal classical limit of quantum mechanics. In this case, the probability peaks sharply at the classical path. Feynman thus provides us with a smooth transition from quantum to classical physics, while uniquely breaking with the Hamiltonian tradition and automatically providing a perspective in which space and time are on the same footing.

The basic ideas of the path integral formulation of quantum mechanics were in Feynman's 1942 Ph.D. thesis (Feynman 1942). Unfortunately, at this pregnant moment in scientific history, war intervened and Feynman joined the scientists at Los Alamos who were building the nuclear bomb. He made important contributions, but would not continue his basic research and publish his new formulation of quantum mechanics until after the war was won.

POSTWAR QED

In just a few years immediately following World War II, quantum electrodynamics grew from a theory that had left the great founders of quantum mechanics in despair to one that matched general relativity in the exquisite precision of its predictions. Furthermore, these predictions could now be tested against data of equal precision.

Two postwar developments made this possible. First, fresh-faced, well-trained, self-confident, and well-supported young theorists and experimentalists joined the attack. Second, the technologies developed in the war, particularly the microwave electronics used in radar, made available new data of unprecedented accuracy to guide theory in the right direction. As the history of physics has amply demonstrated, physical theories makes little progress without adequate experiments to test their predictions. This was emphatically the case with QED.

Two experimental results from Columbia University, reported at the 1948 Shelter Island conference, provided the spark for the flashes of insight that followed. First, Willis Lamb and Robert Retherford found that one of the energy levels of hydrogen is split by an energy corresponding to an emitted photon frequency of approximately 1,000 megacycles per second.[2] The Dirac equation had indicated that there were two different quantum states at the same, "degenerate" energy level. The Lamb-Retherford results indicated that this level was split in energy by 6 parts in a million; they were seeing the low energy, radio frequency photons emitted in the transition between the states. This is now called the **Lamb shift**.

The second Columbia experiment also involved the splitting of hydrogen spectral lines at microwave frequencies. At Shelter Island, theorist Gregory Breit of Yale reported his calculations on an anomaly that had been observed by Isadore Rabi and his students J. E. Nafe and E. B. Nelson at Columbia (Nafe

1947). The main contribution to the magnetism of matter comes from electrons in atoms. However, nuclei of atoms have also have magnetic fields that are thousands of times weaker and produce very small effects such as the "hyperfine" splitting of spectral lines. Today's very powerful medical diagnostic tool, *Magnetic Resonance Imaging* (MRI) is based on nuclear magnetism.

Rabi and his students had measured a splitting of 1421.3 megacycles per second in the ground state of hydrogen that disagreed with the theoretical prediction of 1416.9. Breit suggested that the electron might have a slightly different **magnetic dipole moment**, the measure of its magnetic field strength, from that predicted by the Dirac equation.

Four main figures were involved in successfully calculating these small effects and thereby resolving the infinities problem in QED, although several others, notably Bethe, also made major contributions. Perhaps the most remarkable of the four, because of his circumstances, was Sin-itiro Tomonaga. Amid the postwar rubble of Tokyo, Tomonaga led a group of young theorists in working out details of his "super-many-time" formalism that guaranteed the compatibility of the theory with relativity. Tomonaga had recognized the intrinsically nonrelativistic nature of the conventional quantization procedure for fields. As I have noted, the field was being treated a having a value at each point in space at a single, absolute time— a meaningless concept according to Einstein. Tomonaga modified the formalism so that each degree of freedom had its own time as well as its own position (Tomonaga 1946).

By 1947, Tomonaga and his associates had solved the infinities problem through a process he called "readjustment." In this process, the mass and charge of the electron are calculated with compensating infinities. Following Kramers, Tomonaga viewed the electron as having a bare mass and charge that are themselves unobservable. In the calculations involving electrons interacting with photons, the divergent terms are shown to exactly cancel similar terms that appear in the self-mass and self-charge of the electron (Tomonaga 1948).

Tomonaga's results were not immediately known in the West. There the focus of physics had moved from Europe to the United States, where the scientific war effort had been centered and the greatest minds in European and American physics had gathered. The important role played by physicists in winning the war had gained them great prestige. The opportunities and funding for basic research had boomed to unprecedented levels (no longer matched today). Two gifted young American physicists, who had participated in the war effort in small but useful roles, gained almost instant fame with their independent solutions to the QED infinities problem: Feynman and Julian Schwinger.

Schwinger was so talented that he published his first physics paper at sixteen and completed his Ph.D. thesis before getting his bachelor's degree at Columbia. Since he rarely went to class and studied from dusk to dawn when no one else was around, he did this pretty much on his own. He had worked at the MIT Radiation Laboratory during the war and in 1946 was a (rare) twenty-eight-year-old associate professor at Harvard. By the end of the next near, he had computed corrections to the hyperfine structure of the spectral lines of hydrogen and deuterium that agreed precisely with the new data.

In a way similar to Tomonaga's super-many-time procedure, of which he was not initially aware, Schwinger replaced the time parameter with a space-time surface when computing the evolution of a system. Like Tomonaga and Feynman, Schwinger had recognized that singling out time as a special coordinate is an inherently nonrelativistic procedure. The new relativistic/quantum replacement for the classical equation of motion, which was arrived at independently, is now called the Tomonaga-Schwinger equation.[3]

Schwinger succeeded in removing the QED infinities by a series of mathematical manipulations ("canonical transformations"), all done in a completely relativistic way. The electron mass was renormalized in the manner suggested by Kramers. By a similar procedure, Schwinger renormalized the electron's charge. He wound up with an equation for the "effective external potential energy " of the electron, from which the energy level splittings could be calculated, expressed as a power series. We call this the "perturbation series," since it is similar to what is done in celestial mechanics when the orbits of planets are calculated in two-body approximation, with the effects of other planets treated as small perturbations. However, Schwinger's perturbation series was horrendously difficult to calculate, in at least the form presented, beyond the third term ("second order radiative correction").

Fortunately, Schwinger found that the perturbation series converged rapidly and the first three terms gave results that agreed beautifully with the experiments reported at Shelter Island. He found that the magnetic dipole moment of electron was larger than the Dirac prediction by the fraction 0.00118 (Schwinger 1948a). This accounted for the observed hyperfine splitting for both hydrogen and deuterium. Schwinger also calculated the Lamb shift, obtaining 1,051 megacycles per second compared to the observed 1,062 (Schwinger 1949b). When these results were presented, Schwinger overnight became the brightest star in the physics firmament. Soon, however, another brilliant young physicist from New York would also light the skies.

As we have seen, Dick Feynman also had made an early name for himself, first as an undergraduate at MIT, and then as a graduate student at Princeton. However, he was very different from Schwinger in temperament and approach to physics. Where Schwinger read almost everything, Feynman read almost nothing. Where Schwinger said little until presenting his results in elegant, formal lectures and papers, Feynman talked continuously, entertaining more than lecturing, and writing in the same informal, often careless way he spoke. Since his death in 1988, Feynman has become a scientific legend matched only by Einstein in the twentieth century. As someone who first heard Feynman lecture in 1956 and witnessed him in action many times over three decades, I can testify that most of the legend is true.[4]

Schwinger, on the other hand, faded into relative obscurity—or at least as much obscurity as a Nobel laureate can ever experience. Few had the ability or stamina to follow his logic and mathematics, impeccable though they were. To extend his methods appeared beyond mortal ability. On the other hand, Feynman produced a set of rules so intuitively reasonable that even those of limited mathematical talent could use them to calculate QED processes. Users of the Feynman algorithm not only obtained correct answers, they also felt that they really "understood" what was going on in the physical processes. Schwinger was not intentionally paying tribute when he remarked that "the Feynman diagram was bringing computation to the masses" (Gleick 1992, 276).

All too often, intuitive comprehension gets lost when grinding though some complicated mathematical calculation. This does not happen with the analysis of Feynman diagrams. Although the mathematical rules are still complicated and require some years of training to use, they make intuitive sense every step of the way. For all these reasons, I submit that they are telling us something profound about reality.

FEYNMAN'S QED

The first step in the development of Feynman's version of QED, the machinery for calculating QED processes, was his alternate explanation for the negative energy states that Dirac had interpreted in terms of an infinite sea of electrons filling the universe. In Dirac's theory, these electrons are not observed until a gamma-ray photon with over a million electron-volts of energy comes along and kicks an electron out, leaving behind a hole that is associated with the positron. As we saw above, this was never an appealing idea, but it was the only game in town for two decades of QED calculations.

Feynman, in his Nobel lecture, tells about the call he received from Wheeler in the fall of 1940. Wheeler said, "Feynman, I know why all electrons have the same charge and the same mass." "Why?" "Because, they are all the same electron" (Feynman 1965b, 163).

This seemingly bizarre notion grew out of the work that Wheeler and Feynman had done on classical electrodynamics, described above, in which signals from the future were included along with those from the past. Wheeler proposed that a single electron zigzagging back and forth in space-time would appear at a given time as many electrons at different positions.

Wheeler had the idea that the backward electrons were in fact the few observed positrons moving forward in time. Feynman asked where all those extra positrons were, and Wheeler suggested that they might be hidden in protons. Feynman explains that he did not take the idea that all electrons are the same electron as seriously as he did the observation that positrons could be viewed as electrons moving from the future to the past: "That I stole!" (Feynman 1965b, 163).

Nothing much has ever been made of Wheeler's notion that all electrons are the same electron, which may have been proposed half in jest—or even fully in jest. Nevertheless the idea can still serve as a useful pedagogical tool. It cannot but puzzle the layperson why no two snowflakes are alike while all electrons are completely indistinguishable.

Martin Gardner (1979, 269) has described Wheeler's scenario, but commented that "there is an enormous catch to all of this" since it demands there is an equal number of electrons and positrons in the universe, which is apparently not the case now. However, we can construct a space-time path in which all the positrons occur at small distances or early times, where current theories in particle physics and cosmology predict particle-antiparticle symmetry. For example, as illustrated in figure 7.2, we can have a single electron appearing as two electrons and two positrons "now," with only the electrons being observed "here," the positrons materializing deep inside an atomic nucleus.

This idea may still be shown to have ontological significance. Perhaps all particles are really a single particle circulating in space-time. Note how this provides a simple, if crazy, explanation for the puzzle of the multiple slit experiment. We recall that the puzzle is how a single particle can pass through several slits simultaneously. In the figure, we can imagine the two electron paths shown in the vicinity of here and now passing through two slits.

It was not until after the war, in 1949, that Feynman finally published his theory of positrons (Feynman 1949a). Ernst Stückelberg had published the concept earlier (Stückelberg 1942). While Feynman acknowledges this in his paper, he probably did not know of Stückelberg's work when formulating his own ideas, which were jelling already in the prewar period. Still, Stückelberg should be credited as the first to publish a "Feynman diagram."[5]

Recall that many of the difficulties encountered by QED in the 1930s were the result of the need to make everything into a many body problem. The number of electrons in a reaction was no longer conserved. Feynman now showed us that we can still work with a single electron and simply follow its path, its worldline, in space-time. What appears to us at a given time as an electron accompanied by an electron-positron pair is a single electron when viewed from overall space-time perspective.

Feynman showed that his positron theory simplifies many problems. In the example of electron scattering by an external field, the effects of virtual

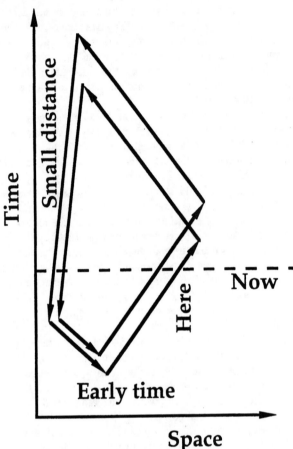

Fig. 7.2. An electron circles around in space-time, appearing four places at the time labeled "now." In conventional forward time, it is interpreted as two electrons appearing simultaneously "here," and two positrons appearing "now" at a very small distance, say inside a nucleus. This describes a situation in which we have very few positrons "here" and yet have complete symmetry between positrons and electrons. Note that the two electron paths in the vicinity of here and now could be passing through the slits of a double slit experiment.

pair production are automatically taken into account in the motion of a *single* electron. Pair production simply represents one of the possible paths, one switchback, that an electron can follow when it scatters back in time and then forward again.

Feynman contrasts this point of view with the Hamiltonian method in which the future develops continuously out of the past. In the case of a scattering problem, "we imagine the entire space-time history laid out." He further asserts: "The temporal order of events during the scattering, which is analyzed in such detail by the Hamiltonian differential equation, is irrelevant."

Feynman followed his paper on positron theory with "Space-Time Approach to Quantum Electrodynamics" (Feynman 1949b). This paper, he says, is to be regarded as a continuation of the positron paper which described the motion of electrons forward and backward in time. Feynman now considers their mutual interaction. Once again, he explains the advantages his approach has over the Hamiltonian method in which the future is viewed as developing from the past and present:

> If the values of a complete set of quantities are known now, their values can be computed at the next instant in time. If particles interact through a delayed interaction, however, one cannot predict the future by simply knowing the present motion of the particles. One would also have to know what the motions of the particles were in the past in view of the interaction this may have on the future motions.

In Hamiltonian electrodynamics, this is accomplished by specifying the values of a "host of new variables," namely, the coordinates of the field oscillators.

In Feynman's words, the field is introduced in the Hamilton picture as a "bookkeeping device" to keep track of all those photons that the electron might scatter from in the future. You need to know their positions and momenta at the present instant, and you need to determine their future positions by solving their equations of motion along with the electron's.

To Feynman, this was a lot of unnecessary baggage to have to carry along as we tiptoe through space-time. He proposed instead that we take an overall space-time view that not only eliminates the need for fields, but is intrinsically relativistic. Like Schwinger and Tomonaga, he recognized the inherent nonrelativistic nature of the Hamiltonian method. But whereas they stayed within that framework and instead reformulated it by replacing absolute time as the running parameter with one that was suitably relativistic, Feynman proposed to discard the whole method: "By forsaking the Hamiltonian method, the wedding of relativity and quantum mechanics can be accomplished naturally."

The Feynman picture, then, is one of particle-particle interactions. He showed how to calculate the probability amplitude for electron-electron scattering in which a photon is exchanged, as was first suggested by Fermi and Bethe in 1932. He also computed the amplitudes for "radiative corrections" in which additional photons are emitted and absorbed. Here again the infinities associated with electron self energy came in.

Feynman had hoped to solve the self energy problem with his space-time approach. As we have seen, this had been a major interest of his since his undergraduate days and motivated much of what he had done since then. Yet, Feynman did not really achieve this goal in his paper on QED; he just swept it under the rug as others had done many times before by introducing a finite cutoff in the calculation. With the competition in Tomonaga and Schwinger already out ahead, he had rushed his incomplete paper into print. As it turned out, Feynman's great contribution was not so much to solve the infinity problems of QED as to provide a method by which QED calculations, and those beyond QED, could be performed more easily. The infinities problem proved ultimately solvable within that framework.

Feynman admitted that he never really derived the Feynman diagram rules using his path integral formulation of quantum mechanics, although that was eventually accomplished by others. He explained that he simply found the path integral approach useful in the guesswork that he employed in inferring those methods. The path integral formulation has never been widely applied in theoretical physics. However, it suggests the viability of a purely particulate model of reality in which fields are simply a mathematical convenience with no reality of their own. As for least action, Schwinger (1951) showed how the "particle aspect" of quantum fields can be derived from an action principle.

UNIFYING QED

The fourth individual who rightfully shares the limelight along with Tomonaga, Schwinger, and Feynman for the success of postwar QED was a Cambridge educated Brit doing graduate studies in the United States at the time. Freeman Dyson put it all together, although he did not share the Nobel prize since it is limited to three recipients. In fact, he never even got a Ph.D. for all his efforts! Nevertheless, he has received many honors (including the million dollar 2000 Templeton prize for contributions to religion) and earned a well-deserved pedestal in the museum of science history.

In the spring of 1948, Dyson was a graduate student at Cornell. Feynman was there, still developing his ideas, and Tomonaga's and Schwinger's results were just becoming public. Dyson saw that the Tomonaga and Schwinger approaches were equivalent, with the Japanese physicist's simpler and less shrouded in esoteric mathematics. Both had shown how to formulate a quantum field theory in a relativistic manner within the Hamiltonian framework.

That summer, Dyson drove across country with Feynman, who was going to visit a girlfriend in New Mexico. Dyson took the bus back to Michigan for a summer school where Schwinger was presenting lectures, bussed across country once more to Berkeley, and then back again to Princeton where he was to continue his studies in the fall. During his long, lonely bus rides, Dyson worked out in his head the equivalence of Feynman's approach with that of Tomonaga and Schwinger (Dyson 1979; Schweber 1994, 502–505).

Dyson realized that he could rewrite the Schwinger perturbation series by including certain "chronological ordering operators" that had been used by Feynman. During a Chicago stop, Dyson was able to derive the entire Feynman theory from Schwinger's. Back at Princeton, he soon worked out the details and realized the far greater power of the Feynman method, especially the possibility to go to higher orders in the perturbation expansion that was prohibitively difficult in the Schwinger approach.

Feynman had developed his calculational machinery largely by intuitive guesswork. However, he always insisted that his guesses were based on ideas that he had already worked out in detail, like path integrals. Furthermore, these guesses were always accompanied by long and arduous mathematical work, checking each step for consistency. Feynman's equations were not just pulled out of thin air, as it sometimes appeared to casual observers.

By October 1948, Dyson had submitted a paper entitled "The Radiation Theories of Tomonaga, Schwinger, and Feynman" to the *Physical Review* (Dyson 1949). He started by outlining the Tomonaga-Schwinger equation, which we recall is basically the time-dependent Schrödinger equation, that is, the equation of motion of quantum mechanics, expressed in terms of relativistically invariant variables. Dyson then derived Schwinger's perturbation series expansion for the effective external potential energy. As we saw above, Schwinger had just computed the first three terms, but these were sufficient to give an estimate of the electron's magnetic moment and the Lamb shift that triumphantly agreed with experiment. Fortunately, the perturbation series converged rapidly enough for Schwinger's immortality.

Dyson recognized that the basic principle of the Feynman theory was "to preserve symmetry between past and future." By implementing this principle, Dyson showed that nasty mathematical elements in the Schwinger calculation can be avoided, leading to a much simpler expression of the perturbation series. He next related the expression directly to the "S-matrix element" that appears in scattering problems. He then demonstrated that the terms in the perturbation expansion can be represented by Feynman diagrams (sometimes called "Feynman-Dyson diagrams"). Dyson further asserted that the diagrams are not to be regarded "merely as an aid to calculation, but as a picture of the physical processes which gives rise to the matrix element."

After all this, Dyson was still not finished. He proved that the various QED processes like self mass and vacuum polarization that lead to infinities can be handled by absorbing the corresponding diagrams into a smaller number of "irreducible" diagrams. For example, the electron self energy dia-

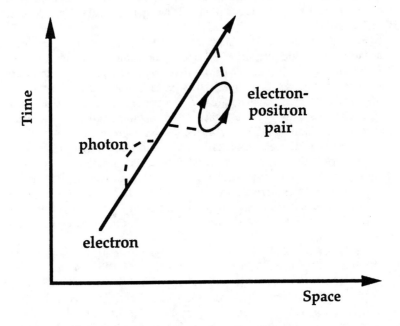

Fig. 7.3. An electron is shown emitting and reabsorbing a photon and then emitting of photon that creates an electron-positron pair. The pair then annihilates into a photon that is reabsorbed by the electron. Processes of this type all contribute to the self energy of a physical electron and can be subsumed into a single electron line.

grams, as illustrated in figure 7.3, in which an electron emits and then absorbs one, two, or any number of photons all get subsumed in a single electron line that represents the electron that is observed in experiments.

Vacuum polarization was another one of the ghostlike QED processes that had haunted physicists since the 1930s. Because the uncertainty principle allows small violations of energy conservation for short times, an electron-positron pair can briefly appear and then disappear in the vacuum. During this brief period, the pair acts as an electric dipole and produces, in continuum language, an electric field. In the Feynman-Dyson picture, no electric or magnetic fields exist—only photons and pairs. Vacuum polarization results when any photon in a diagram briefly creates a pair that then quickly annihilates back into a photon. This is a process similar to electron self energy, and Dyson eliminated it in the same way—by subsuming it into a single, irreducible, photon line on the diagram.

The Feynman-Dyson methods were readily extended to more complex situations. It was still a lot of work, but at least feasible with sufficient effort and knowhow. Indeed, a whole generation of theoretical physicists was soon trained, in physicist graduate programs worldwide, who could perform with ease calculations that had stymied the greatest minds of earlier generations—Einstein, Bohr, Heisenberg, Schrödinger, Dirac, Jordan, Born, Pauli, and the rest. Even Schwinger's Harvard students, now at a competitive disadvantage, could be found drawing Feynman diagrams on the board when their mentor was not looking. Today, much of the drudgery of the Feynman machinery can be handled by computer programs, leaving a tool of unprecedented power literally at the fingertips of physicists.

NOTES

1. For a good discussion and full bibliography, see Schweber 1994, 33–38.
2. The cycle per second is the unit we now call the Hertz (Hz).
3. For reprints of most of the important papers on QED during the period from 1927 to 1949, see Schwinger 1958.
4. A number of biographies of Feynman are now available. I have relied on Gleick 1993 and the extensive material on Feynman in Schweber 1994.
5. While Stückelberg's original paper may be hard to find, his "Feynman diagram" is reproduced in Gleik, 1993, 273.

8 THE TIMELESS QUANTUM

Every particle in Nature has an amplitude to move backwards in time. . . .
Richard Feynman (1986, 98)

SEEKING A QUANTUM ARROW OF TIME

In chapter 4, we learned that three arrows of time are inferred from macroscopic phenomena: the thermodynamic, cosmological, and radiation arrows. We found that these three are essentially the same, and can be understood in terms of the inflationary big bang model of cosmology in a way that maintains overall time symmetry in the universe. In this scenario, we live in the expanding space of the big bang where the entropy at one end of the time scale is much greater than the other, time's arrow then being defined to point in the direction of increased entropy. With t = 0 defined as the point at which inflation was triggered by a quantum fluctuation, then the negative t side of the time axis (negative, that is, from our perspective) undergoes an initially identical expansion in which time's arrow points along the –t direction. Thus, an overall time symmetry is maintained at the cosmic scale, as required by all the known fundamental principles of physics.

One of the strong proponents of a time-asymmetric universe is Oxford mathematician and cosmologist Roger Penrose. He has sought an arrow of time in cosmology, and also looked for one in quantum mechanics (Penrose 1989, 250). His cosmological arrow was discussed in chapter 4. Here let us focus on Penrose's quantum arrow.

As I have noted previously, Penrose separates quantum processes into two categories that act in sequence. Firstly, a **U** process describes the ("unitary") evolution of the state of the system as governed by the time-dependent Schrödinger equation. This is completely time-symmetric in all except certain rare processes that we can ignore for the present discussion and, as we will see, can be made time symmetric anyway by including other symmetry operations. The Penrose **R** process provides for the collapse or reduction of the wave function, or state vector, by the act of measurement. As we have noted, state vector collapse is a separate, unwelcome, and probably unnecessary axiom of quantum mechanics. Penrose points out that this part of quantum mechanics is inherently time-asymmetric and the possible source of a fundamental quantum arrow of time.

He illustrates this with a simple example that I have redrawn in figure 8.1. A lamp L emits photons in the direction of a photocell (small photon detector) P. Halfway between is a half-silvered mirror M arranged at forty-five degrees so that the photon, with 50 percent probability, will go either straight through to P or be reflected toward the laboratory wall in the direction W. The paths are indicated by the solid lines in the figure. The lamp and photocell also contain registers that count the photons emitted. Penrose then asks: "Given that L registers, what is the probability that P registers?" His answer follows from quantum mechanics: "one-half."

Penrose then considers what he calls the "reverse-time procedure" in which a backward-time wave function represents a photon that eventually reaches P. This wave function is traced backward in time to the mirror where it bifurcates, implying a fifty-fifty chance of reaching the lamp L. He then asks: "Given that P registers, what is the probability that L registers?" His answer: "one." His reasoning is as follows: "If the photo-cell indeed registers, then it is virtually certain that the photon came from L and not the wall!" The conventional application of quantum mechanics thus gives the wrong answer in the time-reversed direction and so, Penrose concludes, the **R** process must be time-asymmetric.

While this is correct for the process Penrose describes, note that this is not what you would actually see watching a film of the event being run backward through the projector. This more literal reverse-time process is indicated by the dashed lines in figure 8.1. There the photons that are absorbed in the lamp come from two sources, the detector P and the wall. All the photons registered at P also register at L, as in Penrose's example, but the lamp also receives additional photons from the wall.

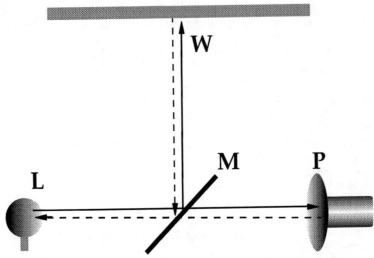

Fig. 8.1. Photons from lamp L have a fifty-fifty chance of either going through the mirror M to the photodetector P or reflecting from M and being absorbed by the wall at W. In the time-reversed process, as imagined in a film run backward through the projector, half the photons are emitted by P and half by the wall, all returning to the lamp at L. At the quantum level, sources can act as detectors or absorbers, and detectors or absorbers as sources. Irreversibility occurs at the macroscopic level because macroscopic sources and detectors are not in general reversible. This figure is based on figure 8.3 in Penrose (1989, 357).

If the experiment were a purely quantum one, with single atoms as sources and detectors, then it would be completely time reversible. The irreversibility that Penrose sees is exactly the same irreversibility I talked about earlier, an effect of the large amount of randomness in macroscopic many body systems.

Now, Penrose is arguing that the actual experiment is done with a macroscopic lamp L, macroscopic detector P, and macroscopic walls and so it is irreversible. That is, L is not normally a detector and P or the wall is rarely a source. But this is exactly the situation I described in chapter 4 while discussing the arrow of radiation. There I related Price's conclusion that the radiation arrow results from the asymmetry between macroscopic sources and detectors in our particular world. That is, they are part of the "boundary conditions" of the world of our experiences and nothing fundamental. Most lamps are not reversible as detectors, although lasers, being quantum devices, are in principle reversible. As mentioned, the first maser (the precursor of the laser, which worked in the microwave region of the spectrum) was designed as a detector. Most macroscopic detectors, like photomultiplier tubes, cannot be reversed into sources, or, at least, very efficient ones. Furthermore, walls do not emit the kind of narrow beam of visible photons we are assuming in this experiment, although they do emit thermal radiation.

Lamps, detectors, and absorbing walls do not look the same in the mirror of time. However, like the face of the person you see when you look in the bathroom mirror, which is strangely different from the one you see in a photo, what is seen is not strictly impossible.

Let us peer more deeply into the lamp, detector, and wall. Consider the primary emission and absorption processes that take place at the point where the photon is emitted or absorbed. This was illustrated in figure 4.3 (a), for both time directions. In "forward time," an electron in an excited energy level of an atom in the lamp drops down to a lower energy level, emitting a photon. This photon is absorbed in either the wall at W or the detector P in figure 8.1, with an electron being excited from a lower to a higher level in an atom in wall or detector. In "backward time," either the wall or the detector emits a photon by the same process as the lamp in forward time while the lamp absorbs the photon.

True that, in actual practice, the wall and detector will irreversibly absorb photons, converting their energy to heat and gaining entropy. But if we could do the experiment with a purely quantum lamp, detector, and wall composed of a single atom each with the same energy levels, say a hydrogen atom, then the process would be completely reversible.

As Penrose (1989, 359) says, "the **R** procedure, *as it is actually used,* is *not* time-symmetric." From this he concludes that we have a unique quantum arrow. The first statement is true. But the irreversibility is still statistical and the result of asymmetric boundary conditions imposed on the experiment, not the result of any fundamental principle of physics. This irreversibility does not actually occur for the primary quantum event at the microscopic level. Rather, the arrow appears where the coherent, usually few-body processes of quantum interactions are replaced by the incoherent many-body processes that interface the quantum world with us denizens of the classical world. As with his cosmological arrow, Penrose has not demonstrated any need for a new fundamental law to explain the undoubted time-asymmetry we observe in many body phenomena. The quantum arrow is again the same as the thermodynamic/cosmological/radiation one.

The **U** process as we understand it, is fundamentally time-symmetric. As for **R**, we saw in chapter 6 that this process can be understood in terms of the decoherence that takes place when a quantum system interacts with macroscopic detectors

Robert Griffiths (1984), Yakir Aharonov and Lev Vaidman (1990), Roland Omnès (1994), and Giuseppe Castagnoli (1995), among others, have shown that conventional quantum mechanics can be precisely formulated in a manifestly time-symmetric fashion. Also, Murray Gell-Mann and James Hartle (1992) have demonstrated that a quantum cosmology can be developed using a time-neutral, generalized quantum mechanics of closed systems in which initial and final boundary conditions are related by time reflection symmetry. At the quantum level, we have no evidence for an arrow of time and no unique entropy gradient, other than those discussed previously, to require us to nevertheless adopt an arrow.

ZIGZAGGING IN SPACE AND TIME

In their book *The Arrow of Time*, Peter Coveney and Roger Highfield (1991, 288) suggest that the macroscopic arrow of time can be used to explain the paradoxes of quantum mechanics. They base this on the attempts by chemist Ilya Prigogine and his collaborators to derive irreversibility at the microscopic level from dynamical principles that they apply first to the macroscopic level. However, Vassilios Karakostas (1996) has shown that these derivations tacitly assume temporal irreversibility. Here we find yet another example of the use of what Price (1996) calls a double standard, in which investigators fail to realize that they have not treated time symmetrically and so fool themselves into thinking they have discovered the source of the arrow of time. Furthermore, contrary to the claims of Prigogine and followers, the recognition that time in fact has no arrow at the quantum level actually helps to eliminate the so-called paradoxes of quantum mechanics. So why should we want to force an arrow back in?

Time symmetry remains deeply embedded in both classical and quantum physics and can be expunged only by making the added, uneconomical assumption that the arrow of normal experience must be applied to both of these areas of basic physics. Why uneconomical? Because it requires an additional assumption not required by the data.

As we saw in the last chapter, Feynman utilized the inherent time symmetry of physics in developing his methods for solving problems in quantum electrodynamics (Feynman 1949a,b). Now, fifty years later, time reversibility still seems indicated by both data and theory. Specifically, it furnishes a way of viewing one of the great puzzles of quantum mechanics: how particles can follow definite paths in space-time and still appear many places at once. Examples include the instantaneous quantum jump of an electron in an atom from one energy level to another, among others.

To see how time reversibility allows for a particle to follow a definite path, and still be several places simultaneously, consider figure 8.2. In (a), an electron zigzags back and forth in space-time, appearing at one place at time A, three places at time B, then two places for an instant at time C, and

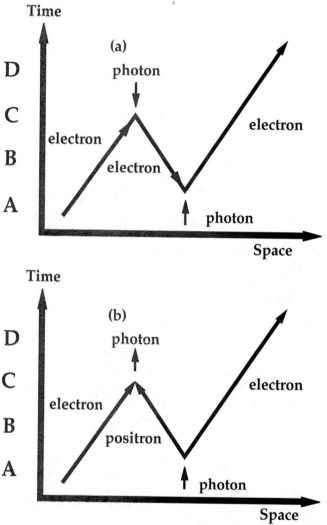

Fig. 8.2 (a) An electron zigzagging back and forth in space-time, thus appearing simultaneously at three different positions at time B and two different positions at time C. (b) The conventional directed-time view in which an electron-positron pair are created by a photon at A, the positron annihilating with an electron into a photon at C. These diagrams can occur "virtually," with zero energy photons in the vacuum, as long as the violation of energy conservation at time C is corrected at time A within a time interval allowed by the uncertainty principle.

back again at one place at time D. The conventional view is shown in (b), where a photon makes an electron-positron pair at A and the positron annihilates with the original electron at C.

This particular phenomenon is observed in experiments with high energy photons (gamma rays), but can also occur "virtually" with the normally undetectable zero energy photons that fill the vacuum (see chapter 7). In this case, the time interval from A to C must be small enough so that the breaking of energy conservation at A is corrected at C. Since an electron-positron pair is being created from "nothing" at A, energy is not conserved by an amount at least equal to the rest energy of the pair, 1.022 MeV. The Heisenberg uncertainty principle applies to measurements of energy and time in the same way it does for measurements of position and momentum. Applying this, the time interval during which an electron zigzag takes place is of the order of 10^{-22} second.[1] The distance over which the zigzag takes place can also be estimated and is easily shown to be of the order of the de Broglie wavelength of the particle.[2] This indicates that the spreading out we associate with the wavelike nature of particles is related to their zigzagging in space-time.

Of course, we do not have direct experience of familiar everyday bodies travelling back and forth in space-time. The time interval for virtual zigzagging is far shorter than anything we can directly measure with our best instruments, and the de Broglie wavelength for the typical macroscopic object is also unmeasurable. Electrons can zigzag over interatomic distances, such as between atoms in a crystal. Particles far less massive than electrons, such as neutrinos and photons, can zigzag over large spatial intervals. Radio photons, for example, can have wavelengths of macroscopic or even planetary dimensions. That is why we don't normally think of them as tiny particles but spread-out waves. But, with space-time zigzagging, we can maintain a single ontological picture of pointlike photons that applies on all scales, from the subnuclear to the supergalactic.

This scheme can also be used to visualize the quantum jump between energy levels in an atom. In the case of excitation to a higher energy level, one photon in figure 8.2 comes in from the outside, the other from interaction with the rest of the atom, as the electron is viewed as jumping instantaneously from one orbit to another at time C.

Feynman's methods for calculating the probabilities of processes such as that shown in figure 8.2 have became basic tools for the in particle physicist. They were heavily utilized in the development of the current standard model of quarks and leptons that agrees with all observations to-date. In applying these methods, most physicists think in terms of the familiar unidirectional time, picture (b) rather than picture (a) in figure 8.2. However, the calculation is the same either way and this simple fact alone demonstrates that time symmetry is evident at the quantum level. Eliminating any preferred arrow of time from our thinking then opens up a new perspective on several of the strange features of quantum mechanics that seem so bizarre in the normal, time-directed convention. Whether or not all the interpretational problems disappear with such a simple expedient will require a more detailed logical and mathematical analysis than I propose to attempt here.

The idea of having particles at different places "at the same time" is not as revolutionary, or as new, as it may seem. Any model of reality we may utilize must take into account the view of time promulgated by Einstein in his theory of relativity that even now, a century later, has not penetrated conventional thinking. As we have seen, Einstein showed that time is not absolute, and so the term "simultaneous" is a relative one. Saying that two events are simultaneous can only refer to the reference frame in which all observers are at rest with respect to one another.[3] Since simultaneity is not absolute, no logical inconsistency exists in describing a particle to be several places at "the same time" as observed in some reference frame.

Note that in the rest frame or "proper" frame of the electron in figure 8.2, the electron is always at a single place in space: "here." This is the same as the understanding that you are always "here" in your own personal reference frame, even when you are hurtling across the sky at 5 miles a minute in a jetliner that places you at well-separated places at different times to an observer on the ground.

If you can be the same place at two different times, why can't you be at two different places at the same time? Only our primitive intuition of absolute, directed time makes this seem impossible. And this intuition is the result of our personal existence as massive, many-particle objects. The zigzagging we do in spacetime is over such small intervals as to be undetectable.

Notice that although the electron seems to jump instantaneously from one point in space to another in figure 8.2, it does not do so by travelling through space at infinite speed. Whether going forward or backward in time, it still moves at less than the speed of light in all reference frames. The electron appears to undergo an infinite acceleration at time A, but that acceleration is not measurable for time intervals smaller than the interval over which the zigzag takes place and is balanced out, just as energy conservation was balanced out at C. Another way to say this is that at one moment the electron is going at a subluminal speed in one time direction, and at the next measurable moment it is moving at a subluminal speed in the opposite time direction.

Still, the violation of Bell's inequality in the Bohm EPR experiment has led to the widespread conclusion that the universe is necessarily "nonlocal," that is, it requires superluminal connections of some type—if not superluminal motion. The conventional wisdom holds that quantum mechanics is complete but nonlocal. But this also has a problem. As I have mentioned, superluminal signals are provably impossible in any theory compatible with the axioms of relativistic quantum field theory. So nonlocality, at least if we read this to mean superluminal motion or signaling, is inconsistent with the most modern application of quantum mechanics. Surely something unfamiliar is going on at the quantum level, but it is not necessarily superluminality.

At the risk of overwhelming the reader with too many unfamiliar terms, I would like to suggest yet another one that I think better describes the situation. David Lewis (according to Redhead 1995) has proposed the term **bilocal** to refer to an object being in two places in space at once. Extending this definition to allow more than two places at once, let me refer to the situation in the quantum world as **multilocal**. While not at one place at one time, a quantum particle is still not everywhere at once but at several well-localized places in space-time. For example, in figure 8.2, the electron is at three localized positions at time B. Again, no superluminal motion is implied because the object goes back in time and then forward again to reach a new position at the same original time. Since we require only localized interactions, no holistic fields need be introduced to provide instantaneous action at a distance.

THE EPR PARADOX IN REVERSE TIME

The experimental violation of Bell's theorem, discussed in chapter 6 (Aspect 1982), demonstrated that when a system of two particles is initially prepared in a pure quantum state, then those particles retain a greater correlation when they become separated than is expected from either classical physics or more general, commonsense notions of objective reality. In particular, the results of a measurement of some property of one particle are found to depend on the results of a measurement of that property for the other particle, even after they have separated to such a distance that any signal between them would have had to be superluminal.

As mentioned above, no superluminal signal is possible within the framework of standard quantum mechanics and relativistic quantum field theory. The observers cannot determine, with absolute certainty, the outcome of either measurement—if these foundational theories are valid. The observers simply set their detection instruments to a particular configuration and take what nature provides.

Olivier Costa de Beauregard (1953, 1977, 1979) was perhaps the first to suggest that the EPR paradox can be resolved by including the action of signals from the future. Unfortunately, he did not view this idea as a way to economically eliminate the paradox. On the contrary, he claimed a connection with "what parapsychologists call *precognition* and/or *psychokinesis*" (Costa de Beauregard 1978). Others have proposed quantum "theories" of psychic phenomena (Jahn and Dunne 1986, 1987; Goswami 1993). Perhaps as a result of his unnecessary foray into the occult, Costa de Beauregard's original suggestion was not taken as seriously as it should have been.

Nevertheless, time symmetry makes the EPR paradox go away—without occult consequences (see also Sutherland 1983 and Anderson 1988). In the original implementation of the EPR experiment, proposed by Bohm, the two electrons are prepared in an initial state of total spin (angular momentum) zero, the **singlet** state. However, any random pair of electrons will be in either the singlet or the spin one **triplet** state, in statistical proportion of three triplets for each singlet, as the names suggest. The triplet pairs are discarded in the preparation of the experiment, but we must not forget that this operation is part of the experimental procedure.

The electrons go off in opposite directions to detectors that measure their individual spin components along axes that can be set arbitrarily by the observers at the end of each beam line. As noted, measured correlations seem to imply some sort of superluminal connection.

Now, imagine the whole process in a time-reversed reference frame (see figure 8.3). Note: I am not assuming that the experimenters go backward in time or anything of that sort. Just imagine taking a movie of the experiment, and then running it backward through the projector. This is simply another equally legitimate way of observing the experiment. What we see are two electrons emitted from the "detectors" at the beam ends and travelling toward what was the source in the original time direction.

The emitted electrons can be expected to have random spin components perpendicular to the beam axis. When they come together, the combined two electron system will be either spin one, the triplet state, or spin zero, the singlet state. Again, the statistical ratio will be three-to-one.

In the conventional experiment, only singlet states are accepted and triplet states discarded. Viewing this selection from the time-reversed perspective, we watch the triplet two electron states being discarded while the singlet states are accepted. This action occurs locally, that is, at a single place and time (in the center-of-mass reference frame). Of course, a correlation is then enforced on the system; only the singlet state of the two electrons are allowed in the sample. However, it should be obvious that no superluminal connections of any kind are required.

All other explanations of the EPR paradox require some type of superluminal connections and considerable logic twisting to argue that these are not in fact "signals," which are impossible within the current paradigms of relativity and quantum mechanics, but "influences." Theorems (Stapp 1985) that are purported to prove that quantum mechanics is required to be non-local in the light of the Aspect experiment admittedly assume time-directed causality. If they are right, then only time reversibility can rescue us from spooky instantaneous action at a distance in quantum mechanics.

It is important to emphasize that time-reversed pictures are every bit as

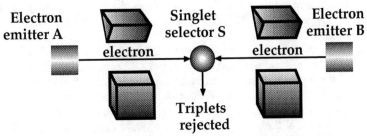

Fig. 8.3. The Bohm EPR experiment viewed in a time-reversed perspective. Electrons with random spin components are emitted by two sources with the triplet states being discarded locally by the singlet selector S.

valid as those drawn with a conventional time direction. That direction is arbitrary. No one viewing the results of an experiment in which electrons of arbitrary polarization are fired at one another and their triplet two-electron states locally discarded would get excited about "nonlocality." So why are so many people excited about the conventional-time EPR experiment, which is indistinguishable from this?

TIME SYMMETRY AND SCHRÖDINGER'S CAT

As we have seen, Schrödinger's cat exemplifies the problem of how a single object in quantum mechanics can exist simultaneously in a mysterious mixture of two states. Unobserved, the cat in the box is somehow a superposition of dead and alive, which I have called "limbo." However, note that no cat is ever seen in limbo—only dead or alive, so no paradox concerning actual observations ever occurs.

Similarly, a photon passed through a left circular polarizer[4] is always found in a single polarized state L. You can verify this by passing the photon through a second left circular polarizer, taking care before doing so not to interact with the photon in any way and possibly change its state.

Let us recall figure 6.7, which shows how the intermediate state of a horizontally polarized photon H (analogous to the cat in limbo) is a superposition of circularly polarized photons L (live) and R (dead). Now suppose we bring time symmetry into the picture. Consider figure 8.4. There I have turned the arrow of the R photon at the bottom around so that it goes into the polarizer instead of out. The L photon at the top is unchanged, still going into the polarizer. The experiment can now be viewed as one in which we have two incoming photons, one spinning along its direction of motion (L)

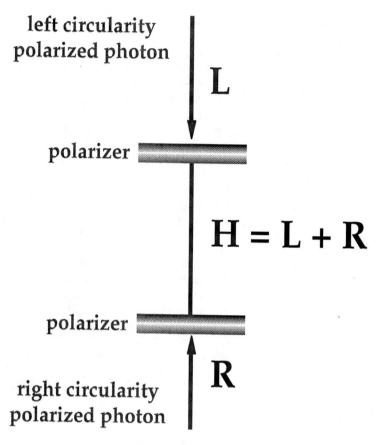

Fig. 8.4. The time-reversible Schrödinger cat experiment as done with photons. The R photon at the bottom is viewed as going into the polarizer rather than coming out. The state in between the two polarizers is then composed of two photons, one coming in and one going out. In the cat experiment, a live cat goes forward in time and a dead one backward in time.

and the other (R) spinning opposite. They combine to form a composite two photon state H.

Alternatively, I could have drawn the two arrows as pointing away from the polarizer. This would be equivalent to the situation where an object disintegrates into two photons. In fact, we know of several such objects, for example, the π^0 meson.

As I mentioned in chapter 6, the quantum formalism contains single photon state vectors for H and V photons, even though these cannot be simply pictured as a single, spinning particle in the same way L and R photons can. Recall that an L photon can be viewed as a particle spinning along its direction of motion while an R photon spins opposite. Because of this easy visualization, I suggested that we might regard circularly polarized photons as the "real," objects and linearly polarized photons as superpositions of two "real" L and R photons. Now bringing time symmetry into the picture, when you send a photon through a linear polarizer that is horizontally oriented, the H photon that comes out is ontologically two photons, an L from the past and an R from the future. That is, it is a "timeless photon."

The point is that at least some of the quantum states that occur in nature can be understood in terms of a superposition of forward and backward time states. This contrasts with the many worlds view in which the superposition is two forward time states in two different worlds. I am suggesting that, with time symmetry, we can get at least these two different worlds into one.

As with the EPR experiment, time symmetry can be utilized to recast our observations into a more familiar form. Based on other observations we have made that exhibit precisely the same phenomenon, nothing is logically inconsistent. The only difference is in the placing of the arrows that specify the time direction. When we are thinking ontologically, we can leave the arrows off. When we must think operationally, as when we consider the results of an experiment, we must place arrow in the direction that is specified by the entropy-generating, or entropy-absorbing, process of our many body experimental setup. This is determined by the boundary conditions of that experiment.

DETECTION AND THE QUANTUM-TO-CLASSICAL TRANSITION

So, we now see how a time-symmetric reality can provide for a particle existing in a superposition of two quantum states. How, then, do we go from that reality to the one we more directly experience in which time has an unambiguous time asymmetry? We manage this by the same process that takes us from the quantum world to the classical world—the act of observation by a macroscopic system, such as a human eye or a particle detector.

As we have seen, the measurement problem has been at the center of much of the controversy over interpretations of quantum mechanics. In the conventional Copenhagen view, the detector is taken as a classical system and the "collapse of the wave function" occurs during the act of detection. Some interpret this to happen under the control of human consciousness.

For good reasons, Einstein called wave function collapse "spooky." It also implies superluminal action at a distance. Bohm and his followers have provided a simple answer to this conundrum: they say that we really have a spooky action at a distance—a mystical, holistic universe in which a quantum potential reaches out instantaneously throughout the universe. Maybe. But we would be wise to rule out less fantastic possibilities first.

In the decoherence interpretation, which has had a number of different formulations, the interaction of the quantum system with macroscopic detectors or the environment itself results in a random smoothing out of interference effects, yielding classical-like observations that do not exhibit these effects. Decoherence provides for wave function collapse in the Copenhagen interpretation, rescuing it from mysticism. It also rescues the many worlds interpretation from incompleteness, by providing a mechanism for selecting out which parallel world is inhabited by a particular observer.

I propose that the quantum-to-classical transition is one and the same with the time symmetric-to-asymmetric transition. As we saw in chapter 4, the arrow of radiation can be traced to the asymmetry between macroscopic sources and detectors, and this in turn is connected to the arrows of thermodynamics and cosmology.

Let us now examine in detail how this asymmetry of detection comes about in an important, specific example.

One of the prime detectors in physics is the **photomultiplier tube**. It provides and excellent prototypical example of a quantum detector since it is sensitive to single photon. Indeed, the photon was first introduced by Einstein in 1905 to explain the **photoelectric effect**, in which light induces an electric current. This played a crucial role in the development of quantum mechanics. The modern photomultiplier tube is an important element of many modern experiments. For example, over 13,000 such tubes provided the primary detection elements in the Super-Kamiokande experiment, on which I worked, that provided the first solid evidence that the neutrino has nonzero mass (see chapter 9).

The detection process in a photomultiplier tube is illustrated in figure 8.5. A photon hits a **photocathode**, which emits an electron into the interior of the evacuated glass chamber that constitutes the tube. The electron is accelerated by the high negative voltage on the cathode to the first **dynode** where it kicks out two or more electrons. This process is repeated, typically ten or so times (only three dynodes are illustrated in the figure), with a multiplication factor of the order of one million. This produces a measurable current pulse that is used to trigger counters and other electronic recording devices.

In figure 8.6, the Feynman diagram for the basic detection process is shown, again in simplified fashion with only single photon exchange interactions indicated. Photons are denoted by dashed lines and electrons by solid

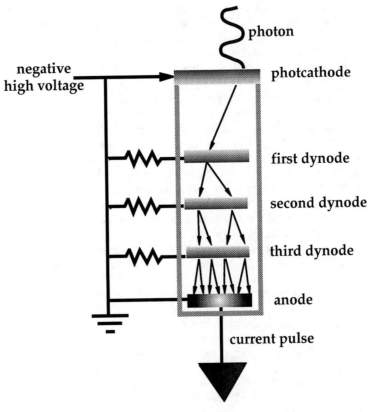

Fig. 8.5. Illustration of the principle of the photomultiplier tube. A photon hits the photocathode which emits an electron into the tube. The electron is accelerated by the high negative voltage to the first dynode, where it kicks out two (or more) electrons. This process is repeated several times (only three dynodes are shown), multiplying the number of electrons and producing a measurable current pulse.

lines. The external photon is shown being absorbed by the cathode electron, which then goes on to interact with a dynode electron by virtual photon exchange, kicking it out of the dynode. The struck electron is accelerated to the next dynode, where it interacts with an electron kicking that electron out of its dynode. Thus, the process of multiplication continues. Note that the cathode and dynode electrons all start at rest and so their worldlines run par-

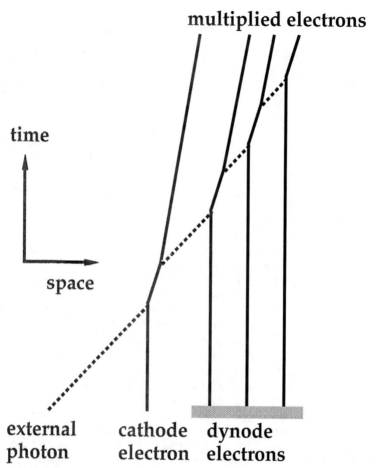

multiplied electrons

time

space

external   cathode   dynode
photon   electron   electrons

Fig. 8.6. The process of photon detection in a photomultiplier tube as a Feynman diagram. While in principle time reversible, it is very unlikely that the multiplier electrons will interact to produce a single photon.

allel prior to interaction. However, once scattered outside the metal of the cathode or dynode, their worldlines are more randomly directed.

Now, each of the lines in the Feynman diagram is in principle reversible (hence, no arrows are indicated). Thus, the struck cathode electron can reverse and emit a photon, settling back to its original state inside the cathode. The process is similar to what was illustrated in figure 4.3 (a).

As we move down the dynode chain, however, precise reversibility becomes increasingly improbable. As noted, the worldlines of the electrons scattered from the cathode and dynode are no longer parallel, meaning they are not at rest relative to one another but flying off in different directions. At the output end where a million or so electrons have produced a macroscopically measurable current, typically a few milliamperes, the probability of them reversing to interact in the exact way needed to result in a single photon being emitted from the cathode is very small. Like a dead cat coming back to life, it is possible but highly unlikely.

Many different Feynman diagrams contribute to the process that converts a photon to a milliampere current. In principle, we would calculate their amplitudes, sum them as complex numbers, and then take the square modulus to get the probability for photon detection. In practice, we simply measure the "detection efficiency" for each tube in the lab and use that to plan our experiments.

In chapter 4, I argued that the three arrows of time—thermodynamic, cosmological, and radiation—were equivalent. In this chapter I have tried to show that no arrow of time exists at the quantum level, but time irreversibility develops during the process of macroscopic measurement. We see that observation is the same decoherent process that makes quantum behavior appear classical, with interference effects being washed out. Indeed, the apparent quantum arrow in what Penrose denotes as the **R** process is just the radiation arrow that results from the built-in asymmetry of macroscopic sources and detectors, and the particular boundary conditions that are imposed on the process of measurement. These boundary conditions are usually those that match the entropy gradient of the apparatus with the cosmological entropy gradient in our part of the universe.

The purpose of measurement is to extract information from a system. Since information is negative entropy, the system under observation must experience an increase in entropy, or disorder. This increase occurs as the entropy of the measuring apparatus decreases, by the very definition of measurement. The reverse process in which a system is made more orderly also can happen, as when you pump up an automobile tire. To measure the pressure with a tire gauge, however, you have to let out a little air and increase the tire's entropy just a bit.

In short, macroscopic measurement processes are irreversible and incoherent by the same mechanism—one that does not apply at the quantum level.

CPT

We have seen how, by changing the time direction of a photon, we can model the superposition of two photon states as a state of two real photons. We need now to generalize this idea to other particles, such as electrons. Let us begin by clearing up something about the process involving photons that was swept under the rug.

Recall that when we time-reflected the L and R photons, we did not change their handedness or chirality. This may seem wrong, because, obviously, the hand of a clock viewed in a film run backward through a projector spins in the opposite direction. But, in fact, we made no blunder. To see why, consider an ordinary clock moving along an axis perpendicular to its front face, as shown in figure 8.7. It is left handed, as can be seen if you place your left thumb in direction of motion of the clock. Your fingers will curl in the direction in which the hand turns on the clock. Now take a film and run in backward through the projector. The direction of motion of the clock will be opposite, but so will be the rotation of the clock hand. The clock is still left handed, as shown on (a). That is, L → L and R → R under time reflection.

Now let us consider space reflection, as shown in figure 8.7 (b). We see that L → R and R → L under space reflection. That is, left goes to right and right to left in a mirror image, a commonsense fact. In physics, we denote time reflection by an operator T and space reflection by an operator P, which stands for **parity**. As we will see, a third operator is also generally involved, denoted by C, which takes a particle into its antiparticle. Photons are their own antiparticles, so this operation did not arise in the above discussion. Later when we deal with electrons in a similar situation, we will combine an electron going forward in time with its antiparticle, the positron, going back in time and viewed in a mirror. The **CPT theorem** has been proved from the most basic axioms of physics; it asserts that every physical process is indistinguishable from the one obtained by the combined operation CPT.

In the case of the photon processes already discussed, time reflection was adequate since those processes are known to be invariant to the T operation. As we will see later, certain fundamental processes, namely, the weak interactions, are not invariant to the P operation. And, a few rare cases of these are not invariant to the combined operation CP. This implies, by the CPT theorem, that these processes violate T. However, we can avoid such complications by being careful to change particles to antiparticles and reflect space as well as time. This will guarantee that the time reflection process will not violate any known principles of physics. That is to say, we can preserve the symmetry of all physical process to time reversal, provided we also change particle to antiparticle and reverse its handedness.

To give a specific example, the CPT theorem says that the interaction between a negatively charged electron and an electron neutrino is indistinguishable from the interaction of a positively charged antielectron, or positron, and an antielectron neutrino, filmed through a mirror with the film run backward through the projector. I realize this is mind-bending, but that's what this book is all about! Please, stick with it. A bent mind is a terrible thing to waste.

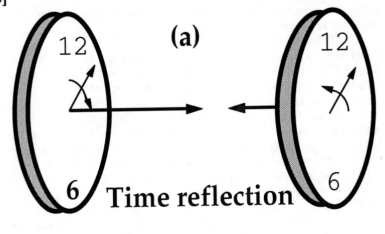

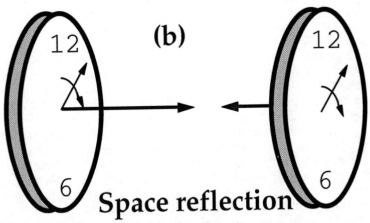

Fig. 8.7. In (a), a clock moving along its axis is left handed in the normal convention and remains left handed upon time reflection. However, it becomes right handed under space reflection (b).

TIME SYMMETRY AND INTERFERENCE

Time symmetry can also be used to remove at least some of the mystery of quantum interference. Recall from chapter 6 that photons and other particles produce interference effects that can be easily explained as wave phenomena but become deeply puzzling when we attempt to visualize the particle as following definite paths in space. This was profoundly demonstrated by Wheeler, who showed how the light from a distant galaxy bent by the powerful gravity of an intervening supermassive black hole can be used to do a cosmic double slit experiment in which a choice made by the experimenter seems, somehow, to act hundreds of millions of years back in time.

Interference effects are seen when photons and other particles pass through any number of apertures, or bend around corners. These are conventionally explained in terms of the wave particle duality. Particles have wavelike properties as exhibited by interference and diffraction. In the Copenhagen interpretation, as promulgated by Bohr, this is a consequence of the complementarity of nature, as explained in chapter 6.

In the de Broglie-Bohm pilot wave interpretations, a particle carries along with it a superluminal quantum field that produces these effects. In the many worlds interpretation, the particle picture is retained but the various paths the particle can follow, for example, through two slits, occur in different worlds that can interact with one another to produce the observed interference.

We also found that the consistent histories formulation of quantum mechanics, due to Griffiths, provided a scheme by which a particle can follow a specific path, in a single world, and still exhibit interference effects. This was discussed in terms of the Mach-Zehnder interferometer, which was illustrated in figure 6.1. Instead of the particle being regarded as a superposition of states corresponding to the two paths, the state of the particle, when it is on one path or the other, is a superposition of the two possible detection states. This leads to the same interference effects as the conventional scheme.

Time symmetry can perhaps help us formulate a realistic model of two-path interference that is consistent with this picture. As seen in figure 8.8, we have one photon emitted from S going forward in conventional time and another photon emitted from D1 which goes backward in conventional time, following, as one possibility, the alternative path back to the detector. (If we had used electrons, then a positron would go backward.) Of course, the photon can also follow the same path back, but this is indistinguishable from the forward photon. Time symmetry then simply gives us a way to have a photon on each path.

Let us take another look at the double slit experiment within the framework of time symmetry (see figure 8.9). A single hit is registered at both the source and detector, within a time interval that would allow the particle or antiparticle to propagate the distance between the two, through either slit. This is conceptually the same as the interferometer example above. Included in the possible paths will be pairs in which the particle passes through one slit and the antiparticle passes simultaneously (in the reference frame of the screen) through the other. And so, a correlation exists between

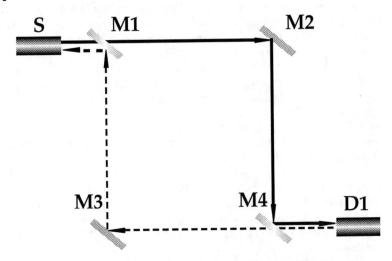

Fig. 8.8. A time-symmetric view of the interferometer. S emits a photon forward in time. When it reaches D1, another photon is emitted by D1 backward in time. It reaches S at the time that S emitted its photon, so the situation is time symmetric.

the two slits in a given event, without the need of a superluminal signal being sent between the two. (Recall, also, the discussion in chapter 7 with regard to figure 7.2.)

Note that the charge balances, as it should. Suppose the particle is an electron. The source loses a negative charge that is gained by the detector some time later, when viewed in conventional time. In reverse-time, the detector loses a positive charge that was *earlier* gained by the source. In either case the net result is the same.

Again, we can see the contrast with the many worlds view, in which the paths through the two slits occur in different worlds. Time symmetry seems to offer an alternative interpretation to the many worlds interpretation of quantum mechanics. This can be extended to multiple slits, or multiple paths. The principle is simply that for every particle following one path in one time direction from point A to point B in space-time, an antiparticle follows the same or another allowed path from B to A in the opposite time direction. In this way all the possible paths between A and B are followed and, as the rules of quantum mechanics require, interfere with one another.

As to "why" the paths interfere, this is basically an axiom of quantum mechanics no matter how it may be formulated or interpreted. I will try to give some idea of its ontological source in the next chapter, but I make no

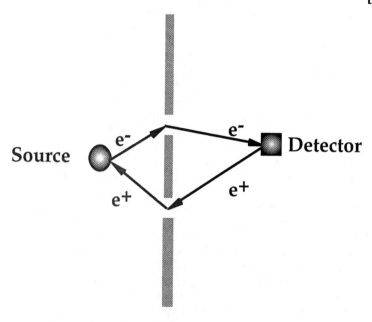

Fig. 8.9. The double slit experiment from a time-symmetric perspective. In conventional time, an electron passes through the top slit. In the time symmetric view, detectors also act as sources and sources as detectors. Here the detector is shown sending a positron back in time through the bottom slit. This is equivalent to an electron going forward in time through that slit. The two electrons pass simultaneously through both slits. Thus, a correlation between the two slits exist that does not require superluminal signals. All the possible paths between source and detector are not shown.

claim to "derive" this very fundamental feature of quantum phenomena. Nor does any interpretation of quantum mechanics.

SOLVING THE TIME-TRAVEL PARADOX

The time-reversible picture presented here should not be interpreted as implying that macroscopic objects, such as human beings, will ever be able to travel back in time. As with a dead cat being resuscitated, it can happen. But we should not expect to see it in the lifetime of the universe. The familiar arrow of time applies to incoherent many body systems, including the human body, and is defined by the direction of most probable occurrences. Because of the large number of particles involved, this probability is very highly peaked in one direction, which is then defined as the direction of time. The time travel implied by the very unlikely but in principle possible processes that can happen in the opposite direction is not to be confused with some of the other types of macroscopic time travel about which people speculate. For example, cosmological time travel might occur when the time axis is bent back on itself in the vicinity of strong gravitational singularities, or cosmic wormholes are used as time machines (Thorne 1994).

The quantum fluctuations that result in the zigzagging in space-time illustrated in figure 8.2 normally occur only over small distances. Deep inside the nuclei of our atoms, elementary particles are zigging and zagging in space-time, but the aggregate moves in the direction that chance, macroscopic convention, and boundary conditions select as the arrow of time.

Large-scale quantum correlations also exist in the EPR experiment, which has now been performed over kilometer distances. However, we have seen that these do not require superluminal connections in the time-reversible view. Macroscopic, many particle coherent systems such as laser beams, superconductors, and bose condensates also exist. Since these are quantum systems, they should also exhibit time reversibility.

Suggestions have been made that the human brain is a coherent quantum system, and that consciousness is somehow related (Penrose 1989, Stapp 1993, Penrose 1994). These ideas remain highly speculative and highly unlikely, although this has not stopped paranormalists from claiming backward causality in quantum mechanics as a basis for the human mind to reach back and affect the past (Jahn 1986, 1987). Again, I will not repeat what I covered in detail on this subject in *The Unconscious Quantum* (Stenger 1995b).

Still, at least one loose thread needs to be tied up in this chapter. The idea that time reversibility solves the quantum paradoxes has been floating around for about half a century, so no parenthood is being claimed here for that notion. However, time reversibility carries what many people think is paradoxical baggage of its own. This has dissuaded influential figures, like Bell, from incorporating time reversal into interpretations of quantum mechanics.[5] If the time travel paradox can be resolved at the quantum level, then the way is open for time symmetry to be a part of any quantum interpretation.

Human time travel would appear to allow you to go back in time and kill your grandfather when he was a child, thereby eliminating the possibility of your existence. This is the famous time-travel **grandfather paradox**. Philosophers technically refer to the problem of backward or "advanced" causation as **the bilking problem**. As Price (1996,128) explains it:

> In order to disprove a claim of advanced causation, we need only to arrange things so that the claimed later cause occurs when the claimed earlier effect has been observed not to occur, and vice versa.

Price (1996, 173) discusses in great detail arguments given by philosopher Michael Dummett (1954, 1964) to avoid the bilking problem. As he summarizes Dummett's view:

> Claims to affect the past are coherent, so long as it is not possible to find out whether the claimed earlier effect has occurred, before the occurrence of the alleged later cause.

In other words, if you go back in time you cannot take an action to prevent some effect if you have no way of discovering the causes of that effect.

I will leave it to the philosophers to debate whether or not bilking can be avoided and macroscopic time travel made logically consistent. I kind of doubt it, without recourse to parallel universes, where you end up in a different world from the one in which you started. Let us instead examine the problem on the quantum scale.

The quantum situation is as if you went back in time to look for your grandfather and kill him, but found that he was indistinguishable from all the other men in the world (only two, in the photon example). All you can do is shoot one at random, which you do not have to be from the future to do. Since no information from the future is used, no paradox exists.

Once again, let us illustrate the effect by considering an experiment with coherent photons rather than macroscopic bodies like cats or grandfathers (see figure 8.10). Suppose we pass initially unpolarized photons through a horizontal linear polarizer H, so that we know their exact state, and thence into a circular polarizing beam splitter. Two beams emerge from the latter, one with left polarization L and the other with right polarization R. We set up the apparatus so that when a photon is detected in the L beam, a signal is transmitted back in conventional time (it need only be nanosecond or so) to insert an absorber that blocks the photon before it reaches the L polarizer. This is analogous to going back in time and killing your grandfather, or yourself for that matter.

Now, each photon that has passed through the H polarizer is in a coherent superposition of L and R states. As a result, we will sometimes block photons that would normally end up in the R beam. That is, we cannot identify an L photon to absorb since that L photon is part of the pure state H. If we act to absorb the photon anyway, we will be killing an H state, not an L state. This action is as if the absorber were randomly inserted in the beam, killing off on average half the photons that will end up in the L and R beams. Since we can achieve the same result without the signal from the future, we do not have a causal paradox. That is, the killing of the photon was not necessarily "caused" by an event in the future.

You might wonder what would happen if, instead of the H polarizer we had an L one. Then we always have L photons and kill them all off. Or, if we

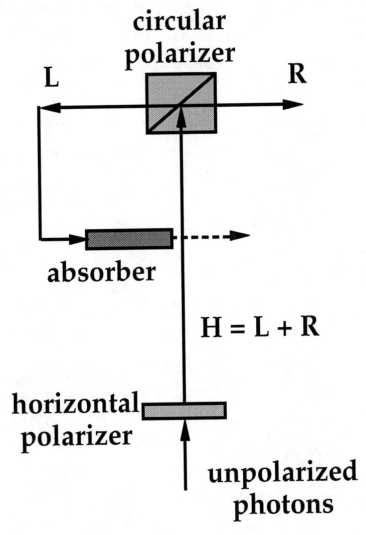

Fig. 8.10. Unpolarized photons are placed in a pure quantum state by the horizontal polarizer. The circular polarizer separates the beam into two circularly polarized beams L and R. When an L is detected, a signal is sent back in time to place an absorber in the beam and prevent the photon from proceeding, thus producing a causal paradox. However, the paradox is not present when the state H is a coherent superposition of L and R, only when it is an incoherent mixture.

have R photons we don't kill any. Again, this can be done without a signal from the future and we have no paradox. The same situation obtains if we try to measure the circular polarization before taking action. To do this we would have to insert a right circular polarizer in the beam, which would produce a beam of pure R. But this can also be done with no information provided from the future.

To appreciate the role quantum mechanics plays in avoiding paradox, consider what happens when H is removed. Then we have an incoherent mixture of L and R photons from the original source, analogous to a macroscopic system. The absorber will only block the L photons that are destined to be in the L beam. The R photons pass unhindered to the right beam. Now we indeed have a causal paradox, with the signal produced by a particle in the future going back in time and killing the particle. While this example certainly does not exhaust all the possibilities, it strikes me that the causal paradox will be hard to defeat for macroscopic time travel, unless that travel is to a parallel universe.

In the clever *Back to the Future* films, the young time traveller Marty (played by Michael J. Fox) had to take certain actions to make sure his father became his father. He had to watch over him, protect him from bullies, and make sure that he dated Marty's future mother (who found Marty strangely attractive). Time travel may be impossible in the classical world without the parallel universes that these and other films and science fiction tales assume, but time travel in the quantum world appears to remain logically possible.

In this discussion, I have implicitly assumed a direction of time as defined by convention and spoken of "backward" or "advanced" causality from that perspective. However, we have seen that the conventional direction of time is set by the direction of increased entropy, and we might wonder where that enters in these examples.

Actually, it enters in an interesting way. Backward causality involves the use of information from the future that is not available in the past. But this implies that the future has *lower* entropy—more information—than the past, a contradiction. The past is by definition the state of lower entropy. On the other hand, the flow of information from the past into the future is allowed. So no logical paradox occurs when a signal is sent to the future (as it is in everyday experience—far in the future of we send it third class mail), just when it is sent into the past.

In the quantum case, since the same result as backward causality can be obtained without the signal from the future, no information flows into the past and no causal paradox ensues. Indeed, this is exactly what we have discovered; quantum mechanics seems to imply time-symmetric causality, but can be formulated without it. My thesis is not that time reversibility is required to understand the universe, Rather, as happened with the Copernican solar system, once we get over our anthropocentric prejudices time reversibility provides us with a simpler and more economical picture of that universe. And it is on that basis, not proof, that we can rationally conclude that time is reversible in reality.

NOTES

1. $\Delta E \Delta t \geq \hbar/2$. This can easily be seen using four-vector scheme of relativity, where time and energy are the fourth components of the four-vector position and momentum. Also, in classical mechanics, energy is the generalized "momentum" conjugate to the generalized "position" that is time. For $\Delta E = 2mc^2 = 1.022$ MeV, $\Delta t = 10^{-22}$ second.

2. Distance $\Delta x \approx h/\Delta p \approx h/p = \lambda$, the de Broglie wavelength.

3. In cosmology, we still often talk as if everything across the universe is simultaneous. This works, as long as we limit ourselves to those parts that are moving with respect to us at nonrelativistic speeds.

4. Two kinds of circular polarizers exist. The usual circular polarizer is a quarter-wave plate with an incident linear polarization at $45°$ from the principal axes, the output will give you either a right circular polarization R or a left circular polarization L, but not both; it essentially depends on the direction of the incident linear polarization with respect to the principal axes. On the other hand, circular polarizing beam splitters give two components, R and L, propagating at right angles.

5. See Price (1996, chapter 9) for Price's attempt to convince Bell of the merits of time symmetry.

MODELING MATTER

The establishment of the Standard Model was one of the great achievements of the human intellect—one that rivals the genesis of quantum mechanics. It will be remembered—together with general relativity, quantum mechanics, and the unraveling of the genetic code—as one of the outstanding intellectual advances of the twentieth century.

Sylvan Schweber (1997)

THE NUCLEAR FORCES

By 1950, quantum electrodynamics could be used to compute all that had been measured about electricity and magnetism, at least in the fundamental interactions of photons and electrons. Even if many physicists were unhappy with the way the theory handled infinities, the renormalization procedure nevertheless worked, giving answers that agreed with the data to great precision. Increasingly precise gravitational experiments and cosmological observations also continued to be described very well by general relativity, in the simplest form originally proposed by Einstein.

However, by this time it had already been long clear that forces beside electromagnetism and gravity exist. Inside the atomic nucleus, protons and neutrons are being held together by the strong nuclear force. A weak nuclear force is evident in the beta decays of nuclei.

It hardly needs mentioning that nuclear physics became a major area of research during World War II. When the hot war was immediately replaced by a cold one, governments recognized basic physics as a critical element of national security. With the kind of support that only comes from the fear of conquest, greatly expanded efforts were undertaken to explore the structure of nuclear particles and forces. Particle physics, for a while at least, benefitted from this largesse stimulated by fear (Schweber 1997).

In the typical nuclear experiment of the postwar period, protons and neutrons were scattered off one another to see what could be learned about the strong force. It was found that this force did not depend on the charge of the particles, although charge remained conserved. Instead, the strong nuclear force depended on another quantity, called isotopic spin, or **isospin**.

Isospin is analogous to ordinary spin, though more abstract. Recall that a spin $\frac{1}{2}$ electron has two possible components along any spatial direction, either "up" or "down" along that direction. We do not normally regard a spin up electron as a different particle from one with spin down. Analogously, the proton and neutron, which have almost the same mass, can be thought of as two different isospin states of a single **nucleon** of isospin $\frac{1}{2}$. Isospin up is associated with the proton state of the nucleon and isospin down with the neutron state.

Just as total angular momentum is conserved in collisions between particles, experiments indicated that total isospin is conserved in strong nucleon-nucleon interactions, and that this interaction is different for different isospins. Two nucleons scattering off each other can exist in total isospin 1 or 0 states, the isospins of the nucleons lining up in the same direction in the first case and opposite direction in the second.[1] The nuclear force in each case does not depend on whether the nucleons are protons or neutrons. For example, the elastic nuclear scattering of two neutrons with total isotopic spin 1 is indistinguishable from the elastic nuclear scattering of two protons with total isospin 1—despite the fact that the charges are different. The electrical forces, by contrast, are not the same in these two cases, being zero between two neutrons and repulsive between two protons. So the strong nuclear force is clearly something different from electromagnetism, at least in the energy regime where experiments are being done in this era.

As we will see in chapter 12, angular momentum conservation is an expression of the rotational symmetry of a system in familiar three dimensional space. Likewise, isospin conservation is an expression of rotational symmetry in this more abstract three dimensional *isospin space*. The same mathematics, the theory of transformations called *group theory*, can be used to describe both concepts. The transformation group applying to spin and isospin is called SU(2), for reasons that need not concern us here.

Early attempts to understand the internulceon forces focussed on meson exchange, in analogy with the photon exchange of QED. I have mentioned Yukawa's theory of meson exchange and his 1935 proposal that a particle should exist with a mass intermediate between the masses of the electron and proton. Specifically, the mass of the meson was predicted to be about 140 MeV, compared to 0.511 Mev for the electron and 938 MeV for the proton. (One MeV = one million electron-volts, one GeV = one billion electron-volts; I will follow the convention of expressing the mass m in terms of its equivalent rest energy, mc^2.)

The predicted mass of the Yukawa meson followed from the tiny range of the nuclear forces, about a proton diameter, one femtometer (10^{-15} meter). The uncertainty principle allows for a particle to be emitted by another particle, temporarily violating energy conservation in an amount equal to the rest energy of the emitted particle, as long as this particle is reabsorbed somewhere else in a sufficiently small time.[2] The range of the force is then inversely proportional to the mass of the exchanged particle.

For example, a neutron might emit a meson that is then absorbed by a nearby proton. Momentum and energy (and possibly even charge) are carried from the neutron to the proton, as illustrated in figure 9.1. By this means, according to Yukawa, the nuclear force is generated. The range of the nuclear force, about 10^{-15} meter, implies (by a more exact calculation) a meson mass of 140 MeV.

When a particle of mass 106 MeV was found in cosmic rays in 1937, many thought it was Yukawa's meson. The mass was a bit off, but no particle having a mass between the electron and proton had previously been observed, so finding anything in this range was a triumph.

However, the new particle was being produced high in the atmosphere and penetrated to the surface of the earth, indicating that it interacted with matter far too weakly to be the quantum of the strong nuclear force. Instead, the new particle turned out to be a heavier version of the electron called the **muon** (symbol μ).

Muons form the primary component of cosmic rays at sea level.[3] In 1946, another particle called the pi-meson (symbol π), or **pion**, was discovered in high altitude cosmic rays. Its mass is 140 MeV, in line with Yukawa's prediction. The scarcity of pions at sea level testified to its strong interaction and made it a far better candidate than the muon for Yukawa's meson.

This discovery encouraged the search for a quantitative theory analogous to QED in which the strong nuclear force results from pion exchange. However, because of the much greater strength of the nuclear force, the perturbation series diverged rather than converged. The higher order radiative corrections, where in this case the "radiation" was composed of pions, were more important than the single pion exchange process! No useful numbers could be obtained with the first few terms in the series, and no one knew how to sum the series.

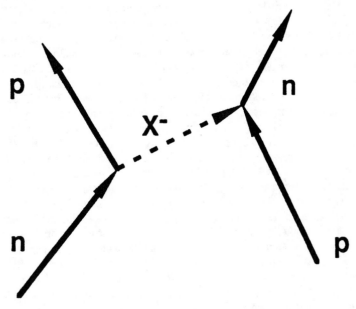

Fig. 9.1. Neutron-proton scattering, n + p → p + n, by the exchange of a negative charged meson X⁻.

During the 1950s, other attempts were made to develop a theory of the strong interaction. Some were quite innovative. Particularly notable was *S-Matrix theory*, also known as *nuclear democracy* and *bootstrap theory*. Pursued by Geoffrey Chew and others, including the now-famous New Age guru Fritjov Capra, S-Matrix theory sought to derive the properties of the strong nuclear interaction and nuclear particles from a few basic mathematical principles. Particles were regarded as being, in some sense, made up of combinations of one another. This approach, which was considered "holistic" rather than "reductionistic," was very appealing in the revolutionary atmosphere of 1960s Berkeley. However, it proved unsuccessful with the advent of the eminently reductionistic quark model.[4] Some of the mathematics, however, still lives on in M-theories, which will be discussed briefly later in this chapter.

THE RISE OF PARTICLE PHYSICS

Despite a theoretical impasse in describing the strong interaction, the 1950s ushered in two decades of immensely productive experimental effort in nuclear and subnuclear physics. Particle accelerators of ever increasing energy produced beams of protons, electrons, and other particles of ever-increasing intensity. These were used in the kind of carefully controlled experiments impossible with cosmic rays. Grand new particle detectors, notably the bubble chamber, visually displayed the interactions produced in these beams and enabled their detailed analysis.

At first, physicists thought that by producing higher energy beams from accelerators they could probe deeper into the structure of the nucleus to learn more about the nuclear force. They imagined themselves as repeating what Ernest Rutherford had done early in the century for chemical atoms, observing the inner structure of matter. Higher energy meant smaller wavelengths, and thus the ability to look to smaller distances.

This probing took place to some extent, but inner structure was not immediately evident. Instead, a new phenomenon captured the attention of most investigators. High energy collisions resulted in the production of many new particles that had never before been seen and were not, like the chemical elements, composite structures of electrons, protons, and neutrons. These new particles were too massive to be produced in chemical reactions or lower energy nuclear collisions, but now appeared as collision energies moved above a billion electron-volts (1 GeV). They were highly unstable, with short lifetimes, disintegrating into lower mass objects in tiny fractions of a second.

Most of the particles are not seen in nature, at least not at the level of typical human observation. Only the cosmic rays bombarding the earth from outer space had sufficient energy to produce these short-lived particles, and this was where several were first seen. However, cosmic ray physicists have no control over their "beam," so observations are difficult to interpret. The controlled beams from accelerators made more extensive particle searches possible. And new detectors, tuned to these beams, enabled more detailed observations of the properties of these previously unimagined components of the physical universe.

By the time I started my Ph.D. research at the University of California at Los Angeles (UCLA) in fall 1959, the new field of particle physics had just split off from nuclear physics. Perhaps a thousand scientists were now working at particle accelerators worldwide, rapidly accumulating vast amounts of data. High-tech electronics and the very new transistorized computers greatly aided this search, as particle physics pioneered the scientific application of computers and other new technologies. I wrote my first computer program in 1959, on punch cards for the IBM 7090. For my thesis, completed in 1963, I studied the interaction of "strange particles," K-mesons (symbol K), or **kaons**, with protons and neutrons as observed in a bubble chamber at the Lawrence Radiation Laboratory (now Lawrence Berkeley Laboratory) in Berkeley, California.

Located in the beautiful hills above the University of California at Berkeley campus, an accelerator dubbed the *Bevatron* had been built a few years earlier that accelerated protons to an unprecedented energy of six billion electron-volts (6 GeV—or 6 BeV in those days). This energy was chosen so that the accelerator would produce *antiprotons*, the antiparticle of the proton, which everyone by then was sure existed. In 1955, soon after the Bevatron was turned on, antiprotons (symbol $\bar{p}$) were observed exactly as anticipated. A bit later, *antineutrons* ($\bar{n}$) were also detected. Many other antiparticles would be included in the tables of new particles that began to take shape as the result of work at the Bevatron and the other new multi-GeV accelerators in the United States and Europe that rapidly appeared. Although only one part in a billion of the normal matter occurring naturally in the known universe, antimatter became a common occurrence in the particle physics laboratory.

Experiments measured the various properties of all the new particles: mass, charge, spin, and other features that were described in terms of newly invented quantities given names such as **strangeness, baryon number**, and **lepton number**. These properties helped classify the particles in a kind of particle botany. Theorists thought long and hard on how to extract any underlying, global principles that might be poking though the data. The great breakthrough occurred around 1964 when Murray Gell-Mann and George Zweig "independently" (they were both at the California Institute of Technology) arrived at the idea of **quarks**. Thus appears the first element of what is now called the **standard model**.

THE QUARKS

Quarks were originally introduced as a means for categorizing the various observations. The great majority of the new particles, eventually numbering in the hundreds, were strongly interacting. Designated as **hadrons**, they seemed to interact with one another by the same force that held the nucleus together. Gell-Mann and Zweig found that the properties of hadrons could be classified by assuming they were composed of more fundamental objects of *fractional* charge. Unlike all previously known objects, from molecules to electrons, the quarks did not carry electric charge in multiples of the unit electric charge. By the early 1970s, experiments with electron beams had revealed pointlike structures inside the proton and neutron that were identified as fractionally charged quarks.

First there were three, then four, five, and eventually six quarks (and their antiparticles). The six quarks of the current standard model are classified in terms of three generations, or families, of two quarks each. Each generation has one quark of charge $+2/3$ and one quark of charge $-1/3$, in fractions the unit electric charge. Quarks were a bold proposal, since no fractionally charged particles had ever been seen.

In the current standard model, the quarks are accompanied in each generation by **antiquarks** with opposite charge. In the first generation, the "up" quark u and the "down" quark d combine to form the nucleons that comprise the nuclei of the chemical elements. The proton is uud, total charge $+1$, and the neutron is udd, total charge zero. The antiproton is $\bar{u}\,\bar{u}\,\bar{d}$ charge 1, and the antineutron $\bar{u}\,\bar{d}\,\bar{d}$ charge zero. In time-symmetric physics, the antiparticles correspond to particles going backward in time; but for now, let us describe things in the conventional way where antiparticles are assumed to be separate objects.

The designations "up" and "down" for the u and d quarks implemented the isospin concept that was earlier applied to the proton and neutron. However, this older "strong" isospin evolved in a different direction and what now appears in the standard model is called **weak isospin**. As we will see below, it also applies to weakly interacting (that is, nonhadronic) particles such as electrons and neutrinos.

The quarks are spin $1/2$ fermions. Putting three together, as in the proton, we still get fermions, particles with spin $1/2$ or $3/2$, and sometimes higher half-integer values. The proton and neutron have spin $1/2$, but other three quark states of higher spin exist. One of the first of the new hadrons discovered was the Δ, a highly unstable particle with mass 1236 MeV and spin $3/2$. It was found in four charge varieties, $+2$, $+1$, 0, and -1, as expected from the quark combinations uuu, uud, udd, and ddd.

Particles formed from three quarks are called **baryons**. Antibaryons made from three antiquarks are also observed. Particles formed from a quark and antiquark comprise the mesons. For example, the pi meson or pion, the particle predicted by Yukawa, comes in varieties with charges ±1, as expected from the quark combinations $u\bar{d}$, and $d\bar{u}$, and a neutral pion that is some quantum superposition, $u\bar{u} + d\bar{d}$. Note that two quarks will form a boson, that is, a particle with integer spin. The pion has spin zero. A particle made from the same quark combinations as the pion, the rho meson, ρ, has spin 1. In "quark spectroscopy," these particles are all viewed as different states of quark-antiquark pairs, analogous to how the hydrogen atom is viewed as different states, or energy levels, of an electron and proton.

Similarly, the different charges of particles correspond to different states of the same quark pair. The masses, like the energy levels in atoms, can be split. For example, the mass of the neutral pion, π^0, is about 3 percent less than the charged pions $\pi^\pm$. In a similar way, we refer to the nucleon as a particle that exists in two charged states, +1 for the proton and zero for the neutron, though the neutron is a bit heavier.

Quark-antiquark pairs can also appear momentarily in and around hadrons, analogous to the electron-positron pairs that populate the vacuum in the vicinity of atoms. However, we assign a *baryon number* $B = \frac{1}{3}$ to quarks and $-\frac{1}{3}$ to antiquarks, so baryons always have $B = +1$, antibaryons $B = -1$, and mesons $B = 0$. Baryon number is a generalization of the nucleon number which is conserved in nuclear reactions. In the higher energies of particle physics, the number of nucleons need not be conserved, but the number of baryons (so far) is found to be conserved in particle interactions. One example is the decay $\Delta^{++} \rightarrow \pi^+ + p$, where the nucleon number is 0 initially and 1 finally, but the baryon number is 1 on both sides of the reaction.

The second generation quarks in the standard model include the c with charge $+\frac{2}{3}$ and the s with charge $-\frac{1}{3}$, and their antiquarks. The "s" stands for **strangeness** and the "c" for **charm**, whimsical names that were assigned to the strange and charming properties that originally led to their discovery. The positive kaon (of my thesis work) is composed of a u and $\bar{s}$. The negative kaon is its antiparticle, $\bar{u}s$. Note that two different neutral kaons exist, $K^o = \bar{s}d$ and $\bar{K}^o = \bar{d}s$. They have opposite strangeness. Neutral mesons made of quarks within their own generation, such as the neutral pion and the phi meson, $\phi = \bar{s}s$, are indistinguishable from their antiparticles.

Prior to the introduction of the quark model in 1964, the hadron classification scheme introduced by Gell-Mann predicted the existence of a negatively charged, triply-strange, spin $\frac{3}{2}$ baryon, the Ω^-. It was observed the same year in the bubble chamber at the Brookhaven National Laboratory, with exactly these properties and at the mass predicted. This was the kind of discovery that convinced particle physicists that this quark business must have something to do with reality. The Ω^- was immediately identified as the system of three strange quarks, sss. Only a single charged state can occur with this quark combination. The charge +1 combination $\bar{s}\bar{s}\bar{s}$ giving the antiparticle Ω^+ was also soon observed.

When the J/ψ particle was discovered at the Stanford Linear Accelerator in California and Brookhaven in New York in 1974, it was hypothesized to be a $\bar{c}c$ meson, a particle with made of a charmed quark and charmed antiquark. With a mass of 3,100 MeV, more than three times that of the proton, the term "meson" had lost its original meaning as intermediate between the electron and proton. The $\bar{c}c$ system proved a rich one, with further exited states and a whole new "charmonium" chemistry was studied at positron-electron collider accelerators. Other particles containing single c quarks were also seen during this period of intense activity.

The c quark was in fact theoretically predicted to exist before evidence for it was found, although most experimentalists (including myself) were skeptical. Theorists discovered that the symmetry provided by having a c, s pair along with the u, d pair helped cancel some of the infinities that were again cropping up in their calculations. With the great success of renormalization in QED, described in the last chapter, physicists were no longer shy about hypothesizing new phenomena that helped make their calculations come out finite. As we will see shortly, they were richly rewarded in this case. The c quark also served to repress certain processes that were not observed, as things quickly began to fall into place.

An amazing dance of theorist and experimentalist occurred during these times, as each supported the other in their findings and suggested new lines of study. Both were somewhat surprised and flattered to receive such skillful support. It was like John Travolta in *Saturday Night Fever*, picking up a girl at the disco and finding that she smoothly followed his every move—and showed him a few of her own. Social scientists would be mistaken to conclude from this, however, that the two groups combined in some conspiracy to simply invent the model out of whole cloth. Enough healthy competition existed, and not every crazy idea that was proposed survived the test of experiment. We just hear about the ones that did.

The quark model was more than a particle classification scheme. Although Gell-Mann originally referred to the notion as "mathematical," he later claimed he always thought of quarks as real particles and had said so (Gell-Mann 1997). Experiments with the Stanford Linear Accelerator electron beam in 1968 revealed a pointlike substructure inside protons and neutrons that could be identified with fractionally charged quarks. Experiments with high energy neutrino beams at the Fermi National Accelerator Lab (Fermilab) and the (now named) European Center for Particle Physics (CERN), further confirmed the quark scheme. I had the good fortune of being a collaborator on experiments at each laboratory in the 1970s, where evidence strongly supporting the quark picture, including the charmed quark, was gathered in great quantities.

The third and probably final generation of quarks began to appear in 1977, when evidence was found for the upsilon particle at 9.7 GeV by a team led by Leon Lederman at Fermilab. Theorists had reason to suspect a third generation existed, and again experimentalists complied. The upsilon was interpreted as a $b\bar{b}$ state, where b (for "bottom" or "beauty") is the $-\frac{1}{3}$ charged quark of the third generation. Evidence for its partner, the $+\frac{2}{3}$ charged t (for "top" or "truth") quark was not found until 1995, in two independent experiments at Fermilab. The top quark took a while to be confirmed because it is so heavy, 174 GeV, almost 200 times as massive as a proton, and its discovery had to await the construction of a beam of sufficient energy.

The following table lists the masses and charges of the three generations of quarks. The antiquarks have the same mass and opposite charge.

Quark	Mass (Mev)	Charge
u	1.5–5	$+^2/_3$
d	3–9	$-^1/_3$
c	1,100–1,400	$+^2/_3$
s	60–170	$-^1/_3$
t	173,800±5,200	$+^2/_3$
b	4,100–4,400	$-^1/_3$

A conceptual problem remained as the quark model developed. No quark has ever been observed as a free particle, the way the electron, proton, kaon, and even the Ω^- and its antiparticle have been seen as tracks in bubble chambers and other particle detectors. Quarks bound inside protons and nucleons are seen by scattering electrons and neutrinos off them. You might ask how we can regard as real objects that are never observed in a free state. The short answer is that we never observe the moon as a free object, since it is bound to the earth by gravity, yet only solipsists and mystics would deny that the moon is real. Later in this chapter, I will give the explanation for the absence of free quarks that is provided by the standard model.

This brief summary of the quark picture cannot convey the profound way that it naturally explained many of the particle properties that seemed very puzzling when first observed. Almost overnight, the quark model brought order to what many of us particle physicists at the time despaired was an incomprehensible mess. For example, why is there no neutral Ω? Because this object is composed of three s quarks that yield only a single state with charge –1, or +1 for the antiparticle. Similarly, the quark model explains why we have a doubly positive charged Δ^{++} but no doubly negative charged Δ^{--}. You can get charge +2 with uuu, but only –1 with ddd. Of the literally hundreds of hadrons that have been observed in experiments, none contradict the quark picture. Many searches have been conducted for hadrons with properties not allowed in the quark model, and none have been found.

LEPTONS AND THE WEAK INTERACTION

A spare three generations of pairs of spin $\frac{1}{2}$ quarks with electric charge $+\frac{2}{3}$ and $-\frac{1}{3}$, along with their antiparticles of opposite charge, form all the known hadrons. In the standard model, the quarks of each generation are accompanied by pairs of non-strongly interacting spin $\frac{1}{2}$ particles called **leptons** with electric charges 0 and –1, and their antiparticles (see figure 9.2). The electrically charged leptons are the electron e, the muon μ, and the tauon τ. All have antiparticles with charge +1. The three neutrinos, ν_e, ν_μ and ν_τ comprise the neutral leptons for each generation, respectively.

The muon, we have seen, is a heavier version of the electron. The tauon

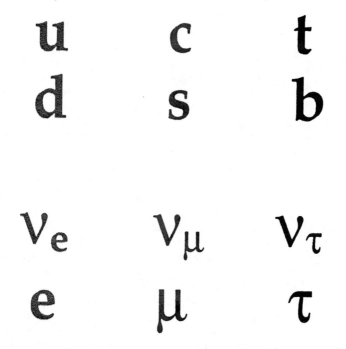

Fig. 9.2. The standard model showing the three generations of quarks and leptons. Familiar matter is composed of just u and d quarks, and electrons.

is heavier yet. The electron neutrino, v_e, was first proposed as an additional particle produced in the beta decay of nuclei. The muon neutrino, v_μ, was found to accompany the decay of muons. Direct evidence for the tau neutrino, v_τ, has just been reported as I finalize this book.

The neutrinos were long regarded as massless. For forty years since their first observation, no significant mass could be measured and the standard model, as of early 1998, assumed zero mass. However, evidence for nonzero neutrino masses has now been strongly confirmed by an experiment called "Super-Kamiokande," in a Zinc mine in Japan using a technique I first suggested in 1980. Again I seem to have had the good fortune to be in the right place at the right time, as one of fifty American scientists who joined seventy Japanese colleagues to collaborate on this important experiment, widely considered one of the most important scientific results of 1998.

Leptons are characterized by not being involved in the strong interactions. They are not hadrons. Yet they interact with each other, and with quarks, by means of the electromagnetic and weak interactions. The weak interaction was originally identified as a short-range nuclear interaction separate from the strong interaction that holds nuclei together. As the name implies, this interaction is much weaker, first appearing in the beta decay of nuclei in which an electron and antineutrino are emitted. The neutrino is only weak interacting, which is why we can observe them passing straight up though the earth. For the same reason, we can only detect neutrinos with highly massive detectors where the chance of some neutrino colliding with some nucleus is high enough to produce reactions that can be observed and studied. The detector in Super-Kamiokande is 30,000 tons of water.

Once nuclei were recognized as being composed of protons and neutrons, beta decay was associated with the decay of the slightly heavier neutron into a proton, electron, and an antielectron neutrino, $\overline{\nu}_e$. This is the antiparticle of the electron neutrino ν_e. In the standard model, the fundamental beta decay process occurs when a d quark inside the neutron disintegrates into a u quark, electron, and antielectron neutrino. In the other generations, similar quark decays occur. Quark and lepton decays are also seen across generations, as the higher masses cascade down to the lowest masses in the first generation.

The weak interaction acts to link the various generations. For example, the μ^- in the second generation decays into $e^- + \nu_\mu + \overline{\nu}_e$. Heavier quarks also decay into lighter ones across generations, leading to hadronic decays such as $K^+ \rightarrow \pi^+ + \pi^0$. This occurs with an average lifetime in the kaon rest frame of about 10^{-8} second. While a very small time interval by human standards, when produced at high energy the kaon can go meters before decaying, leaving a clear track in detectors such as the (now obsolete) bubble chamber. I studied hundreds of such tracks in my thesis experiment, so many years ago.

The relative "stability" of particles like the kaons is to be contrasted with other hadrons such as the rho meson. The decay of the positive rho meson into the same final state of two pions described above for the K^+ also happens here: $\rho^+ \rightarrow \pi^+ + \pi^0$. This occurs with a lifetime of the order of the 10^{-24} second, hardly long enough for the rho to travel the diameter of a nucleus before disintegrating. Of course, the rho leaves no measurable track in any detector; its existence is inferred by indirect means, which are nevertheless highly reliable.

What is the difference between the decays of the rho and positive kaon? In the first case the quarks involved are all within the same generation; in the second case the decay, viewed at the quark level, is across generations. Strong interactions do not occur across generations, but weak interactions do. The kaon decay is a weak interaction, occurring with much lower probability, and thus longer lifetime, the rho decay, which is a strong interaction. Once again, a puzzle succumbs to the quark model.

Although differing greatly in strength, both the strong and weak interactions are short range, occurring at distances no greater than the diameter of a proton, 10^{-15} meter. Indeed Yukawa had originally proposed a unified theory for both the weak and strong forces. However, as we have seen, the pion exchange model for the strong interaction was not fruitful. In the case of the weak interactions, a theory proposed by Fermi in 1934 and later extended by Feynman, Gell-Mann, and others was reasonably successful at low energies. It was not renormalizable, that is, the perturbation series gave infinities in the higher order terms, but because of the weakness of the interaction, perturbation calculations to first order gave usable results and the infinities were swept under the rug.

Fermi's theory assumed that the weak interaction had zero range. It gave a reasonable description for beta decays of neutrons and other particles. We can visualize the Fermi picture for neutron decay, in **Feynman diagram** terms, as a neutron line ending at a vertex from which a proton, electron, and antielectron neutrino emerge (see figure 9.3 [a]). In the particle exchange picture, which we recall Fermi had invented in collaboration with Bethe, we can imagine the vertex under a magnifying glass (see figure 9.3 [b]). The neutron changes into a proton by the emission of a particle W, which then decays into the electron and antineutrino. That is, the weak interaction proceeds by the exchange of a particle, a weak quantum W, that must be very massive so that the interaction range is very small, appearing as a point at the comparatively low energy of beta decay.

The weak quantum, called the **W boson**, was required to exist in at least two charge states, ±1, to allow for the beta decay in which positrons are emitted, which was also observed. Unsuccessful searches for the W± were performed in the 1950s and 1960s, with ever-increasing limits being set on its minimum mass and consequent ever-decreasing upper limits on its maximum range. The W boson was eventually found in 1983 in two gigantic experiments at CERN, at a mass of 80 GeV. The next year, an accompanying neutral particle, the **Z boson** at 90 GeV was seen. These observations implied that the range of the weak nuclear force was much smaller than the size of a nucleon, on the order of 10^{-18} meter. The masses were exactly as predicted by a new theory that had been developed in the previous decade, **electroweak unification**.

ELECTROWEAK UNIFICATION

Ever since Newton showed that earthly and celestial gravity were the same, the unification of the forces of nature has been a major goal of science. We have seen how Michael Faraday and James Clerk Maxwell unified electricity and magnetism in the nineteenth century. However, the unification of the

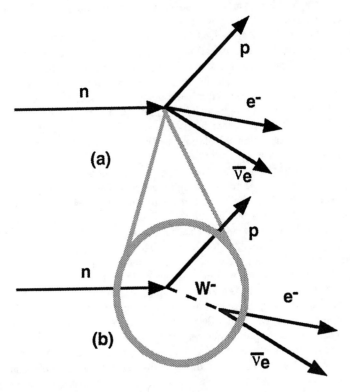

Fig. 9.3. The Fermi model of neutron beta decay (a) with the vertex magnified in (b) to show the role of the W boson in the fundamental process.

electromagnetic and weak forces in the 1970s was highly unexpected. Most of us working in the field at that time saw little connection between electromagnetism and the weak force. The first had an infinite range, while the second had a range smaller than a proton diameter. If any unification was to happen, it seemed it should be between the weak and strong forces, both being short range and nuclear. Gravity was a far more likely candidate for unification with electromagnetism than the weak force, since both fell off with the square of distance. But things did not happen in the expected sequence.

The first step on the road to electroweak unification was taken in 1954 with the development of a mathematical model by Chen Ning Yang and Robert Mills that at first seemed to have little to do with reality. Yang and Mills looked at QED and asked what would happen if they extended it to include the exchange of charged particles, that is, "photons" with charge ±1. They were motivated by the strong interaction, where pion exchange was not giving fruitful results. The pion is spinless. Perhaps the exchanged particle should have spin 1, like the photon. Since charge is exchanged in strong interactions, the strong quantum would have to occur in charge +1, 0, and –1 varieties. The rho meson, discovered a few years later, emerged as a possible candidate.

Yang and Mills wrote down a model that described electromagnetic-like interactions in which the exchanged particle had spin 1 and three charged states. But a problem occurred when they tried to give these bosons mass, which they presumably needed to make the force short range. Then the theory was no longer gauge invariant, which I will now explain.

Gauge invariance is a property possessed by the classical electromagnetic fields that is maintained in quantum electrodynamics. As we will explore in detail in chapter 12, invariance principles, or symmetries, are deeply connected to conservation principles. For example, in both classical and quantum physics conservation of energy is found to result from the invariance of a system to translations along the time axis, that is, from the absence of any special moment in time. The laws of physics are the same today as they were four hundred years ago or four hundred million years ago.

Linear momentum conservation similarly results from invariance under translations of the three spatial axes. The laws of physics are the same on earth as in the most distant galaxy. Likewise, angular momentum conservation results from invariance under rotations about the three spatial axes. No special direction in space exists, and the laws of physics are the same in New York as in Sydney. In classical physics language, the corresponding time, space, and angular coordinates are said to be "ignorable" when a system possesses the symmetry that corresponds to that coordinate. And, in both classical and quantum physics, the momentum "conjugate" to that coordinate is conserved.

Charge conservation is analogously viewed as a consequence of the invariance of the system to gauge transformations. In classical electrodynamics, the gauge transformation takes one set of potentials, the scalar and vector potentials (the scalar potential is just what is called "voltage" in an electrical circuit), to a different set that give the same electric and magnetic field, and thus the same observed forces on charged particles and currents. This symmetry— the same observations for different sets of potentials—is **gauge invariance** and leads to charge conservation. In the language of group theory mentioned earlier, electromagnetic theory is invariant to transformations in the group U(1). The Yang-Mills paper greatly clarified this connection between gauge invariance and charge conservation, and extended it to the group SU(2). As we saw above, the internulceon force conserves isospin, which is described by SU(2), and this symmetry is built into the Yang-Mills scheme.

In quantum electrodynamics, gauge symmetry is maintained—as it must be to agree with the observed fact of charge conservation. In wave function language, if the phase of the wave function is allowed to vary arbitrarily at each position and time, the resulting quantum mechanical equations are invariant to gauge transformations of the potentials and electric charge is conserved. Later, when we interpret the laws of physics in terms of unbroken global symmetries, and broken local symmetries, we will see how profound this notion is.

One of the consequences of U(1) symmetry is that the photons must have zero rest mass. Fortunately, this agrees with experiment, where the current experimental upper limit on the rest mass of the photon is 10^{-48} gram! Recalling that the range of forces depends inversely on the mass of the exchanged particle, the fact that we can see galaxies billions of light years away shows that the photon must have very low mass indeed.

Yang and Mills recognized that maintaining gauge invariance was important in order to extract other conservation principles analogous to charge conservation, such as the isospin conservation of the strong interaction. But this required charged spin 1 bosons with zero mass, and such particles are not observed. Thus, the Yang-Mills theory was initially viewed only as a mathematical exercise, a "toy" model that did not apply to the real world.

A few theorists, however, were sufficiently intrigued to pursue the Yang-Mills idea further. Could a way be found to make the charged "photons" massive and still maintain gauge invariance? Yes, yes, yes! The secret lay in another direction—*superconductivity*.

Superconductors are materials that allow the flow of electric current with zero resistance. John Bardeen, Leon Cooper, and John Schrieffer (BCS) had explained the effect as resulting when electrons become paired by *phonon* (quantized sound wave) interactions with the structure within which they move. Being made of two spin $\frac{1}{2}$ electrons, these "Cooper pairs" must be integer spin (they are actually zero spin), which means they are not restricted by the Pauli exclusion principle. At very low temperatures, all the bosonic pairs can drop to their lowest energy state in what is called a **Bose condensate**. Since no additional energy can be taken from the pairs, they are then able to move freely without resistance. Hence superconductivity.

What does this have to do with particle physics? Physics is physics, and the good physicist keeps his or her ears tuned to developments in all specialties. In 1961, Yoichiro Nambu suggested that the splittings between hadron masses, such as between a neutron and proton or between the neutral and charged pions, might result from the **spontaneous symmetry breaking**, the accidental breaking of the underlying symmetries that unite them in single multiplets. He did this by analogy with the BCS theory of superconductivity, replacing the electron-electron pairing with hadron-antihadron pairing. This was prior to the discovery of quarks. As we will see, spontaneous symmetry breaking has turned out to be the mechanism by which we can understand all structure formation in the universe.

At about the same time, condensed matter physicist Philip Anderson remarked that the massless phonons in a superconductor become massive by way of spontaneous symmetry breaking. In 1964, Peter Higgs and, independently, Francois Englert and Robert Brout produced a model by which spontaneous symmetry breaking gives mass to **gauge bosons**, the name given to the elementary bosons exchanged in the gauge models. They hypothesized that spinless, uncharged elementary particles assemble as a Bose condensate in the vacuum. These particles, called **Higgs bosons**, fill the universe as quanta of what is called the **Higgs field**. They are not directly observable, however, because like the Cooper pairs in a superconductor they have the lowest possible energy and so cannot give up any energy to a detection apparatus unless that energy is provided from the outside.

Now, you might expect that the lowest energy state is one with zero energy, like photons in the vacuum. In that case, all the Higgs bosons would be simply zero mass particles with zero kinetic energy and zero spin, which is about as close to being nothing as you can get. At least a zero energy photon has spin. Suppose, however, that the lowest energy state is not zero energy. In that case, the Higgs boson has an inertial mass (recall that energy and inertial mass are equivalent in relativity) and particles moving through the vacuum will scatter randomly off the Higgs particles that are everywhere. The Higgs bosons resist the motion of other particles, effectively slowing them down the way photons are effectively slowed as they pass through a transparent medium such as glass, although the analogy is not exact. These photons continue to travel at the speed c in space between atoms in the glass, but because of their bouncing around, they take longer to traverse the medium and end up with an effective speed c/n, where n is identified as the index of refraction of the medium. Applied to the Higgs mechanism, the vacuum has an index of refraction.

According to the equations of relativity, the rest mass of a particle moving at the speed of light is zero. At less than the speed of light, the particle has a rest mass greater than zero. So a massless particle moving at the speed of light c through the Higgs field, that is, between collisions with Higgs particles, obtains the operational equivalent of a rest mass greater than zero as it proceeds with an average speed less than c. Note how a particle ontology allows us to picture massless particles moving at the speed of light and then to view nonzero mass as an inertial effect. This is precisely what mass has meant from the time of Newton. In this regard, the Higgs particles themselves are also fundamentally massless and gain their mass by scattering off one another. Thus, the Higgs mechanism, for the first time in history, provides us with an explanation for the origin of mass.[5]

It remained to implement this notion in a practical theory, one that could be directly applied to existing phenomena. Thomas Kibble proposed that the charged Yang-Mills's bosons gain mass by the Higgs mechanism. The resulting theory remained gauge invariant and the way was then cleared for identifying the new heavy "photons" with the weak bosons.

However, a number of questions remained. First, why is the lowest energy level for the Higgs boson nonzero? For the Higgs particles to collide off one another, they can't have zero energy. Another way to ask this, in the field language (which we can still view as a field of pebbles), is: Why isn't the energy density of the Higgs field zero? Here another analogy with condensed matter physics is often used—ferromagnetism.

At high temperatures, above what is called the **Curie temperature**, a ferromagnet has zero magnetization, and thus zero net magnetic field, even though iron is filled with little magnetic domains. Atoms are intrinsically magnetic, with whirling charged particles, but most materials are nonmagnetic because the atomic magnetic fields point in random directions and cancel out. In iron, domains exist whose atoms are locked with their magnetic moments pointing in one direction, making that domain magnetic. Above the Curie temperature, all the little magnetic domains within the iron are being jiggled around so much by thermal motion that the net magnetic field still adds to zero. Below the Curie temperature, however, this motion slows sufficiently for all the domains to lock into place. Think of the domains as little bar magnets. Because of their mutual interaction, they will tend to line up when the thermal motion is small, yielding a net magnetic field.

This process is called spontaneous symmetry breaking, since in the absence of any outside magnetic forces the resulting magnetic field will point in a random direction. Once formed, however, that field breaks the original symmetry of the system by selecting out this random axis as a special direction. The field now has a nonzero energy density. Furthermore, this arrangement of domains will be the ground state of the system. So we have a situation where the lowest energy of the system, the one it will eventually relax into, breaks the rotational symmetry of the underlying physics—in this case, Maxwell's equations.

Theorists conjectured that the Higgs field, at least in our current cold universe, exists in an analogous situation. The interactions between Higgs particles are presumed sufficiently strong so that the ground state is one of broken symmetry, leading to a nonzero energy density and giving mass to the bosons that constitute the field. This was only a guess, not derived from any deeper theory and with no idea what the Higgs mass might be. Nevertheless, a mechanism was developed that provided for a Yang-Mills type gauge theory in which the quanta of the field carried mass and thus mediated short-range forces. The quanta remained intrinsically massless, however, and so the theory was still gauge invariant and thus conserved isospin.

I will get back later to the question of the source of spontaneous symmetry breaking, which relates very deeply to cosmology and the evolution of the early universe. Here we must follow the line of thought that ultimately lead to electroweak unification.

I do not want to leave the impression that when the above ideas were first promulgated everyone jumped on the bandwagon and headed off straight down the yellow brick road to the standard model. Quite the opposite. Most physicists were unaware of the significance of these results, and even those who were remained highly skeptical of what seemed a pretty ad hoc procedure. The theory contained a number of problems that had to be painstakingly grappled with. Still, it was promising enough that a few brave souls pushed on. And, while the names of certain individuals are always mentioned for their particularly important contributions, it should be noted that there is no legendary "Einstein" of the standard model. It was very much a communal effort and I will not be able to mention by name everyone who deserves credit.

In 1967, Abdus Salam and Steven Weinberg independently proposed a Yang-Mills type of gauge invariant theory with spontaneous symmetry breaking that unified the electromagnetic and weak interactions. Sheldon Glashow had published an earlier model in 1961 that had some of the same ingredients but without spontaneous symmetry breaking.

Weinberg's theory was the simplest and most complete of the three, although he originally limited himself to leptons, that is, the interactions between electrons, muons, and neutrinos. He proposed that in addition to the photon three heavy Yang-Mills bosons of the type discussed above exist. Two were the charged W bosons, long regarded as the quanta of the weak interaction. He discovered that a massive neutral boson, which is called Z, also was needed, as Glashow had realized, too. Weinberg estimated the masses of these particles to be at least 60 GeV, a number that would later be greatly refined by a combination of theory and experiment to prediction of 90 GeV.

While the $W^{\pm}$ and Z masses were not directly predicted, they were expressed in terms of a parameter now called the **Weinberg angle**, or weak mixing angle θ_W. The fundamental bosons in the theory were actually a **weak isospin** triplet (W^+, W^o, W^-) and a singlet B. The neutral bosons mixed together to give the observed photon and Z. If you think of a two-dimensional Hilbert space in which one axis is the W^o and one the B, rotating that coordinate system by θ_W gives the photon and Z as new axes.

The electroweak interaction conserves charge and weak isospin. Each generation of quarks and leptons contains weak isospin doublets. For the first generation, (u, d) and (v_e, e^-) have weak isospin (up, down). The other generations follow similarly.

The Weinberg angle also relates the electric charge, which measures the strength of the electromagnetic interaction, to the strength of the weak interaction that occurs in the Fermi theory. Weinberg's theory did not predict the value of θ_W. It had to be determined by experiment. However, once θ_W was known, many other quantitative predictions could be made, notably the masses of the W and Z bosons.

One prediction of Weinberg's theory was the existence of "weak neutral current" reactions such as $v_\mu + e \rightarrow v_\mu + e$ which were not previously thought to exist. These result from Z exchange. This prediction was confirmed in 1973 in a bubble chamber experiment at CERN. This and a wide range of other experiments lead to accurate estimates of the Weinberg angle and from that to predictions of the W mass to be 80 GeV and the Z mass to be 90 GeV. These predictions would be confirmed in two independent huge colliding beam experiments in 1983 at CERN.

At this writing, weak bosons now are produced by the thousands in colliding beam accelerators and interest is focussed on exactly what they are telling us. For example, e^+e^- colliders at the Stanford Linear Accelerator, CERN, and elsewhere have been used to produce Z bosons in large numbers. By a careful study of the Z decay rates into all possible observable particles, the number of generations of quarks and leptons has been shown to be no more than three. This was done by eliminating the possibility that Z decays into any additional neutrinos by the reaction $Z \rightarrow v\overline{v}$ for any but the neutrinos of the three known generations. Interestingly, cosmologists had also found evidence for a limit of three light neutrino generations.

What is now generally referred to as the Weinberg-Salam-Glashow electroweak theory has more than met the minimal requirements of a successful scientific paradigm. It made highly risky, quantitative predictions whose failure to be confirmed by experiment would have falsified the theory. The triumphant confirmation of its predictions should leave us with little doubt that this paradigm is telling us something very deep about objective reality. It is internally consistent, gauge invariant, and, as proved by Gerard 't Hooft in 1971, renormalizable! The earlier Fermi theory of the weak interactions was not renormalizable, so the proof of renormalizability was a crucial step in the development of the electroweak theory that turned many skeptics into true believers. Finally we had a theory that was both gauge invariant and renormalizable, like QED but going beyond and subsuming it.

The electroweak theory was extended to quarks, which require additional parameters that, like the Weinberg angle, must be determined by experiment. Not all of these have yet been measured since they appear only in the rarest processes, such as b-quark decays. However, experiments are gradually filling in the details, some of which remain important for the full understanding of fundamental processes.

QCD

The standard model also includes the modern theory of the strong interaction called **quantum chromodynamics**, or QCD by analogy with QED. QCD was developed in parallel to the unified theory of electromagnetic and weak interactions and, like them, is a renormalizable gauge theory of the Yang-Mills genre.

In QCD, the role of the electric charge in the electroweak theory is taken by another kind of charge called **color charge**; hence the "chromo" in quantum chromodynamics. The use of "color" in this context is not simply whimsical, like "strangeness" and "charm." It comes from the fact that the QCD color charge has properties analogous to those of the primary colors of light: red, green, and blue.

Recall the Δ^{++} baryon, which is composed of three u quarks, uuu, or the Ω^-, which is the combination sss. Other baryons, such as the proton and neutron, contain two seemingly identical quarks, and analyses indicated that these are often in the same quantum state. This presented a problem. Quarks are spin $\frac{1}{2}$ fermions and as such must obey the Pauli exclusion principle, which says that two or more fermions cannot exist in the same quantum state. The solution given in the standard model is that quarks have another property that comes in three primary varieties and combinations of those three. The analogy is drawn with the familiar concept of color.

Think in terms of the familiar primary colors red, green, and blue. All other colors can be formed from a mixture of these three and an equal mixture gives white. Imagine the three primary colors as defining axes 120 degrees apart in a plane we will call "color space." Any arbitrary direction in that plane represents a unique color. Call an arrow pointing in that direction the "color vector." Take it to have unit length. The projections of the color vector along each axis then will give the relative amounts of the three primary colors.

Following this analogy, quarks are assumed to come in three colors, that is, color charges, which we will call red, green and blue. We have red, green, and blue u quarks, d quarks, s quarks, and so on. Thus the Δ^{++} is made of one red, one green, and one blue u quark. The Ω^- is a similar combination of three s quarks. The Pauli principle is okay because the particles are different, or at least in three different color charge states.

The antiquarks have "anticolors," $\bar{r}$, $\bar{g}$, and $\bar{b}$, which we can think of as turquoise, violet, and yellow the colors you get when you filter red, green, or blue light respectively from a beam of white light. The anticolor vectors point in the opposite direction to the color vectors in color space. And so, $\bar{r}r$, $\bar{g}g$, and $\bar{b}b$ all give white. Other quark-antiquark combinations exist, such as $\bar{r}b$, that are not colorless, that is, not white.

QCD indicates, as we will see below, that only white particles can exist as free objects. This explains why we always must have systems of three quarks or three antiquarks in the case of baryons, or an antiquark and quark in the case the mesons. We see no states such as uu or ud$\overline{s}$ because you can't make them white. Note that this also explains why we only observe zero or integral charge in matter that is in fact composed of fractionally charged elementary particles.

This leads us to the picture of the strong interaction that developed with the standard model and replaced the original, unsuccessful pion-exchange picture and other aborted attempts to develop a theory of the strong interactions. In QCD, the strong force acts between quarks and is mediated by the exchange of eight massless, spin 1 particles called **gluons**. That is, the gluon is the quantum of the strong interaction in the same way that the photon is the quantum of the electromagnetic interaction. However, unlike the photon which carries no electric charge, the gluons carry color charge. They are the Yang-Mills bosons of the strong interaction, only we have eight rather than three as in the electroweak interaction. Recall that the isospin symmetry of the original Yang-Mills theory was provided by the transformation group SU(2). In the gluon case, the symmetry group becomes SU(3). In SU(2) we have two possible orientations of the isospin vector, up or down, or two electric charges in each doublet quark or lepton. In SU(3), as applied here, we have three color charge vector orientations or three color charges in each triplet quark, red, green, and blue. Since leptons are not strongly interacting, they presumably lack color. Think of them as white.

It helps to view each gluon as a composite state of a quark and anti-quark of different colors. As mentioned, not all combinations are white. Six gluon pairs will give nonwhite colors, with interesting tints such as $\overline{r}$g. While three white combinations also occur, only two of these combinations are independent. Thus we have eight gluons, six colored and two colorless, being exchanged to provide the strong force between quarks. Like the charged bosons in the original Yang-Mills theory, most of the gluons mediating the strong interaction carry color charge. This, as we will see, has profound consequences for the color force.

In QED, the electromagnetic force extends to infinite range because the photon is massless. A massless particle need not violate energy conservation when it is exchanged in an interaction, and so can travel a long distance without having to be reabsorbed to restore energy conservation. The gluon is also massless in QCD, guaranteeing gauge invariance and thus color charge conservation. Note especially that the Higgs mechanism is not invoked in QCD in order to give the gluons mass. They remain massless.

Why, then, does the strong interaction have such short range? The QCD answer requires some explanation, but this explanation also neatly ties up a number of other loose strings I have left dangling. Since gluons carry color charge, the strong force behaves very differently from the electromagnetic force where the force carrier, the photon, has no electric charge. Recall how an electrically charged particle, such as an electron, is surrounded by a (pebble) field of photons and electron-positron pairs continually being created and destroyed in the vacuum by quantum fluctuations. The charges in the pairs tend to screen off the bare charge of the particle, so that the effective charge we measure for the particle is less than the bare charge. This is taken into account in the charge renormalization procedure of QED, which is carried over to the electroweak theory.

Now, if you probe the region around a charged particle to smaller and smaller distances, using higher and higher energy probes, you would expect to see an increase in its effective charge. In the case of QED, however, this increase occurs very slowly until you reach energies far higher than we have available today at accelerators. Thus, the electric charge is constant to a good approximation at the scale of current experiments. Nevertheless, the effective charge ultimately goes to infinity in the theory. This remains a logical difficulty with QED that led some notable theorists in the past, such as Feynman and Lev Landau, to question whether the theory could ever be made logically consistent. It can, by unifying electromagnetism with the other forces.

QCD is different from QED by virtue of the fact that its quanta, the gluons, carry color charge. A bare colored quark is surrounded by a field of gluons and quark-antiquark pairs. Like the electric case, the color charge of the bare quark is screened off by the color charges of the quarks in the pairs, which tends to increase the effective color charge as you probe deeper. In 1973, David Gross and Frank Wilczek, and, independently, H. David Politzer (the latter two were graduate students at the time, and Gross was only thirty-two) discovered, by rather complicated calculations, that Yang-Mills theories like QCD have the opposite property. Apparently 't Hooft (also a very young man at the time) had arrived at the same conclusion. Color-charged Yang-Mills bosons screen the bare color charge in such a way that the effective color charge gets smaller rather than larger at shorter distances. This was called "asymptotic freedom," but let me refer to it less technically and more descriptively as *short-range freedom*.

Short-range freedom has a number of profound consequences. First, it implies that when quarks collide at very high energies (again, high energies mean small distances) the force between them is weak and the perturbation series converges. A small number of Feynman diagrams can then be used to calculate reaction rates. This also implies that quarks move around comparatively freely at the small distances inside hadrons. This neatly explained another puzzle that had bothered people when the experiments probing protons and neutrons with electrons and neutrinos indicated pointlike substructure. If the quarks being scattered off were tightly bound to other quarks the probing particle would be expected to scatter off the whole rather than the individual parts. Again we find that the whole is not much more than the sum of its particles.

Short-range freedom implies *long-range slavery:* the strong interaction gets stronger as you increase the separation between quarks. When the distance between quarks is about a nuclear diameter, the force is so great that a color discharge of the vacuum occurs that neutralizes the space in between the quarks and cuts off the force. This situation is analogous to the electrical breakdown in a gas that occurs when the voltage gets too high, as in a lightning bolt. This occurs because the gas becomes ionized and the freed electrons conduct a current that neutralizes the voltage.

This also explains why no free quarks and only colorless hadrons are seen as free particles. As is well known to anyone who has worked with electricity, an electrical conductor, like ionized gas, cannot maintain separate electrical charges. Similarly, a color charge conductor cannot maintain free color charges. Because of colored gluons, the vacuum is a color charge conductor. Only color neutral objects can exist separated from others by more than a proton diameter. At high enough energy, however, a "quark-gluon plasma" might be seen. Indeed, some evidence has been found for "glueballs," colorless objects made of gluons and quark pairs.

Despite this beautiful conceptual scheme that offers a logically consistent, gauge invariant, renormalizable theory of strong interactions, QCD has not achieved the level of predictivity of QED. The calculations are straightforward at the quark-gluon level, but become more ambiguous when they have to be compared with experimental data, which are limited to measurements on composite, colorless objects. Since no free quarks are seen, we must study their interactions by slamming composite quark systems like mesons and baryons together, or by scattering leptons from hadrons. It is like trying to study electron-electron interactions by scattering atoms from atoms. It can be done, but the data are not as clean as obtained from observing the primitive processes directly. The calculations are more laborious and uncertain, requiring assumptions about the detailed structure of the atom. Similarly, the structure of hadrons must be modeled in order to compare theory with data.

For example, assumptions have to be made on how quarks are "dressed" into hadrons and at what minimum energy to calculate "gluon radiation." Still, few doubt that the gluon exchange picture of strong interactions is correct, and many useful calculations have been performed that agree reasonably with the data. Since the theory is renormalizable, infinities can be subtracted. The self energy of quarks and vacuum color-polarization effects are incorporated as are the analogous processes in QED. Furthermore, QCD fits well together with the other critical element of the standard model, the unified electroweak force. Both are Yang-Mills gauge invariant theories with spin 1 boson exchange.

BEYOND THE STANDARD MODEL

The standard model contains about twenty-four parameters, depending on how you count, that must be set by experiment. They include the masses of particles, the force strengths at a given energy, and various mixing parameters like the Weinberg angle. However, it should be understood that these constants are basically needed to connect the theory to the data in current "low energy" experiments. The fundamental underlying theory is simpler at very high energies beyond what we can currently achieve in accelerators, where all masses are at least approximately zero and all force strengths are the same. At low energies, where we make our measurements, the basic symmetries of the theory are broken, leading to a more complex array of particles and forces.

In any case, no measurement has been made that is inconsistent with the paradigm and many more than twenty-four experimental measurements have been made, some to exquisite precision, that are fully consistent with calculations. The detection of apparent neutrino mass, mentioned above, requires at least some modification of the standard model as it has existed now for over a decade, but this does not negate the successes so far obtained.

A big unfilled gap remains, however, in the failure, so far, to detect the Higgs boson. As we saw, the introduction of the Higgs boson in the electroweak theory was ad hoc, with no real estimate of its mass. Searches at the largest particle accelerators have placed a lower limit on it mass of about 60 GeV at this writing. It could be more massive, or perhaps not even exist. However, it cannot be too massive or else the whole theory becomes inconsistent, with an upper limit somewhere in the vicinity of 300 GeV.

As the 1990s began, U.S. high energy physicists had pinned their hopes for finding the Higgs boson and pursuing physics beyond the standard model on a giant particle accelerator, the *Superconducting Supercollider* (SSC), that was to be built in Texas. The SSC was designed to reach into the energy regime where either the Higgs boson would be produced or alternative new physics would almost certainly be found. In 1993, the U.S. Congress canceled the project, though a billion dollars or so had already been spent. A slightly less ambitious project, the *Large Hadron Collider*, is going ahead at CERN, with substantial U.S. support and participation. The energy of the Large Hadron Collider should be high enough to produce the Higgs boson, if it exists and current estimates on its highest possible mass are reliable. However, the experiments may not be able to disentangle the new physics that would be implied should the Higgs not appear on schedule—a possibility that cannot be ignored.

What might that new physics be? I do not want to indulge in too much speculation here, since my object is to draw ontological conclusions based on what we now can assert, with high reliability, about the structure of matter and the universe. However, several ideas merit mention. Should the standard model, as described above, turn out to be incorrect in one detail or another, that will not necessarily imply that the particulate picture of reality I have described must be wrong or drastically modified. Furthermore, the standard model is regarded as an "effective" theory that must grow out of a more fundamental, more highly unified theory with far fewer parameters— perhaps none.

One alternative to the Higgs mechanism, called *technicolor*, draws more heavily on the analogy with the BCS theory of superconductivity, mentioned earlier, in which pairs of electrons form a Bose condensate. Recall that in the standard model the Higgs boson is assumed to be a real particle, the quantum of a field, that forms a Bose condensate. In the technicolor scenario, pairs of quarks form the condensate and no Higgs particles, as such, exist. However, little can be said about this model without some data. A new force is implied, with technicolor gluons, presumably. The Higgs picture is more developed theoretically, but we can only wait and see.

Most ideas for new physics beyond the standard model center on further unification of the forces and particles. Soon after the success of electroweak unification became apparent and quantum chromodynamics was developed along the same gauge theory lines, physicists naturally moved to try to unify the two. This was called **Grand Unification**, and the theories called grand unified theories or **GUTs**.

The simplest GUT first proposed was called *minimal SU(5)*, the simplest symmetry group (nonuniquely) encompassing the symmetries of the standard model. We will avoid the mathematical technicalities here; I only mention the group names so the reader can connect our discussion to other literature on the subject where these terms might be encountered. Furthermore, this grand unified theory has no other descriptive name.

We have seen how SU(2) was originally applied to the nucleon system. The proton and neutron were treated as if they were states of a single particle of "strong" isospin 1. The neutron has slightly more mass than the proton, but at sufficiently high energies this difference can be ignored. The original notion was that the nuclear interaction is fundamentally invariant to rotations in "strong" isospin space and the mass difference arises when that symmetry is broken at lower energies.

With the knowledge that the proton and neutron are actually composite, the isospin idea was carried over to the electroweak unification scheme. The first generation quarks (u, d) are treated as if they are two charged states of a single quark while the leptons (v_e, e) are taken as two charged states of a single lepton. The generic quark and lepton each have "weak" isospin $\frac{1}{2}$. The photon, W and Z bosons are treated as mixtures of two bosons of weak isospin 1 (three charge states +1, 0, and –1), and 0 (one neutral charge state), as described above.

At sufficiently high energy, the electroweak interaction is fundamentally invariant to rotations in weak isospin space. The differences between the masses of the various charged states of the quarks, leptons, and bosons result when the symmetry spontaneously breaks at lower energy. All the particles, except the photon, gain mass and the weak force appears as a force with different properties than the electromagnetic force.

In the case of quantum chromodynamics, no symmetry breaking takes place. Even at low energies, the strong interaction is invariant to rotations in color charge space. The three quarks of each color are indistinguishable except for their color charge, and the gluons are massless. QCD does not care about the electric charge of quarks or to what generation they belong. And so the theory is one of a single generic quark that comes in three colors. These generic quarks interact with other generic quarks by means of exchanging generic gluons that come in six colors plus white. For example, a red t quark will interact strongly with a blue s quark just as a red c quark will interact with a blue b quark.

In SU(5) and other GUTs, the quarks and leptons are combined into a single fermion, called a **leptoquark** or, less conventionally, a **quarton**. The electroweak and strong forces are combined into a single unified force above some unification energy. This energy has been estimated by calculating, in the standard model, how the interaction strengths vary with energy, or equivalently, distance. Recall that the electric charge and color charge behave oppositely as you go to smaller distances or higher energies, the first increasing and the second decreasing. In the electroweak theory two force strengths exist that are related to the electric charge and weak interaction strength by a relationship involving the Weinberg angle. These also increase with energy.

Some have argued that the electroweak theory is not truly a unified theory since it still has these two different strength factors. However, while this is true at low energies, the notion is that they become equal at some "unification energy." At that energy, the weak and electromagnetic interactions have the same strength.

When we take the current values of the three strengths and plot them as a function of energy, using the theory to give the energy dependence, we find that they indeed eventually come together in the vicinity of 10^{16} GeV. The original estimates were rough, but with time the data and calculations have both improved. In a particular version of grand unification in which another symmetry, **supersymmetry**, is invoked, the three curves intersect in very small region. While two converging lines will always intersect at some point, it seems remarkable that three lines, projected twelve or thirteen orders of magnitude in energy, come together at a single point. This is taken as good evidence for what is called *SUSY-GUTs*, which I will get back to in a moment.

The unification energy here is far above anything that can be reached in the laboratory. Nevertheless, the particular GUT, minimal SU(5), had the virtue being a "good" theory, namely it made a risky prediction that could be tested by experiment in a reasonable time with current technology. In general, since they unite quarks with leptons, GUTs allow for the possibility of transitions between them. For example, the reaction $u + u \rightarrow X \rightarrow e^+ + \bar{d}$ is allowed, where X is one of the twelve new gauge bosons of the theory. This makes possible the decay of the proton: $p \rightarrow e^+ + \pi^0$, where the d in the proton combines with the $\bar{d}$ produced to give the π^0. Note that baryon number is violated in this process, since there is no baryon in the final state.

Up to this point, it had always been assumed that the proton was stable. The fact that the visible matter in the universe, which is mostly protons and electrons, has been around for ten billion years or so seemed to indicate a stable proton and electron. However, minimal SU(5) predicted that the proton ultimately decays to lighter particles, with an average lifetime of 10^{30} years. While this is 10^{20} times the age of the universe, there are about 10^{30} protons in a ton of water. Thus, if you watch a ton of water for a few years, you have a chance of see a few proton decays, if their lifetime is in the predicted range.

Experiments were soon mounted to look for proton decay. No proton decays were reported. The Super-Kamiokande experiment has the best current limit, which at this writing is 10^{33} years. These and previous results have strictly falsified minimal SU(5) and several other proposed GUTs. Incidentally, minimal SU(5) required neutrinos of zero mass, so the discovery of neutrino mass is another nail in its coffin. Other GUTs remain viable, however, both in terms of the proton decay lifetime experimental limit and neutrino mass. However, few have much to commend them, lacking experimental consequences that can be tested.

SUSY-GUTs, on the other hand, still has enough promise to justify the continued search for evidence to support or reject it. It gives (after the fact) a Weinberg angle that agrees with the measured value. Recall that this parameter is undetermined in the standard model. The calculation that the three interaction strengths come together at a single point in the supersymmetry calculation is also very promising. This could be an accident, but that is unlikely—although it could be a result not unique to supersymmetry. In short, supersymmetry solves a number of theoretical problems and seems to provide the only way to ultimately bring gravity into the unification scheme. Keeping with my intention not to speculate too boldly, I will only briefly outline the idea.

As I have noted, GUTs have the feature of carrying forward the unification scheme that started with the nucleon and continued with electroweak theory and QCD. GUTs generally treat quarks and leptons as a single fermion object, the leptoquark. Quarks and leptons appear out of leptoquarks when the GUT symmetry breaks. Similarly, the gluons and electroweak bosons emerge from a single boson object, the photon emerging from the weak bosons when the electroweak symmetry breaks—at perhaps the same energy, if supersymmetry is correct.

In supersymmetry, we go the next step and unify fermions and bosons. It may be difficult to imagine a super particle that is both integer and half integer spin, but basically these concepts also emerge from the breaking of a symmetry. In this case, however, it seems that the symmetry breaking is not "spontaneous" but rather "dynamical." That is, it is not accidental but determined by other principles.

Supersymmetry makes many predictions, but none have so far proved possible to either verify or falsify the notion. Particularly, SUSY predicts that every known particle has a supersymmetry partner. For example, the spin $\frac{1}{2}$ electron has a spin 0 supersymmetry partner called the *selectron*. The supersymmetry partner the spin $\frac{1}{2}$ quark is the spin 0 *squark*, and the partner of the spin 1 photon is the spin $\frac{1}{2}$ *photino*. You get the idea. Other particles are also predicted that must be searched for, including more Higgs bosons.

Searches for the supersymmetry partners have so far been negative, but the theory unfortunately cannot predict their masses and is thus not easy to falsify. However, like the Higgs particle (supersymmetry partner: the *Higgsino*), if supersymmetry partners do not show up at the Large Hadron Collider or another next generation accelerator, the theory is likely to be wrong.

Serious candidates for the *dark matter* of the universe are the supersymmetry partners of the weak bosons, the *wino* and the *zino*, sometime referred to as *wimps*, "weakly interacting massive particles." These would fit in well with cosmological ideas that seem to require relatively massive particles ("cold dark matter") rather than low mass neutrinos ("hot dark matter") to account for the structure observed in the cosmic microwave background, although perhaps both are present.

M-BRANES

Finally I will also just briefly mention **superstring theories**, or what have now been generalized as **M-theories**. Often called "theories of everything," these have for some time now been the only known, viable candidates for theories that can ultimately unify gravity with the other known forces. They include supersymmetry within their framework, so any failure of supersymmetry would also strike a death blow to M-theories. Like GUTs, a whole class of such theories currently exist. Much more will have to be learned before any one, if any, is declared the ultimate theory of everything. Enough progress has been made, however, to keep people working enthusiastically on the subject (see Greene 1999 for a recent popular exposition).

A few features are common to M-theories. They envisage that the universe has at least ten dimensions and the basic objects are not zero dimensional point particles but one dimensional strings or m-dimensional "m-branes," where a particle is a "0-brane" and a string is a "1-brane." Whatever the basic object, all but four of the 10+m dimensions are "rolled up" so the objects appears as a particle moving in three dimensions of space and one of time. Everything I have to say about "primal objects" in this book will apply to m-branes, should they be confirmed, and my use of the term "particle" in that context should be recognized as to include these as a possibility. That is, m can be greater than zero.

To envisage a rolled-up dimension, just take a two-dimensional piece of paper and roll it into a tight tube. That tube will look like a one-dimensional line. The idea of compatified dimensions was first suggested by Theodor Kaluza in 1919 and elaborated shortly thereafter by Oscar Klein. They suggested that electromagnetism might be combined with gravity in a generalization of Einstein's general relativity, with a fifth dimension providing the extra degree of freedom of electric charge. In the process they discovered that the component of the gravitational field in the direction of the rolled-up spatial dimension obeyed Maxwell's equations. Einstein, who spent most of his life after relativity looking for a unified field theory of electromagnetism and gravity within the framework of general relativity, was keen on the idea. But it never worked out. The nuclear forces were found, and it soon became clear than any unified theory would have to be a quantum theory.

M-theories bring that early effort considerably up-to-date, with quantum theory and general relativity combined as no other theory has been able to do. The theorists who work on m-branes have proved a number of theorems that encourage them to carry on in what has become a very esoteric mathematical exercise that only true aficionados with high mathematical skills can follow. These workers are still far from nearing the goal of a final theory, in what is largely a mathematical exercise that proceeds with essentially zero experimental guidance. This is a recipe that has never worked before in the history of physics, and the odds still would have to be against it succeeding now in this fashion.

What is not clear is whether the theory will be able to make unique predictions that are testable in a practical experiment in the foreseeable future. Quantum gravitational effects are not thought to be important until you reach distances as small as 10^{-35} meter, the Planck length, where even space and time become impossible to operationally define. This regime is so distant from what is achievable with current technology that we cannot not conceive how it might be studied experimentally. The compact dimensions of m-branes curl up at the Planck scale, although attempts are being see if the scale is in fact much greater and within experimental reach in the new millennium.

It seems highly unlikely that our generation will know the answer to the question of the precise nature of the fundamental objects of reality. Humankind simply will have to wait to see how all this turns out. But, in the meantime, we have quarks, leptons, and bosons—or perhaps leptoquarks and bosons—possibly with supersymmetric partners, pointlike particles that constitute all known matter.

NOTES

1. This is an oversimplified description of the quantum situation, but good enough for our purposes.

2. The time-energy uncertainty principle is $\Delta E \Delta t \geq \hbar/2$. Thus, energy conservation can be violated by an amount equal to the rest energy of the exchanged particle, $\Delta E = mc^2$, as long as this particle is reabsorbed somewhere else in a time interval Δt less than $\hbar/2mc^2$.

3. Ignoring neutrinos, which outnumber the rest by orders of magnitude but are almost impossible to detect.

4. Capra's book, *The Tao of Physics* (Capra 1975) can still be found on bookstore shelves and remains a major reference for those who promote mystical interpretations of physics and use this as support for paranormal claims. Thus, it is important to point out that the S-matrix theory on which much of his case for a new, holistic physics was based is no longer considered a serious proposition in physics. (See *The Unconscious Quantum* for more details.) None of this is meant to detract from the serious efforts of Chew and other physicists to develop S-matrix theory, whose work is still valuable.

5. I hope my physicist colleagues will not object too strongly to my attempt here to give a simplified explanation for a mechanism that is deeply buried in esoteric mathematics.

10 DREAMS OF FIELDS

Since Maxwell's time, physical reality has been thought of as represented by continuous fields. . . . This change in the conception of reality is the most profound and the most fruitful that physics has experienced since the time of Newton.

Albert Einstein

CONTINUOUS FIELDS

The field in both classical and quantum theory is a mathematical entity that has a set of values at every point in space and time. These mathematical entities can be one-component *scalars*, three-component spatial *vectors*, four-component space-time vectors, or n-component *tensors*. Each of these components can be *complex numbers*; that is, they have a real and imaginary part, thus doubling the number of real numbers you need to specify the mathematical object at each point in space and time. No wonder Feynman despaired of the immense "bookkeeping" that is involved in field theories and sought an alternative.

In a simple example from classical physics, the (scalar) mass density field specifies the mass per unit volume at each point in space within a material object. If matter were a continuous fluid, as once was thought, then the density field would be continuous. Indeed, many practical applications successfully treat matter in this way, even though it is ultimately discrete, because the approximation of continuity is very often good enough and greatly simplifies calculations. Water and air, our two most common fluids, can be assumed continuous for most everyday situations.

By the late nineteenth century, most physicists were convinced that matter occurs in localized chunks or *atoms*. However, they also thought they saw evidence for an additional, continuum component to nature. Certainly light seemed continuous, as was the medium responsible for gravity.

I mentioned in chapter 2 how Newton had puzzled over gravity, which seems to act instantaneously over great distances. By the nineteenth century, gravity was pictured as an invisible field that surrounds a mass, just as an electric field surrounds an electric charge and a magnetic field surrounds an electric current. Today's science teachers still draw "lines of forces" to illustrate these fields.

With the development of the theory of electromagnetism, electromagnetic fields described the stresses and strains in a continuous, elastic medium that was believed to pervade the universe, the *aether*. Laboratory observations of the invisible transmission of energy from one point to another were interpreted as "aether waves" that travelled at the finite speed of light, implying light was the same basic phenomenon.

Electromagnetic waves are produced by jiggling charges and currents, but gravity is apparently too weak for a detectable gravitational wave to be produced by jiggling a mass—at least on the laboratory scale. As mentioned in chapter 3, gravitational waves have not been seen to-date. Physicists continue to search for the gravitational waves expected to be produced by the jiggling of huge masses on cosmic scales, such as in a binary star system where one or both objects is a compact neutron star in which the mass of a star is confined to the volume of a planet. We await these observations.

In the aether model, fields are not just mathematical abstractions but a physically real medium. The stresses, strains, and vibrations of these aetheric fields lead to the transmission of energy from point to point, without an actual movement of matter between the points. This is by analogy to familiar sound waves or water waves in normal matter (when assumed continuous).

We have seen how developments at the dawn of the twentieth century led to the demise of the classical aether as a viable concept. First, Michelson and Morley failed to detect its presence. Second, Einstein developed the special theory of relativity, which pointedly assumed the nonexistence of an aether. The existence of an aether implies an absolute reference frame to the universe, in contradiction to Galilean relativity. Third, the final nail in the aether's coffin was provided by the photon theory of light. Like matter, light was found to occur in localized chunks; it was not smooth, continuous aether waves after all. It was particles.

Actually, if we look back at this discussion we will see that the philosophical position of positivism was being enforced in the assumption that something undetectable, the aether in this case, does not exist. While philosophers no longer insist that we can only talk about that which is observable, such talk still must have some rational basis that rests on observational data. Otherwise we have nothing to constrain our imaginations, nothing to limit the number of concepts on which to focus our attention. Even in metaphysics we must rely on observation as a guide, if we wish to remain rational. And, the fact is that observations show no need to hypothesize even an undetectable aether.

Some physicists argue that the aether remains in physics, in quantum field theory. I will discuss this viewpoint later in this chapter. For now, I know of no serious ontological or epistemological proposal to restore the classical aether—although this is implied in much of pseudoscience which talks about "mind waves" and "bioenergetic fields" (Stenger 1990a, 1999c). A far more serious candidate for a field ontology, because it has survived unscathed for over eighty years, can be found in Einstein's theory of gravity.

With general relativity we have neither an aether nor a gravitational field. As we saw in chapter 3, the gravitational force, as such, does not appear explicitly within the framework of general relativity. Rather, four-dimensional space-time forms a plenum in which what we interpret as gravitation occurs by means of objects following their natural paths, the paths of least action, in non-Euclidean space. The metric of space-time and the energy-momentum tensor are the sixteen-component fields of general relativity, and Einstein's equations relate the two.

We are thus confronted with two ontological possibilities: either space and time are abstract relations between objectively real material objects, or space-time is an objectively real *substance* in its own right. Einstein is generally regarded as holding to the latter view, but the former seems more plausible in the light of those developments in quantum mechanics that have dominated twentieth-century physics.

As elegant as general relativity indisputably is, and as successful as it has been as a physical theory, all attempts to extend it to forces other than gravity or to bring it along into the world of quantum mechanics have so far met with minimal success. Whatever happens in the future, any field ontology that develops almost certainly will have to be fundamentally quantum in nature. Nonquantized general relativity, by itself, does not appear to provide us with a viable model for a reality based on continuous fields. And when a quantum theory of gravity is developed, it will likely contain **gravitons** and possibly other particulate quanta.

QUANTUM FIELD REALITIES

As mentioned in earlier chapters, a one-to-one correspondence between field and particle exists in quantum field theory, with the particle interpreted as the "quantum" of the field. No known observational test distinguishes between a particle and quantum field reality. Physics data do not provide us with a direct, unambiguous message on the particle versus field nature of ultimate reality.

The conventional response to this fact has been particle field, or more familiarly, particle-wave duality. In this view, the reality of an object, particle or field, is not determined until a measurement is performed. This leads, however, to solipsism and quantum mysticism, which are so extraordinary in their implications that Occam's razor requires that we seek alternatives before taking them seriously. Alternatives of a pure field or a pure particle reality need to be examined to see if either is viable.

As I have emphasized, no mathematical or logical proof, no call for conventional trial by data, can be used to determine "true reality." Other criteria must be developed, or else we will be forced to conclude that we are wasting our efforts. Pushing ahead anyway, what might quantum field reality look like?

Clearly the quantum field is something quite different from classical material or ethereal fields. Early attempts by de Broglie and Schrödinger to give a classical-type wave-packet reality to the wave function met with quick failure. The nonrelativistic wave function for two electrons is not a space-time field but rather a field in an abstract space of six spatial dimensions and one time dimension. Furthermore, the two-particle wave function is a complex number that, when made relativistic, becomes a complex function of eight space-time coordinates in which each electron has its own time axis. In the Dirac theory, this gets extended even further, with spin up and spin down and negative as well as positive energy components.

And this is not the end of the story. In the standard model, the quantum fields that replace the wave function have many more components to account for all the other variables that define the particle it represents. For example, the lepton field has two additional components of weak isospin. A quark has all of these, plus three more components to specify color charge.

Obviously, physicists do not attempt to generate the tables that give all the numbers that are needed to describe a quantum field, but seek formulas that allow you to calculate the field at whatever point you wish. Whatever the formula, however, quantum fields are clearly very far from the notion of stresses and strains in an elastic medium. If they are waves in an aether, then that aether permeates a many-dimensional space, not familiar three dimensional space. If that space is real, then it is a Platonic reality we are talking about here.

In the standard model, much of this is now calculable. The mathematics is quite elegant and beautiful and, most important, successful. But how can we extract a model of reality from it? Paul Teller (1990) has examined the "harmonic oscillator interpretation" of quantum field theory from a philosophical perspective. He sees this as an alternative to a reality based on localized particles. However, Teller recognizes that the harmonic oscillator also can be used to describe the behavior of free particles. In his view, the infinite number of quantized states of the oscillator provides a basis for the superposition of states that characterizes quantum systems. That is, Teller uses the many states of the oscillator to provide the means by which a particle can be found simultaneously in many states.

Nick Huggett and Robert Weingard (Huggett 1994) have criticized Teller's proposal on two counts. First, they argue that Teller's ontology applies only to bosonic (integer spin) fields, where many particles (quanta of the fields) can reside in the same state. By contrast, only one fermion (half-integer spin) can be found in a single state. Second, Huggett and Weingard point out that Teller's picture does not account for renormalization. Teller has expanded his discussion to book length (Teller 1995), but his emphasis there is on the logical status of models of quantum field theory, rather than metaphysics.

Both the classical equations of motion and the quantum field equations can be reformulated as equations for a harmonic oscillator such as a mass on a spring. The oscillator equations are often used in solving problems. Both classical and quantum calculations can thus be done *without ever mentioning* the notion of continuous fields. This is not merely practical and pedagogical. Continuous fields are a familiar carryover from nineteenth-century physics that have been a major cause of many of the problems of twentieth-century physics. So we should realize that a viable alternative exists.

HOLISTIC FIELDS

Now, you might be inclined to think that a field ontology offers the great promise of providing an explanation for puzzling quantum coherence effects. Indeed, this was the original idea of Schrödinger when he introduced the wave function. Waves are an easy way to visualize and mathematically describe interference and diffraction. This may be another reason why field ontologies remain popular. They are used pedagogically in this fashion, in giving university sophomores who will be going on to be electrical engineers or chemists some easy way to visualize the quantum mechanics that underlies both electronics and chemistry.

Beyond the familiar coherence effects of interference and diffraction lie more spooky ones associated with the EPR and Schrödinger cat paradoxes.

The inseparability of quantum states strongly suggests to many that there is no escaping the basic holistic nature of reality. This is exemplified by the experiments in which the outcome of a measurement at one point in space is highly correlated with the outcome of a measurement at another point, when the two measurements are on a system that was initially prepared in a pure quantum state.

Note my careful wording, which I emphasized in the earlier discussions of these experiments. The results of a measurement at one point are not *determined* by some action taken by a human or machine at another point, a misstatement you will often hear. In fact, experiments only see a *correlation* larger than the maximum allowed in certain types of theories, thus eliminating these types of theories.

The types of theories eliminated include conventional classical mechanics, which is both local and deterministic. Any deterministic theory, such as the primary version of Bohm's hidden variables theory (he also has an indeterministic version), must necessarily have superluminal effects. Local indeterministic theories are also commonly thought to be ruled out, which has lead to the common belief that quantum mechanics is necessarily superluminal (Stapp 1985). However, local indeterministic theories seem to be possible with time reversibility. Stapp and most others assume the conventional arrow of time in their analyses of the EPR experiment. Their proofs are thus predicated on time-directed causality and, unless reformulated, do not apply to a time-symmetric reality.

We may be forced to a superluminal field ontology if we are to maintain intrinsic time irreversibility, that is, if an arrow of time also exists at the quantum scale. One possibility is that the basic underlying reality is a universal field that reacts instantaneously as a whole to a disturbance at a distant point. Bohm's quantum potential is such a field. However, he readily admitted that this field is not Lorentz invariant, that is, it violates Einstein's relativity (Bohm 1993, 271). No surprise. The whole point is that some kind of superluminal connection must be taking place if his model is correct.

Unlike Bohm's quantum potential field, the quantum fields of the standard model preserve Lorentz invariance and, as we have seen, do not allow superluminal signalling. Both motion and signals moving at greater than the speed of light is provably impossible in any theory consistent with the axioms of relativistic quantum field theory. Thus, quantum fields cannot serve as the superluminal fields some think are necessary to explain the violation of Bell's inequality. These fields do not suffice to represent reality in a scheme in which we preserve the arrow of time. Other, nonlocal fields are still required.

Note that even the nineteenth-century luminiferous aether cannot serve as the holistic cosmic field, since it does not allow for superluminal signalling. Classical aether waves travel at the speed of light! The only model we have in classical physics for a superluminal field is Newtonian gravity. However, this superluminality does not carry over to general relativity, where the gravitational waves predicted by the theory also travel at the speed of light. Similarly, gravitons, the presumed quanta of any quantum field theory of gravity, would be expected to travel at the speed of light. In short, modern physics has no place for superluminal quanta.

And so, we cannot use as an argument in support of a quantum field ontology that it provides for the superluminality that many think is required by the EPR experiments. Quantum fields, at least those in conventional relativistic quantum field theory, do not allow for motion or signals at speeds greater than light. It follows that a quantum field ontology is not compelling, despite its ostensible "holistic" nature. Such an ontology offers no better explanation for the EPR paradox than a particle ontology. As long as a microscopic arrow of time is presumed, both share the same problems in interpreting the paradoxes of quantum mechanics.

Let us next consider whether quantum fields, supplemented by time reversibility or additional, nonaetheric fields such as the quantum potential, provide an attractive alternative to a model of reality based on localized objects moving about in four-dimensional space-time.

ARE FIELDS PARSIMONIOUS?

I claim that the elements of empirically successful modern theories, particularly quantum electrodynamics and the standard model, provide the best candidates for the status of "real" objects, although not every element of the theoretical structure must be real. I propose specifically that the quarks, leptons, and gauge bosons are real. They are made of real stuff that moves about in familiar space-time. By contrast, the corresponding fields exist in an abstract, mathematical world. Any ontology that takes them to be more real than particles is necessarily a Platonic one, where the fields are then Forms.

I leave room for these particles themselves to turn out to be composed of more elementary objects. Composite bodies formed from the basic objects, like the chemical elements and molecules, can still be considered real, just as we consider ourselves real.

As I have already emphasized, I am not suggesting that the term "field" be banished from physics. We still can talk about a discrete field of particles, such as the quantized electromagnetic field of photons, but we have no need to consider continuous fields as real. The quantum field is like a field of pebbles.

Consider the alternative in which the fields in the standard model, rather than the particles, constitute the basic reality. How is this any different from particle reality? We have seen that a one-to-one correspondence exists in the standard model between particle and field. Each particle is viewed as the quantum of a field. You can transform your equations back and forth between particle and field equations motion, so the two points of view are identical, at least in the mathematical theory.

We are led to conclude that nature has not, at least yet, provided us with an empirical way to distinguish between the reality of particles and their corresponding quantized fields. Thus, it becomes a matter of finding other, nonempirical and nondeductive, arguments to prefer one over the other—recognizing that this choice is being made for reasons other than the traditional scientific one of letting experiment or logical deduction decide.

Granting that empirical consistency is the single most important criterion in science, and that quantum fields satisfy this criterion as well as particles, what other criteria might be applied? I suggest that simplicity, or more precisely, *parsimony* provides us with an added condition that prefers the particle to the Platonic field ontology.

Simplicity is not always equivalent to ease of understanding or common sense. A theist may argue that the simplest explanation is that God made it all that way and we poor mortals cannot hope to understand why. A solipsist may argue that the simplest explanation is that it is all in our heads, made up as we go along. A postmodern relativist may argue that the simplest explanation is that, while reality is really there, we still make up our models of reality as we go along, according to our culture, and one culture is as good as another.

This I why I added parsimony to simplicity; it is a more precise concept. Parsimony is applied in the sense of Occam's razor, limiting you to those models or theories that have the fewest independent hypotheses and at the same time make risky predictions. But even parsimony is arguable. For example, how do you count hypotheses? The theist will say he makes only one—God. But the atheist could just as well say the theist makes an infinite number, since everything observed in nature counts as a different event. Absent of any theory of God capable of making risky predictions, the theist hypothesizes, each time in ad hoc fashion, "God did it."

I will address the principles of simplicity and parsimony in a later chapter, where I will apply them to show how the particle ontology, when combined with symmetry principles and chance, provides a deeper understanding of the laws of physics themselves.

The conceptual problem with any field ontology is this: what is doing the waving? Waves in air or water we can understand—even see. But what is waving in the case of quantum fields, including the electromagnetic field? Physicists who talk about quantum fields as the "true reality" do not provide us with any answer to this question, or any way to visualize these fields. None makes any claim that some kind of continuous material medium is still out there, jiggling away to give the events we see in our detectors. Quite the contrary. They acknowledge that such a medium is very unlikely to exist.

The aether died with the Michelson-Morley experiment and Einstein nailed its coffin shut and buried it six feet under.

Clearly, the reality that is imagined for quantum fields is not a reality akin to water waves rippling across the ocean, but a Platonic reality of an abstract many-dimensional space. In the quantum field ontology, fields are mathematical objects that, along with their equations, are taken to be more deeply real than the particles they describe. The same can be said about the universal wave function, or state vector, in many worlds quantum mechanics. It, too, resides in abstract space. To believe literally in the many worlds ontology, as many do, is also to be a Platonist. Since we can readily dismiss the solipsistic ontologies of mystical physics, we are left with only two viable choices: particle reality in which electrons, quarks, and physical space-time are real, or Platonism in which they are an illusion.

PENROSE PLATONISM

The leading spokesman for the modern Platonic view of reality has been the distinguished mathematician and cosmologist Roger Penrose. In several books, starting with *The Emperors's New Mind* in 1989, Penrose has championed the view that the truths of mathematics are not simply there by invention but represent elements of ultimate reality (Penrose 1989, 1994).

Mathematics deals with the logical consequences of an assumed set of axioms. These axioms include the definitions of mathematical objects and the rules for the operations to be performed with those objects. For example, in familiar arithmetic you define numbers as your objects and perform operations on these numbers, such as addition and multiplication.

Now, it would appear that any logical consequences you draw that are "true" within the framework of the particular mathematics within which you are working are the result of those initial axioms. They may not be obvious when the axioms are first written down, but are developed as mathematicians or logicians work them out. Penrose (1989, 96) draws a distinction between "invention" and "discovery" in this process, asking if it is possible that these "truths" have an existence independent of the mathematicians who first wrote the axioms or worked out their consequences. He asks, are these truths elements of objective reality? He suggests that mathematicians may have stumbled on "works of God."

The view that mathematical concepts exist in a "timeless, etherial sense" is called by Penrose (1989, 97), "mathematical Platonism." He bases most of his case on the celebrated theorem of Kurt Gödel (1931). Gödel proved that in any formal system at least as complicated as arithmetic, truths exist than cannot be proven within that system. Penrose (1989, 112) interprets this to mean that "there is something absolute and 'God-given' about mathematical truth." That is, mathematical truth has a Platonic existence that "goes beyond mere man-made constructions."

The main thesis of Penrose's series of books is that the human brain cannot be a simple "computer" carrying out instructions according to a formal computational algorithm. Gödel's theorem, Penrose claims, shows that "the mental procedures whereby mathematicians arrive at their judgements of truth are not simply rooted in the procedures of some specific formal system" (Penrose 1989, 110). He then speculates on what this noncomputational physical process in the brain might be and suggests it may have to do with quantum gravity.

Penrose's ideas have been widely debated by experts in a full range of disciplines, from quantum physics to artificial intelligence to neuroscience. The consensus among experts in these disciplines, so far, is against him (which, of course, does not make him wrong). I have discussed his ideas about a quantum role in human thinking in my previous book (Stenger 1995b, 268–92), where other references can be found. Here my concern is more with the notion of a Platonic reality carried not by the equations of pure mathematics, but by the equations of modern physics.

In mathematics, changing one of the initial axioms of a formalism can change the practical applications that formalism. For example, suppose we eliminate the commutative law of multiplication, AB = BA. You then have a new formalism, like matrix algebra, where multipliers do not commute. The quantities A and B in this algebra cannot then be used to represent many quantities from common experience.

For example, two boxes containing three apples in each contain a total of six apples. Three boxes with two apples in each also contain a total of six apples. You are not free to represent the number of boxes and number of apples with noncommuting mathematical objects. If A = the number of apples, and B = the number of boxes, then AB = BA, six in this case.

We have seen how noncommutative algebra is used to represent observables in quantum mechanics. However, when the time comes to compare calculational results with measurements, the comparisons are made with commutable real numbers.

The mathematical games applied in physics must agree with observations. Reality provides constraints that mathematicians like Penrose are not bound to obey when playing games in their own yards, but must accept when they are playing on the physicist's ball field or the Oxford cricket pitch just around the corner from Penrose's office.

In neither case, with or without the commutative law, are you making any statement about ultimate reality. You are only stating a rule of a game that has been created for the amusement of humans, which is no more a permanent feature of ultimate reality than the infield fly rule in baseball. It can be changed tomorrow.

WEINBERG PLATONISM

Most theoretical physicists think that it is only a matter of time before they or their successors find that ultimate "theory of everything" with which they will be able to calculate all the constants of physics, such as the masses of the elementary particles and the relative strengths of their mutual interactions, from a few basic axioms. They believe that only one logical possibility exists, that the world must be the way it is.

Many theorists have told about how they became enamored with the beauty of the equations of physics and astounded that they described experimental results so well. Furthermore, it seems as if the more beautiful the equations, the better the agreement. Dirac is quoted as saying, "It is more important to have beauty in one's equations than to have them fit experiment . . . because the discrepancy may be due to minor features which are not properly taken into account" (Dirac 1963). Contemporary Nobelist Steven Weinberg has said, "I believe that the general acceptance of general relativity was due in large part to the attractions of the theory itself—in short, to its beauty" (Weinberg 1992, 98).

Weinberg explains how our historical experience teaches us that we find increasing beauty as we look below the surface of things: "Plato and the neo-Platonists taught that the beauty we see in nature is a reflection of the beauty of the ultimate, the *nous*" (1992, 165). He thinks it "very strange that mathematicians are led by their sense of mathematical beauty to develop formal structures that physicists only later find useful, even where the mathematician had no such goal in mind." Weinberg refers to an essay by Wigner entitled "The Unreasonable Effectiveness of Mathematics," adding his own jibe about "the unreasonable ineffectiveness of philosophy" (Weinberg 1992, 169).

Weinberg sees reality as existing beyond the observed world. He says that much of the "angst over quantum electrodynamics" was tinged with a positivist sense of guilt." He writes that "some theorists feared that in speaking of the values of the electric and magnetic fields at a point in space occupied by an electron they were committing the sin of introducing elements into physics that in principle cannot be observed." He indicts positivism for the (unsuccessful) reaction against field theory that was led by Chew in the 1960s at Berkeley, which I described in chapter 9: "The positivist concentration on observables like particle positions and momenta has stood in the way of a 'realist' interpretation of quantum mechanics, in which the wave function is the representation of reality"(Weinberg 1992, 181).

Weinberg makes his Platonic position quite explicit, connecting it to symmetry principles: "Broken symmetry is a very Platonic notion: the reality we observe in our laboratories is only an imperfect reflection of a deeper and more beautiful reality, the reality of the equations that display all the symmetries of the theory" (1992, 195). In other words, the underlying symmetry is the reality and the broken symmetry is the world of experience. This mirrors Plato's notion that the planets follow perfectly circular orbits and that the "wandering" motion that is observed which give them their name is a distorted view—reality viewed through poor quality lenses.

Weinberg sees Platonic reality as the only ontology that can serve as a realistic alternative to positivism. As I have noted, however, many physicists, especially, experimentalists, remain positivists—or, at least, instrumentalists. Actually, most give little thought to philosophy and take it for granted that the real world is what they measure with their instruments. Theoretical physicists, on the other hand, tend to look at their instruments—the mathematical objects and equations of their theories—as the true reality.

Being an experimentalist by training myself, I am afraid that I must break here with Weinberg, whom I admire greatly and agree with 99 percent of the time. It seems to me that the "true reality" is not necessarily the most beautiful. Beauty seems too highly subjective a criterion upon which to judge the nature of reality. A circular orbit for Mercury might be regarded as more beautiful than the rotating ellipse implied by the equations of general relativity. The world is not the most beautiful of all possible worlds according to subjective human criteria. But it may be the simplest in terms of having no more laws governing its behavior than are required by logical consistency. Beyond that, it is random. And constrained randomness can lead to complexity and deviations from symmetry, as I will elaborate upon in chapter 12.

GOD AND STEPHEN HAWKING

Inevitably in discussions of Platonic reality, God comes into the picture. The modern Western notion of God is probably closer to Plato's Form of the Good than the white-bearded Yahweh/Zeus on the Sistine chapel ceiling or the beardless Jesus/Apollo on the wall. Weinberg (1992, 256) raised his atheist flag on the Vatican steps when he said that "the more we refine our understanding of God to make the concept plausible, the more it seems pointless." Stephen Hawking has been slightly more circumspect, ending his stupendous best-seller *The Brief History of Time* by saying that if we ever discover a complete theory of the universe, "we would know the mind of God" (Hawking 1988, 175). This line alone probably added millions to the sales.

Hawking's readers were disappointed if they thought they would find the mind of God in Hawking's book. In fact, he is as much a nonbeliever in a traditional God as Weinberg and at least eighty percent of physicists and astronomers. Hawking was merely doing what many pantheist physicists, notably Einstein, have frequently done, following Baruch Spinoza (d. 1677) in using the term "God" to refer metaphorically to the laws of physics themselves.

Recently we have seen an explosion of attempts to reconcile science with religion by essentially saying that the order of the universe uncovered by sciences is equivalent to the concept of God. Fueled by the largess of wealthy financier John Templeton, the discussion is carried on in books, magazines, and conferences. *Newsweek* has announced on its cover: "Science Finds God" (Begley 1998). This was more than a little misleading, since the story itself admitted that "Science cannot prove the existence of God, let alone spy him at the end of a telescope." But many scientists are quoted giving various statements of the old argument from design. Physicist Charles Townes gave the typical view that "somehow intelligence must have been involved in the laws of the universe."

Kitty Ferguson, a Hawking biographer, has said that Hawking replaced the older pantheist notion of God as the "embodiment of the laws of physics" with a more precise description: "The laws of physics are the embodiment of a more fundamental 'rationality'—to which we could give the name 'God'" (Ferguson 1994, 147). However, even if this conclusion is acknowledged for the purpose of discussion, a huge leap must be taken from this Spinozan God to the personal God of Christianity (see, for example, Polkinghorne 1994).

Whether called "God" or "fundamental rationality," the Platonic ontology rests on a belief that ultimate reality resides in some other realm than the physical one of space, time, mass, and energy. The doctrine subscribed to by many field theorists and cosmologists is that the universe is they way it is because that is the only way it can possibly be. Einstein had wondered, in his metaphorical fashion: "How much choice did God have in constructing the universe?" Hawking proposes, again metaphorically, that God had no choice in setting the "boundary conditions," because none are required, but still had freedom to choose the laws the universe obeyed (Hawking 1988, 174).

A simpler alternative exists to this frankly mystical notion of Platonic reality. The laws and constants of physics did not have to be handed down by God, as theistic Platonists claim. Likewise, they do not have to be the only laws and constants that are logically consistent, as atheistic Platonists like Hawking and Weinberg believe. As we will see, some of the laws and constants appear to be nothing more than definitions. And the rest are probably accidents.

11 PARTICLE REALITY

I want to emphasize that light comes in this form—particles. It is very important to know that light behaves like particles, especially for those of you who have gone to school, where you were probably told something about light behaving like waves. I'm telling you the way it does behave—like particles.

Richard Feynman (1986, 15)

PARTICLES OR FIELDS?

The standard model offers a picture of elementary quarks and leptons, interacting by the exchange of a set of elementary bosons, the gauge bosons. A Higgs boson is included to account for the masses of particles, in particular to give the weak bosons mass in a way that still maintains the symmetry of the theory, a symmetry necessary to account for observed conservation principles.

In this book I am making the unremarkable suggestion that the quarks, leptons, and bosons of the standard model can be safely regarded as elements—perhaps the only elements—of an objective physical reality. This statement does not require that these objects be the ultimate, uncuttable "atoms." Rather, they simply constitute elementary ingredients we have established at our current level of knowledge about the world. Working with established particles, rather than more speculative entities such as strings or m-branes, provides a concrete framework. It can hardly be called speculation to treat electrons and photons as real. At the same time, should m-branes, or any other localized entities ultimately prove to be the constituents of the particles of the current standard model, then little that I say here will have to be drastically changed.

The alternative ontology in which continuous fields are "more real" than particles was discussed in the previous chapter. First, we saw that a dual ontology of fields and particles, as existed in the nineteenth century, contradicts the one-to-one correspondence between particle and field in modern quantum field theory. We can have either a reality of fields or a reality of particles (or other localized objects). We cannot have both without asserting some new physics not described by relativistic quantum mechanics. Such an assertion is uneconomical—not required by the data. Even when gravity is brought into the picture, quantum ideas would seem to require that it have associated quanta, the gravitons, and possible other quanta to provide for the cosmological constant field, or *quintessence*, that seems to be called for in the latest cosmological data (see chapter 13).

Second, we saw that any viable field ontology based on relativistic quantum fields necessarily entails a Platonic view of reality. In Plato's ancient doctrine, the mathematical objects, equations, and laws of physics exist as a more perfect realm than the material, ephemeral objects of our experience. Particles are stuff. Quantum fields are Platonic forms.

One of the reasons ancient philosophers looked to a more abstract, hidden reality beyond observations was the impermanence of the material objects of experience. Today we hear this same point raised with respect to the reality of elementary particles such as electrons, which are "created" and "destroyed" in modern theories. But, in a time symmetric universe, elementary particles have universality and permanence. In place of electron-positron pair production and annihilation we have a single electron zigzagging around in space-time. The electron is permanent! It is neither created nor destroyed.

Let us now examine how the application of time symmetry at the microscale helps to keep open the possibility of a simple non-Platonic reality of real, localized objects moving around in otherwise empty void.

SEEING PARTICLES

Let me begin by arguing why it is reasonable to think that the standard model particles are real—if not as the primitive entities, then at least as real objects at some level. If the electron is not a primary uncuttable atom, then it still is more accurately described as being real than as being a simple figment of physicists' imaginations. Physicists like myself have seen many characteristic trails in particle detectors and identified them as tracing out the paths of electrons, muons, and other particles of the standard model. Our colleagues have measured and calculated the parameters of the electron, and its heavier version, the muon, to many significant figures of precision. Currently, the measured value of the electron's magnetic dipole moment is $1.001159652193 \pm 0.000000010$ Bohr magnetons. The magnetic dipole moment of the muon is $1.001169230 \pm 0.0000000084$ muon magnetons.[1] Both measurements agree with theoretical calculations within the stated measurement error. Surely the electron and muon are as real as anything can get. We know their parameters more precisely than we know those of the earth or the human body.

As we have seen, the leptons, like the electron and muon, remain elementary within the standard model. That is, they are not composed of more elementary objects. The neutral leptons, the neutrinos, have masses many thousands of times less than the mass of the electron. We have seen the tracks and measured the detailed properties of literally hundreds of hadrons, such as protons and kaons, which we do not regard as elementary. Other hadrons, such as the delta and rho, have very short lifetimes and are less directly inferred, but the physics used in that inference is so basic and noncontroversial that the reality of these particles is as secure as their more directly observed cousins.

Similarly, the reality of the individual, discrete photon can be readily demonstrated by the way it produces electron pairs in a bubble chamber or triggers pulses in a photomultiplier tube. The W and Z gauge bosons are observed in more advanced particle detectors. The tauon, the third generation charged lepton after the electron and muon, is also seen. The evidence for the reality of the three neutrinos and their antiparticles is more indirect, but still very difficult to refute without tossing out a good deal of established physics.[2] The quarks and gluons are hidden from direct observation, but once again the evidence for their objective reality is compelling. The fact that they are not seen as free particles does not make them unreal. You and I are not free bodies but are bound to the earth (we have negative potential energy), yet we are (presumably) real.

In short, only the most drastic revision of both commonsense and scientific thinking, a descent into solipsism or Platonism, can lead us conclude that the particles of the standard model are not objectively real. While the Higgs boson is still not observed and alternatives remain at this writing, belief in the reality of quarks, leptons, and gauge bosons is justified by the great success of the standard model—just as belief in the reality of atoms is justified by the great success of the atomic theory. This statement cannot be "proven" conclusively by any known application of mathematics or logic. However, the absence of any alternative over the past two decades argues strongly for its validity.

Some might claim that the absence of alternative theories is simply the result of the inbreeding of physicists, that other proposals are suppressed inside their "closed community." All I can say is that in my forty years as a physicist I have seen popular proposals, such as proton decay, fail to be confirmed—at least, so far. I have also seen initially unpopular notions, such as the charmed quark, gain acceptance as the data came in. I think I reflect the feelings of all practicing physicists when I say that I personally would like nothing better than to be the inventor of a theory, or the observer of a phenomenon, that is disputed by all of science but later confirmed. That's the stuff Nobel Prizes are made of.

Even if some, or all, of the particles of the standard model are eventually found to be composite objects, this will not detract from their reality. The chemical elements are no less real for being composed of nuclei and electrons. Nuclei are no less real for being composed of nucleons. Nucleons are no less real for being composed of quarks. And all these objects—molecules, atoms, nuclei, quarks, leptons, and bosons (except the Higgs) have been "seen," at least in the generic way I define "seeing."

None of these particles is less real because it cannot be seen directly with the naked eye. Observing a particle with modern detectors attached to computers that must go through a complex analysis to provide a signal is no different, in principle, from seeing something with the naked eye. The eye is a photodetector, less sensitive than a photomultiplier tube but remarkable nonetheless. It sends its data to the equally remarkable neural network of the brain where complex analysis is performed before a signal is perceived.

Furthermore, the "theory ladenness" that philosophers of science insist marks all scientific observations is no different, in principle, from the assumptions our brains make in the process of visual and other sensory perception. Both are subject to error, and both require independent corroboration. If I were the only one to see an electron track in a bubble chamber, no one would believe me any more than they would if I were the only one to see a unicorn.

As I have emphasized, most of the methods that are used to make calculations at the elementary particle level are based on relativistic quantum field theory. A one-to-one relation between field and particle exists, the particle being the "quantum" of the quantum field. Since either a particle or field representation describes the same data, a quantum field reality remains possible. That is, we cannot decide between the particle and field ontologies by experiment alone. This is where most physicists would end the discussion. However, I promote the particle ontology on the basis of other criteria that, while not "data" or "proof," are still, I claim, rational. The particle ontology is as close to common sense as we are likely to get. It is simpler and more economical than Platonic field ontologies. And, it can be usefully applied in the thinking processes that are used to draw the **Feynman diagrams** by which we can proceed to calculate much about matter and its observed behavior.

MAKING THE VIRTUAL REAL

Most experimental particle physicists will agree that the quarks, leptons, and bosons of the standard model are objectively real, at least at some level, while theorists will tend to follow their more abstract thinking processes and opt for a reality of Platonic fields. However, even experimentalists will hedge a bit when the subject of *virtual particles* arises. Virtual particles occur as the internal lines in Feynman diagrams. They also appear as particle-antiparticle "loops," such as the electron pairs that are created and destroyed in the vacuum or contribute to the so-called radiative corrections in QED.

Consider the process of elastic scattering A + B → C + D by X exchange shown in figure 11.1(a). Let us now look at figure 11.2(a), which shows what happens when we view the vertex in figure 11.1(a) in the reference frame of A, with X shown as coming in. X collides with A producing C; that is, the basic reaction is X + A → C. Assume, for simplicity, that A and C have the same mass. From momentum conservation, particle X imparts its momentum to C. From energy conservation, it also imparts its energy to C, which goes into kinetic energy. This may be a bit brain-twisting, but one can prove that the velocity of a particle X, whose momentum equals the momentum of another particle C and whose total energy equals the kinetic energy of C, travels faster than the speed of light and has imaginary mass, where imaginary here refers to the class of numbers that contain the factor $\sqrt{-1}$.[3] In other words, the virtual particle in this reaction is a **tachyon** (see chapter 3).

Except for the gluons, the virtual particles in Feynman diagrams appear under other conditions where they are measured to have the normal proper mass of a free particle. For example, the W boson is observed as a free particle with a proper mass of 81 GeV, a real number. A virtual W boson has all the properties of a free W boson, for example, the same spin and charge. The one exception is its imaginary proper mass. Similarly, a virtual electron has all the properties of a free electron except for an imaginary proper mass.

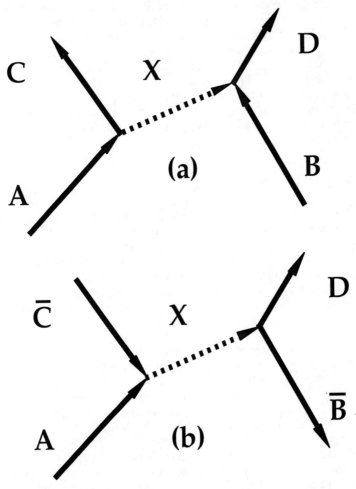

Fig. 11.1. (a) Feynman diagram for A + B → C + D by the exchange of particle X.
(b) Equivalent reaction A + C̄ → X → B̄ + D in which X has a real mass.

Most physicists are not overly concerned with the imaginary mass of virtual particles. You can't "see" these exchanged particles, so we do not have to think of them as real. However, a picture in which particles exist "in reality" would not be complete unless all the particles used in its representations of fundamental processes are understood as objectively real. How can I possibly assert that the particles exchanged in Feynman diagrams, the

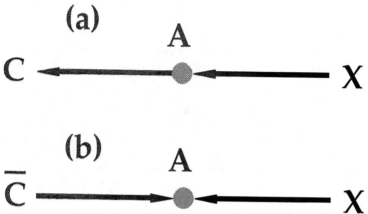

Fig. 11.2. (a) Particle A is at rest. Particle X comes in, interacts with A to produce particle C. (b) The antiparticle of C is seen coming in, interacting with X, to give a real particle A.

objects that are responsible for the forces of nature, are real when they have imaginary mass? Furthermore, an imaginary mass particle moves faster than the speed of light.

This leads to a another puzzle. Recall that the range of an interaction that is mediated by an exchanged particle in a Feynman diagram is given by the maximum distance that particle travels in the time interval over which the exchange takes place. That maximum time interval is given by the uncertainty principle and is determined by the rest energy the particle is observed to have when it is free and the maximum speed is taken to be the speed of light. For example, if the particle exchanged is a W boson of rest energy 81 GeV, the range of the interaction will be of the order of 10^{-18} meter, about a thousand times smaller than the radius of a proton.[4] This is consistent with the data, which verify that the weak interaction is very short range, even shorter than the strong interaction. But, a superluminal speed for the exchanged W would imply much greater, perhaps infinite range.

So, something is clearly amiss with this picture of virtual particles being tachyons, faster-than-light particles. The data indicate they are not, so it must be the theoretical description that is wrong.

The way out of the dilemma, once again, is time symmetry. Recall that when we look down on spacetime itself (figure 3.3), we can watch bodies moving along worldlines that can turn back in time as well as space. As described earlier, Feynman used this notion to develop a manifestly relativistic picture of particles and their interactions. Only when time is placed on the same footing as each of the spatial coordinates does a theory provide intrinsic consistency with the theory of relativity.

Time symmetry, or more precisely, CPT symmetry, provides us with an almost trivial solution to the virtual particle problem. Consider the process A + B → C + D proceeding by generic X exchange, as illustrated in figure 11.1 (a). That is, we have particles A and B coming in and C and D going out. The exchanged particle X, as we have seen, has imaginary mass.

Now, here is the trick: simply turn around the arrows of one particle on each side of the interaction and change it to an antiparticle.[5] As shown in figure 11.1 (b), we now have A + $\overline{C}$ → $\overline{B}$ + D. Figure 11.1 (a) and (b) are indistinguishable. Particle A scattering off particle B with the exchange of particle X is completely equivalent to particle A colliding with the antiparticle $\overline{C}$ to produce particle X, which propagates in spacetime and decays in to the antiparticle $\overline{B}$ and the particle D.

And here is the wonderful part: In figure 11.1 (b) the mass of particle X is arithmetically real! This can be easily seen by looking at figure 11.2 (b), where we see $\overline{C}$ colliding with X to produce a *real* particle A at rest.

In other words, it all depends on your perspective. If you continue to insist, despite all indications to the contrary, that time is unidirectional at the fundamental level, then you have to include superluminal particles with imaginary mass in your scheme. But, if you simply admit that a particle going forward in time is indistinguishable from its antiparticle going backward in time and the opposite direction in space, then you can picture the exchange process in terms of exclusively subluminal particle with real masses.[6]

Now, the mass of the particle X in figure 11.1 (b) is not in general exactly equal to the rest mass of X you will find in particle tables. This is clear in the case of colliding beam accelerators, where a particle and its antiparticle collide head-on with (in most cases) equal and opposite momenta. The particle X is produced at rest. Yet, from energy conservation X must have a rest energy equal to the total energy of the colliding particles, not in general equal to the mass of a free X particle.

Consider the case of a high energy e⁺ e⁻ collider, such as the one at CERN, that has a variable beam energy. Suppose the positron and electrons each have an energy of 100 GeV. Assume they produce a Z boson in the collision. This Z boson will have a mass of 200 GeV, much greater than the published mass of 91.2 GeV.

Is this a problem? Not really. The Z-boson (in the conventional time picture) quickly decays into something else. The possibilities include another e⁺ e⁻ pair, a μ⁺ μ⁻ pair, a whole bunch of pions and other hadrons, or nothing but neutrinos. The uncertainty principle allows for a large range of particle rest energies when time intervals are short. Indeed, if you vary the beam energy you find that the collision probability distribution has a sharp peak

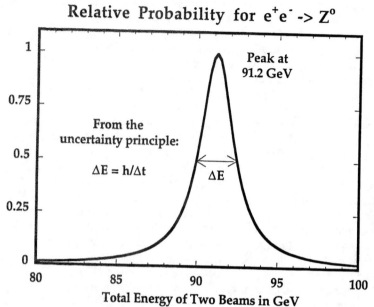

Fig. 11.3. The relative probability for $e^+ + e^- \rightarrow Z^0$ as a function of the total energy of the beams. The peak is at 91.2 GeV, the mass of the free Z^0. The width of the peak ΔE is what is expected from the uncertainty principle.

at the Z mass and a width given exactly by what is expected from the uncertainty principle (see figure 11.3).

Because of the uncertainty principle, unstable particles do not have precise rest masses or, equivalently, rest energies. In order to measure the rest energy of a body with high precision, you must observe it for a long time. When that time is forced to be short by the particle being highly unstable and decaying quickly, then the corresponding energy uncertainty will be large.

If an electron and positron collide with a total energy far from the Z mass, you have very little chance of producing a Z. On the other hand, if the beams are each tuned at 91.2/2 = 45.6 GeV, a great many more Zs are produced. The width of the Z energy distribution is 2.5 GeV, represents the uncertainty in its rest energy. By the uncertainty principle, this implies a mean decay lifetime of about 2.6×10^{-25} second.[7]

Thus, $e^+ e^-$ collisions are equivalent to $e^- e^-$ elastic scattering, the prototypical interaction we have frequently referred to throughout this book. Z-exchange actually makes a small contribution in this case, the main process being photon exchange. However, the situation for photon exchange is no different, in principle, than that for Z exchange. The exchanged photon has imaginary mass when the reaction is viewed as an $e^- e^-$ scattering process, and positive mass when viewed as an $e^+ e^-$ annihilation/creation process. In the latter case, even when the positron and electron are initially at rest the virtual photon has a proper mass at least twice the electron mass. But, again, this is no problem. The photon will recreate the pair in a time interval consistent with the uncertainty principle. While a zero mass photon will have infinite lifetime, because it has nothing it can decay into, massive photons can decay, in this case into $e^+ e^-$ pairs.

Realizing that exchanged photons are massive also provides a metaphysical explanation for the fact that the longitudinal polarization of photons must be included in calculations involving photon exchange. Massless photons can have only transverse polarization. That is, although a photon has unit spin and, in principle, three possible orientations of the spin axis, only two orientations are allowed for particles travelling at the speed of light. In the case of the photon, this accounts for the fact that the electric and magnetic field vectors of a classical electromagnetic wave are always in a plane perpendicular to the direction of propagation of the wave.

A massive unit spin particle, such as the W and Z bosons, can have longitudinal polarization. That is, they can have a spin component along their direction of motion. Physicists have found that they cannot explain the data unless they include longitudinal polarization for exchanged photons as well, as if the photons have mass. Well, they do when they live for finite periods of time, or, at least, they have an uncertainty in mass as given by the uncertainty principle. Time symmetry makes it all so simple.

LINES AND VERTICES

So, at least as far as the single-particle exchange Feynman diagram is concerned, we can view all the particles as real. This can easily be seen to apply to more complicated diagrams as well. With time symmetry, the space-time directions of all particles are arbitrary. Thus, we can take the arrows off the particle lines in Feynman diagrams, as in figure 11.4. Let us refer to those particle lines that represent our incoming/outgoing particles as *external lines*. And, let us refer to the lines for the so-called virtual particles, which we now know are real, as *internal lines*. Completing the terminology, let us call any point where internal or external lines meet a *vertex*.

Fundamental processes, as far as we know, always have three lines meeting at a vertex. This can be viewed as a point in space-time, that is, an *event* at a particular place and time in which two particles annihilate to give

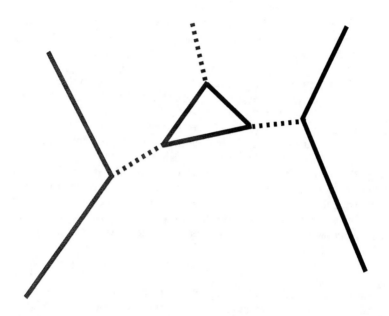

Fig. 11.4. In the time reversible view, the lines in a Feynman diagram that represent particles have no arrows. The basic interactions between particles occurs locally, at the vertices where three lines meet. This diagram shows five vertices, five external lines, and five internal lines. The dashed external line could be a photon that is emitted or absorbed by the vacuum.

the third particle. Or, if you prefer, think of the vertex as the decay of one particle into the other two. Put arrows back on the lines anyway you choose—all the corresponding reactions will be equivalent. We take the vertex to represent the fundamental interaction in nature. It is always local—occurring at a point, or at least within some confined region.

Note that, in this picture, we have no force fields "in reality." The theory may contain mathematical constructs called fields, but no continuous medium is assumed to pervade the space between vertices. This is not meant to exclude the notion of a "field of pebbles," photons, Higgs bosons, electron pairs, or other particles in the space between the particles going in and out of the process we are considering. Indeed these all contribute to a complex, multibody process of interaction in which the primary particles will not travel in simple straight lines but scatter and zigzag around in space-time. Thus, for example, the photon shown in figure 7.3 is being emitted and reabsorbed by an electron, or the electron-positron pair, are shown as curved paths. This can be imagined as the result of multiple scattering from other background particles that are not drawn into the diagram. I have indicated what one such photon would look like in figure 11.4.

Now, some physicists might say that this demonstrates why we cannot live without the idea of fields in the interaction process. I have no problem with this, as long as we recognize that ultimately these fields are not a material continuum, like the aether, but regions of space populated by many background particles—the "particles of the vacuum." The field concept is then used for approximating purposes, replacing a discrete system with a smoothed and averaged continuous one so that we can proceed to make calculations that would otherwise tax the most powerful computer. This is exactly what we do when we replace the discrete atoms of matter with a continuum, using thermodynamics or fluid mechanics to obtain workable solutions to complex problems on the macroscopic scale. But, as we saw in chapter 7, taking this continuum too literally leads to infinities in calculations that must then be subtracted by the complex procedure of renormalization.

Getting back to the Feynman diagram of microscopic events, the mathematical theory associates with each vertex a quantity (not always a simple number) that measures the strength of the interaction. For example, if the vertex represents $e^+ e^-$ annihilation into a photon, the vertex strength factor will be proportional to the square of the electron charge. If instead we have the annihilation of a $2/3$ charged quark with its $-2/3$ charged antiparticle into a photon, the strength factor will be $(2/3)^2 = 4/9$ of the positron-electron annihilation strength factor. For the $1/3$ charged d-quark and its antiparticle, the strength is $1/9$ as much. Observations have confirmed these rules, thereby verifying the fractional charges of the quarks.

Similar rules are obtained for the other interactions, where color charges give the strong interaction strengths and the weak interaction strengths are related to the unit electric charge by the Weinberg angle (see chapter 9) . We need not work out all the details to appreciate the concept: Feynman diagrams are built of unarrowed lines and pointlike vertices, with strength factors determined by properties of the particles that connect to each vertex.

In the next chapter, we will see how this basic process in which a particle splits into two can be understood as symmetry-breaking that occurs spontaneously at random points in space-time. The universe is then viewed as a web of these lines and vertices.

SUMMING DIAGRAMS

We still have another ontological problem. A lone Feynman diagram does not describe most processes observed in the laboratory. The interference effects that mark quantum mechanics, whether you are talking about electron diffraction or electron-positron annihilation, arise, in the theory, from the phase differences of the various amplitudes. The probability amplitudes for the processes described by Feynman diagrams are obtained by summing up the amplitudes for all possible diagrams you can draw that have the same incoming and outgoing particles—however you define incoming and outgoing. The number of diagrams is very large, infinite when you assume continuous space-time (which may not be necessary), but the probabilities remain finite.

So, how do we deal with the sum over all possible internal lines? Compare figure 11.5 (a–c), which shows three ways that electron antineutrinos can elastically scatter from electrons. In (a) we have the (charged) W⁻ indicated as the exchanged particle. In (b) Z exchange is shown. In (c) the electron and antineutrino meet to form a W⁻ which then converts back to the lepton pair. This may seem complicated, but I urge the reader to study these three diagrams. Much fundamental physics is incorporated.

All three diagrams are important and must be included in the probability calculation, and many others as well. How can we imagine them all contributing to a *single* event? Single events are not found to be describable by single Feynman diagrams; rather, the data indicate that each event is somehow a combination of all possible diagrams acting together coherently.

In figure 11.5(d–f) the time arrows on the line have been removed, along with charges and antiparticle signs, to reflect CPT symmetry (the particle chiralities were not indicated to begin with). Notice that diagram (d) and (f) are identical, so we can consider them as the same diagram. Furthermore, within the standard model the W and Z particles are broken-symmetry states of the same underlying gauge boson. Ontologically, the W and Z are identical and we can reduce the three diagrams (a)–(c) to one.

Now, these are not the only Feynman diagrams that must be drawn, in principle, to describe the interaction. For example, we have "radiative corrections" of the type that were illustrated in figure 7.3. There we saw that diagrams of this type can be incorporated in the description of the physical particle. Other diagrams can also occur. With CPT symmetry and some work, they, too, can be simplified. In general, though, many paths must still be considered in describing the outcome of any fundamental process.

Consider a transition between two states of a particle system, a and b. We may call these "initial" and "final" in familiar language, but we now understand that this designation is arbitrary. A given Feynman diagram rep-

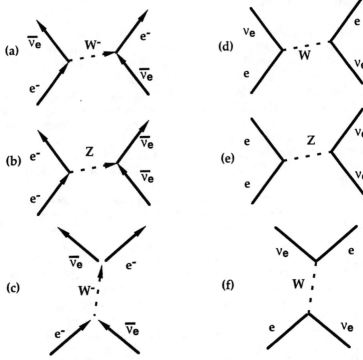

Fig. 11.5. The elastic scattering of an electron antineutrino and electron by (a) W⁻ exchange, (b) Z exchange, and (c) direct W⁻ production. These reactions are shown again in (d), (e), and (f) with the time arrows, charges, and antiparticle signs removed. Note that (d) and (f) are now identical. Furthermore, in the standard model, W and Z are broken-symmetry states of the same fundamental gauge boson. Thus, these all reduce to a single diagram.

resents the specific path of the system analogous to the particle path between "source" and "detector" that passes though a particular slit in the double slit experiment. For a given source and detector, the path that goes through the other slit is likened to a different Feynman diagram for the same "initial" and "final" states. Similarly, we can represent any other quantum transition you wish to consider as a generic process a ↔ b that connects the two states by a specific path. Quantum mechanics, QED, and the standard model tell us that all possible paths contribute to the transition a ↔ b, and that the probability amplitudes for these paths must be added as complex numbers with magnitudes and phases, so that interference effects will occur. The magnitude of the resultant sum is squared to get the transition probability.

Now, note an important point: We do not calculate amplitudes for a ↔ c, a ↔ d, and add them together to get some overall transition probability to a set of "final" states that often includes all possible final states. We are frequently interested in such a probability, such as when we calculate the decay rate of a radioactive nucleus with several different decay products, but we obtain that result by summing probabilities in the usual way. That is, paths between different pairs of states do not interfere. Quantum effects come in only in the sum over paths between the same two states.

In the conventional interpretation of this process, we simply "cannot say" which path the system took between a and b unless we "look." In a strict Copenhagen view, the system takes none of these paths until we look. In the many worlds interpretation, we have a separate world for each diagram.

Within particle formulations of quantum theory, we can use the Feynman path integral scheme to calculate the observed interference patterns for the classic optics experiments. This pattern is measured as the distribution of "hits" on particle detectors placed along the screen. Recall that, in Feynman's method, a complex amplitude is formed for every possible path between two the source and detector. The phase of each amplitude is the action averaged over that particular path, in units of $\hbar$, the quantum of action. The amplitudes for all possible paths are summed as complex numbers, and the resultant magnitude is squared to get the probability for the "transition " that takes the particle from the source to detector.

For the reader who may not be able to follow this description of the mathematical procedure, then the following section should make it much clearer. For the reader who knows the mathematics, this section may still prove interesting—because it describes how Feynman explained all this, without mathematics, to high school students.

FEYNMAN'S ARROWS

Feynman has provided us with a beautifully intuitive description of his path integral method in a little book called *QED*, which presents a series of lectures he gave to a nonscientist audience of mostly high school students in 1984 (Feynman 1985).

In his later years, Feynman did little writing of his own but lectured widely, often to enthralled lay audiences. Many of those lectures were transcribed and edited for publication by his friends and admirers, usually physicists themselves. Indeed, at this writing books are still coming out containing these transcripts—more than a decade after his death.

The *QED* lectures were edited by Ralph Leighton, and Feynman acknowledged that the published version was greatly modified and improved upon by the editor. In any case, the result was a remarkably simple and yet profound exposition of many of the phenomena associated with light and quantum mechanics. This was done using a picture of particles following definite paths that are conventionally described in wave terms. But waves are hardly mentioned in *QED*.

Feynman and Leighton represented the complex probability amplitude (without calling it that) by a little arrow drawn on the board during Feynman's lecture, as illustrated in figure 11.6 (a). The square of the length A of the arrow gives the probability of an event. The arrow also has a directional angle ϕ, which is used to represent the phase of the amplitude. The probability for two processes is then obtained by adding the arrows of all the paths in the usual two-dimensional vector fashion and squaring the resultant.

The next step is to allow the arrow to rotate like the hand on a clock. To follow convention, however, the arrow rotates counterclockwise.

The first problem Feynman and Leighton describe with this technique is ordinary reflection. A light source S sends photons against a mirror that reflects them to a photomultiplier tube P, as seen in figure 11.6. A barrier prevents light from going directly from S to P. A few of the infinite number of possible paths are shown. Each path is viewed as having a little rotating arrow that gives the elapsed time along the path.

The usual result is called the law of reflection, in which the light bounces off the mirror at an angle of reflection equal to the angle of incidence. This is seen as the path for which the elapsed time is minimum. If we add up the arrows of all paths to get a final amplitude, we see that the major contribution comes from the paths E, F, G, H, I. Their arrows can be seen to point in about the same direction and so the corresponding amplitudes "add constructively."

Snell's law, the law of refraction for the path of a light ray from one medium to another, is obtained by having the arrows change speed of rotation as they pass from one medium to another. As with reflection, we have the path of minimum time. This is just what is expected from Fermat's principle, which, as we saw in chapter 2, is the principle of least action applied to optics.

Feynman and Leighton also derive the intensity patterns for diffraction and interference. In these cases, their method is equivalent to what is called the "phasor" method in lower division university physics textbooks. This technique is not normally applied to reflection and refraction, however, but we see that this can be done. These phenomena are usually considered "geometrical optics," where light is assumed to travel in a well-defined beam (the "light ray"), like a beam of particles. Interference and diffraction normally appear in another chapter on "physical optics," where the wavelike behavior of light is introduced and represented by phasors that are simply

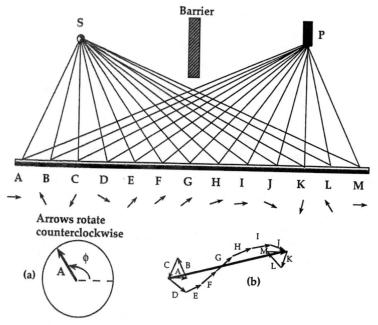

Fig. 11.6. Possible paths of a photon scattered from a mirror. Insert (a) shows the probability amplitude with magnitude A and phase φ, viewed as a clock carried along by the photon (rotating counterclockwise). Insert (b) shows how the amplitudes add, with the major contribution to the final amplitude coming from the shortest time paths E-I. In time-reversible quantum mechanics, all paths occur with the law of refraction resulting as a resonance effect. This figure is based on figure 24 in Feynman (1985).

Feynman's arrows. All the complicated patterns one observes with multiple slits of finite width, where both interference and diffraction come in, can be derived by the phasor method, and so they can also be illustrated with Feynman's arrows.

We can now see how the arrow method also implements Feynman's path integral formulation of quantum mechanics. As mentioned above, the phase of the amplitude is the classical action, in units of Bohr's quantum of action ƛ. Recall from chapter 2 that the action is the average Lagrangian over the path multiplied by the time taken by a particle to travel that path. In most applications, the Lagrangian is the difference between the kinetic energy and potential energy. In the absence of external forces, that is, for a "free" particle, the Lagrangian is simply the constant energy E of the particle (the constant potential energy can be set equal to zero). This applies in the optics examples we have discussed, where the photon travels as a free particle from one point to another such as from its source to the mirror. Furthermore, if we have only localized forces in fundamental processes, all the lines in Feynman diagrams represent the motion of free particles between vertices where the interactions occur. So, for our purposes, the action along a path, and thus the phase of the amplitude, is always proportional to the elapsed time.[8] This justifies Feynman and Leighton's representation of the complex amplitude as the hand on a clock.

As was seen in an earlier chapter, the principle of least action follows as the classical limit of quantum mechanics performed with Feynman's path integral method. In the classical domain, the action is very much larger than $\hbar$. Indeed, this defines the classical domain. When the amplitudes for the paths are added together, all except those paths infinitesimally close to the path of least action cancel out. In the case of free particles, this corresponds to the path of least time. Thus, Newton's mechanical laws of motion and Fermat's optical principle all follow, not as externally imposed laws of nature but as the most probable occurrences in an otherwise random set of processes.

Now, you might complain that all we have explained with Feynman-Leighton arrows so far are very old optics results. This is true, but the method nevertheless illustrates an important point that does not appear in the usual classical (wave) descriptions of these phenomena. Consider, again, reflection. The various paths shown in figure 11.6 are envisaged as *all actually occurring*—not just the particular one that satisfies the law of reflection. Similarly for refraction, and indeed classical particle motion in general which is usually thought of as being "governed" by the principle of least action. Here is where, I think, a time symmetric quantum reality, not even hinted at by Feynman and Leighton in *QED*, comes in to give much additional insight into the very nature of natural law. Indeed, it suggests an ontological model to explain how all these paths contribute simultaneously.

All we have to do is regard every path between S and P as actually traversed. Viewed from a frame of reference looking at the experiment from above, these traverses are seen to happen at once, with all paths contributing simultaneously to the process S ↔ P. No law of nature forbids any path from occurring. The single path of least time does not happen by the action of some special law of physics, engraved in the stone tablets of the universe. Rather, it is seen as the most probable occurrence that results when nearby paths interfere constructively. Indeed, the principle of least action, from which Newton's laws of motion can be derived, is itself derived as the classical limit of underlying quantum processes in which all things happen and some things just happen more frequently than others.

PROPAGATORS

While Feynman and Leighton did not carry it that far in *QED*, which was aimed at high school students, we can easily see how the same picture of rotating arrows can be applied to Feynman diagrams. Each line in a Feynman diagram, whether external or internal, is represented by an amplitude called the **propagator**. That line represents the space-time path, the worldline, of a particle between two vertices where the interactions with other particles take place, If the particle has energy E, the arrow will rotate at a frequency f = E/h as given by Planck's rule.

The interactions at the vertices are pictured as local, contact collisions. This is where the real "physics" comes in, where the graduate student puts in all the factors that have to do with the strength of the interaction and the properties of the particles—their spins, isospins, masses, charges, colors and whatever else affects the probability that the interaction will take place. By now this algorithm is all worked out, even computerized. The well-trained graduate student knows exactly what to do, or learns it quickly. I don't mean to imply that the calculation is trivial. These are pretty smart people who have been intensively trained and often must use innovative techniques to get useful answers.[9]

The paradigmatic rules for the standard model are not all that can be applied. Extensions that go beyond the standard model, such as those mentioned in chapter 9, may also be calculated from the same picture of the underlying reality. For example, the probabilities for interactions involving all the still undiscovered supersymmetric particles can be fully computed within the framework of that theory. The only unknowns are the SUSY particle masses. Indeed, theoretical particle physicists today can make calculations on a wide range of models that go beyond the standard model. The problem is that too many of these models remain perfectly viable and the data are still insufficient to distinguish one from another.

Despite this, we are able to draw ontological conclusions that may reasonably be expected to remain standing as the best routes beyond the standard model become clarified. All the current models based on quantum fields can for our purposes be interpreted as proceeding from the local interactions of particles, the pointlike quanta of the fields, supplemented by time reversibility.

THE PRIMAL OBJECT AS LOCALIZED CLOCK

Many physical objects have some characteristic that can be used to mark time. The earth's rotation has been traditionally used by humans as a clock.

The human heart pulsates. Molecules oscillate and vibrate. Elementary particles, though visualized as points, have spins that can be detected by magnetic interactions, as in magnetic resonance experiments.

As we saw in chapter 7, the photon is the "quantum" associated with the oscillations of the electromagnetic field. All the particles of the standard model are similarly quanta of oscillating fields. The mathematics that describes the behavior of these fields can be expressed in terms of the equations of motion of localized oscillators. A short step from this is all that is needed to view those oscillators as the "real" entities of a non-Platonic metaphysics.

I feel no compulsion to specify the exact nature of the ultimate, primal oscillators and exactly how they interact with each other, for this would be little more than speculation. Whatever the details, however, it seems reasonable to model our elementary objects as oscillators. The vibrating string or m-brane model is as good as any, and concurs with current trends in the (mainly mathematical) search for the primary entities (Greene 1999). As we saw in chapter 9, these hypothetical objects exist in a nonabstract multidimensional space—ten or more dimensions, depending on the detailed model. All but four of these extra dimensions are curled up on the Planck scale. While attempts are being made at this writing to push that scale up to something more within experimental reach in a generation or two, these objects appear as particles in all existing experiments and I will continue to refer to them as such. Should m-brane theory ultimately prove to be wrong, or simply intractable, we should be able to still utilize a model of a localized oscillator, or clock, for the primal objects.

Consider a clock moving parallel to its face. As we see in figure 11.7, under the combined operations of space (P) and time reflection (T), the direction of motion does not change; we get one reversal from space reflection and a second reversal back to the original direction from time reflection. The clock hand, however, reverses only under time reflection. Now suppose that the two clocks are superimposed into a single clock. The new clock will continue to move in the same direction but its hand will remain fixed. That is, time stands still for the superposition of a clock and its PT partner.

To guarantee symmetry even for the rare processes where PT might not be an invariant operation, the transformed clock should be made of antimatter. Our primitive object then is composed of a particle superimposed with its CPT partner. This is the *timeless quantum*.

As we saw in chapter 8, a horizontally polarized photon is a superposition of left circularly polarized photon from one time direction and a right circularly polarized photon from the other time direction. Similarly, an observed electron is a timeless superposition of electron and positron. This should not be interpreted as two particles, which would give a boson rather than a fermion, but a single superposition of quantum states from both the past and the future.

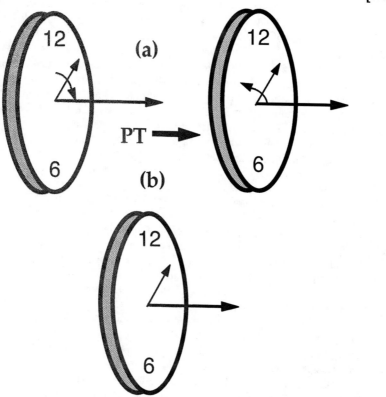

Fig. 11.7. (a) A moving clock undergoing the combined operations of space and time reflection (PT). The direction of motion for the transformed clock is unchanged (two reversals) while the clock hand runs counterclockwise. (b) The hand on the superimposed clock remains fixed. The two clocks in (a) are analogous to L and R photons, while the bottom clock is timeless.

The elementary oscillator and its CPT partner act together to form the timeless quantum. One brings information from the "past," and other from the "future," however these may be defined. At the level of observation, detection devices like photomultiplier tubes enforce a preference for a particular time direction. This is a purely statistical preference, time's arrow being the direction in which entropy, a probabilistic quantity, increases. The reverse processes are not impossible, just very unlikely in most familiar applications. The phenomena we associate with the quantum world arise when the relative probabilities in the two time directions become comparable.

The notion of our primal objects being clocks, or at least carrying along with them some kind of clock, is certainly not original with this book. Recall the operational view of space-time discussed in chapter 3. There I presented Milne's 1935 development of relativity in terms of a set of observers with clocks who communicate with one another by exchanging timed signals. From the timing of the returned signals alone, the observers are able to create a mutually consistent space-time framework with which to describe events—assuming the speed of the signals is the same for all observers. Space is then seen to be defined in terms of time as the primary variable. The Lorentz transformation falls naturally out of the Milne description, and the proper times for each observer are invariant. In this view, each object in the universe keeps its own proper time as it follows a worldline in the arbitrary space-time coordinates of any generic observer.

Leibniz seems to have had a similar insight with his monads, the basic metaphysical objects that formed the basis of his ontology. Monads keep their own private time, while space is the "relationship among all the monads inherent in their contemporaneous mutual perceptions, this general universal ordering throughout time" (Leibniz as interpreted by Rescher 1967, 89). Here you discern a clear intuition of Milne's view of space as just a picture we draw from the timing of our received signals. We also glimpse Einstein's notion of relative time.

In the model of reality presented here, elementary objects can exist in many local regions at once, that is, they are multilocal. The so-called nonlocal reality that we read about in both physics literature and media reports can possibly be recast in this way, most importantly without superluminal signalling. The idea is that we have no instantaneous action at a distance, no continuous fields—only particles, the void, and localized forces. These particles are all free as they move from interaction region to interaction region where they meet other particles, including the short-lived particles that fill the vacuum. For those portions of their paths where they are not interacting with other particles, our elementary objects follow all possible paths, back and forth in time, with their classical laws of motion following in the limit of large phase angles. At the quantum level, the effects of different paths manifest themselves to provide the quantum effects that no longer need be regarded as "spooky.

NOTES

1. The magneton unit is $e\hbar/2mc$, where m is the mass of the particle. For the Bohr magneton m is the mass of the electron. For the muon magneton, m is the mass of the muon. The differences between the measured quantities and unity are the contributions of the radiative corrections that are calculated in QED.

2. In this regard, the standard model still allows for a neutrino interpretation alternative in which neutrinos are identical to their antiparticles. These are called Majorana neutrinos. This approach has come into greater favor among some theorists in light of the observation of neutrino mass, and has experimental consequences that continue to be investigated.

3. In units where $c = 1$, $v = p/T = p/(E - m) = p/[(p^2 + m^2)^{1/2} - m]$. After some algebra, $v = (1+m^2/p^2)^{1/2} > 1$. Here p, E, and T refer to particle B and v to particle X. Note that the square of the proper mass of X will be given by $T^2 - p^2$, which is negative since p > T from the above calculation and so the proper mass is imaginary.

4. $R = c\Delta t = \hbar c/2\Delta E = 10^{-18}$ meter where $\Delta E = 81$ GeV.

5. To be complete, we also do a space reflection on B and C, that is, switch their chiralities. Then we only rely on CPT symmetry and cover those rare processes where P, T, or CP symmetry is violated. In terms of momentum and energy, we reverse the energy and all three components of momentum.

6. Once again I am not introducing any new physics here. Indeed, this idea of changing the directions of the lines in Feynman diagrams without changing the basic physics was part of the 1960s S-matrix theory mentioned in chapter 9. Then it was called *crossing symmetry*. While S-matrix theory itself is no longer fashionable, many of its elements remain valid today. However, since the S-matrix was expressed as a function of four-momenta, it was regarded a nonspace-time description of nature and time reversibility was not considered.

7. If the mean lifetime of the particle in is rest frame is τ, then the rest energy uncertainty is ΔE, obtained from $\Delta E\tau \geq \hbar/2$. This ΔE is the width of the energy distribution that is observed in experiments.

8. The phase is $\phi = Et/\hbar = 2\pi ft$, where the action is Et and the particle energy $E = hf = 2\pi\hbar f$.

9. In actual calculations, a Fourier transform is performed on the amplitude so that the relativistic four-momentum becomes the independent variable in the place of space-time position. But these are technical details that the graduate student works out.

12

GLOBAL SIMPLICITY AND LOCAL COMPLEXITY

According to one mode of expression, the question What are laws of nature? may be stated thus: What are the fewest and simplest assumptions, which being granted, the whole existing order of nature would result?

John Stuart Mill
System of Logic (Earman 1984)

MOSTLY CHAOS

Because of their proximity and pervading influence over our lives, the mountains, rivers, and forests of earth loom large in our minds. When, on the other hand, we view them from space, these features disappear into a smooth, blue and white spheroid that gives far less evidence of structure than a cathedral or a spaceship from *Star Wars*. And the farther away we get, the smaller this spheroid becomes, contracting to a point of light and then disappearing into the vastness of space.

When examined from a cosmic perspective, the universe as a whole possesses a remarkable degree of simplicity and symmetry. It is far smoother and more structureless than most people appreciate. In our era, the observable universe is mostly composed of photons and neutrinos in not quite perfect thermal equilibrium. The temperature of this particle bath is only 2.7 degrees Celsius above absolute zero, 270.3 degrees Celsius below the freezing point of water. It is only one part in a hundred thousand from being purely random.

The chemical elements that form the stars and galaxies represent only a billionth of all the particles of the cosmos. And nine out of ten of these elements are hydrogen atoms, the simplest of all, containing only a single proton and single electron. The carbon atoms that form the basis of earthly life are a tiny, almost negligible, fraction of physical reality.

High precision experiments studying the cosmic photon background measured the tiny temperature deviation from equilibrium across the sky mentioned above. Importantly, it was just the right amount expected from the inflationary cosmological model in which structure development, such as galaxy formation, is the result of small density fluctuations in the otherwise smooth early universe. These observations and others force us to conclude that the universe is mostly random chaos, with a few tiny pockets of order like our solar system scattered around.

The laws of nature are normally thought to be part of nature's organization. Governments institute laws to specify the allowed behavior of the governed, with the avowed purpose of avoiding chaos. By analogy, the laws of nature are assumed to represent the manner in which matter and energy are required to move in orderly ways quite different from the expectations of pure randomness and chance. These laws are regarded as equivalent to statements about the supposed supernatural design of the universe. However, when we recast the most powerful and familiar laws of physics in terms of the structureless symmetries of space, time, and more abstract "inner" dimensions, we find that they appear inevitable—strongly suggesting the *absence* of external design.

Only in recent times has the deep connection between structureless symmetry and the laws and forces of nature been appreciated. Developments in the twentieth century have led to a remarkable conclusion: the great foundational principles of physics—conservation of energy, conservation of linear and angular momentum, Newton's laws of motion, Einstein's special relativity—are simply statements about the natural, global symmetries of space and time, indicative of a structureless universe. And the force laws, such as those that describe gravity and electromagnetism, are statements about the ways that these symmetries can be locally and spontaneously broken.

TIME TRANSLATION AND CONSERVATION OF ENERGY

Let us first consider **conservation of energy**, which is the principle underlying the **first law of thermodynamics**. This, we will see, is equivalent to **time translation symmetry**, expressing the fact that no special moment in time exists—that one time is as good as another.

Time translation symmetry is not to be confused with the symmetry between the two directions of time we have already said much about. To distinguish the two, this type of directional symmetry will henceforth be referred to as **time reflection symmetry**.

We can try to understand the profound connection between time translation symmetry and energy conservation with a simple experiment. Imagine watching a motion picture film in which a pendulum swings monotonously back and forth against a blank background. Assume the pendulum's friction is reduced to negligible proportions, so its maximum swing does not measurably decrease for the duration of the film. By simply watching the film, we cannot judge when it was shot; it could have been yesterday, a month ago, or (if well-preserved) years ago. Indeed, the experiments with pendulums performed 400 years ago by Galileo give the same results when performed today. From watching a pendulum, we cannot single out any particular moment as special. If we rewind the film after someone leaves the screening room, when she returns she will be unaware that the film has been restarted.

A pendulum, even one on film, can be used as a clock (and, of course, often is). We can keep track of the passage of time by counting pendulum swings, and use the count to time events that happen in the viewing room. However, note that the absolute numerical value of the time we assign to any given event is arbitrary. We can say that someone left the room for 20 swings, but it does not matter whether we say she left at time 0 and returned at time 20, or that she left at time 1,000 and returned at time 1,020. The interval of 20 swings is the same in either case. The reference point from which we measure time is arbitrary.

When certain physical processes, such as the motion of an undamped, undriven pendulum are observed not to single out some special moment in time, they are also observed to conserve total mechanical energy. Note that the system, in this example, is not stationary in time and its internal energies are constantly changing. At the top of its swing, the bob is at rest and its kinetic energy (energy of motion) is zero. At the bottom of its swing, the potential energy (stored capacity for doing work) is zero and the kinetic energy is maximum. The total energy of the bob at any instant is the sum of kinetic and potential energy, and that is constant. However, neither of the two forms of energy of a pendulum, kinetic and potential, are conserved. As the bob falls, it loses potential energy and gains kinetic energy. As it rises, it gains potential and loses kinetic. But, as long as no energy is input from the outside, and no friction causes the loss of energy to heat, the total energy remains constant.

All isolated systems show this effect. Their energies are distributed among their various particles and components, and are transformed from one type of energy to another as these parts interact, but the total energy remains constant as long as the system remains isolated from its environment. That is what we call energy conservation.

Suppose, after watching our film for a while, we observe a sudden increase in the maximum swing of the pendulum. Now the situation has changed, and we can specify one particular time—the one special moment when the increase in swing took place. Apparently, in this case, time translation symmetry is violated. Someone who leaves the room can now come back and identify that moment as distinct from all others. That time can be labeled "absolute t = 0," and other times referred back to it, with positive or negative signs.

What has happened to ruin the nice symmetry in time in the previous example? Running the film once more through the projector and examining the screen closely, we may notice a slight shaking of the pendulum bob at one end of its swing—just before its enhanced swing occurs. This is a hint that something interfered from the outside. Perhaps a whiff of a breeze gave the pendulum bob a little boost, or someone off camera blew on it (God?). In either case, the pendulum was, for an instant, no longer isolated from its surroundings.

As this example shows, the interaction of a physical system with the outside world can result in apparent violation of time translation symmetry. The puff of air increased the amplitude of the pendulum's swing by adding energy to the system. The apparent violation of energy conservation, indicated by the pendulum's greater swing, resulted from the system interacting with something outside itself.

Similarly, as happens in the practical situation, friction will eventually bring the pendulum to a halt. The moment when it stops is a special moment that can be identified in any viewing of the film, regardless of when the projector was started or whether it was rewound when we were out of the room. And the fact that this special moment is possible to define is the direct result of the fact that the energy of the pendulum is no longer conserved but lost to friction. We see this scenario repeated many times in physics; when a symmetry is broken, some quantity associated with that symmetry is no longer conserved.

Thus, a direct connection exists between time translation symmetry and energy conservation. This connection is found in both classical and quantum mechanics. The observation of energy conservation in a system is equivalent to the observation that the system does not single out a special moment in time. The breaking of energy conservation is equivalent to the breaking of time translation symmetry and the establishment of an absolute time scale.

The law of conservation of energy, as usually stated, applies to isolated systems. The universe itself is presumably such an isolated system, so it must contain time translation symmetry, when viewed as a whole. Now let me turn this argument around. The law of conservation of energy is merely a statement about the invariance of the universe as a whole, and other isolated systems within that universe, to the operation of time translation. Conservation of energy results simply from the absence of absolute time:

Time Translation Symmetry $\Rightarrow$ Energy Conservation

Now, you may protest that a special moment in time did once occur—the start of the big bang, the moment of "creation." Energy conservation was indeed violated at this point. The spontaneous appearance of the order of 10^{28} V electron volts of energy, the equivalent of 2×10^{-5} grams of matter, not much by cosmic standards, was needed to initiate the inflation of space-time.

However, we have no inconsistency. A breakdown of energy conservation is allowed at a singular time, without the necessity to classify the event as a miracle. This provision is automatically built into quantum mechanics, with Heisenberg's uncertainty principle. Another way to think of this is in terms of the **zero point energy**, the lowest energy any particle can have when confined to a finite region of space. The uncertainty principle provides a well-defined exception to energy conservation, allowing small violations for short enough time intervals. As we have seen, this helps us to understand the "virtual" processes in fundamental interactions. In this case, the allowed time interval was a mere 10^{-43} second. Beyond that time, energy is conserved and no further special moments exist.

Note that the 10^{-43} second time interval around the conventional $t = 0$, the start of the big bang, is not unique. The same violation of energy conservation will be seen in all the similar time intervals from $t = 0$ to now, and on the negative side of the time axis as well (recall chapter 4). So we still have time translation symmetry all along the time axis and $t = 0$ is no special point in time.

The first law of thermodynamics amounts to a statement about energy conservation. In its usual expression, the heat into a system equals the increase in the internal energy of the system plus any work done by the system. Note how it allows for a nonisolated system. For example, the heat entering a cylinder of gas with a piston at one end can go into increasing the temperature of the gas (that is, increasing the kinetic energy of the gas molecules) or into work done as the gas expands against the piston, or a combination of both that adds up to the total heat in. The total internal energy of the cylinder need not be conserved, but the heat came from somewhere, and the work went somewhere, and when the cylinder and its environment are taken to be an isolated system as a whole, the total energy is again constant.

SPACE TRANSLATION SYMMETRY
AND CONSERVATION OF MOMENTUM

By similar reasoning, no call upon external agency is required to explain the second great conservation principle of physics: **conservation of momentum**. Descartes and Newton had used as the "quantity of motion" of mechanics, what we now call **momentum**, the product of the mass m and velocity v of a body: p = mv. Newton's three laws of motion are equivalent to conservation of momentum for an isolated system, and the definition of force as the time rate of change of momentum for an interacting system. An isolated system maintains constant total momentum, the same way that a system insulated from any energy flow to or from the outside maintains constant total energy. Force is quantitatively the rate at which the momentum of a system is changed when the system is not isolated, just as power is technically defined in physics as the rate of change of the energy of a system—the rate at which work is done on or by the system.

And just as energy conservation follows from time translation symmetry, we can show that momentum conservation follows:

Space Translation Symmetry ⇒ Momentum Conservation

Unlike energy, which is a single number, momentum is a vector with three components. That is, it is a set of three numbers. Space has three dimensions, or axes, and space translation symmetry along each axis corresponds to the conservation of the component of momentum in that dimension, what I have previously noted is called the conjugate momentum. For example, space symmetry along the x-axis of a Cartesian coordinate system leads to the conservation of the x-component of momentum, where the latter is simply the mass times the x-component of velocity, $p_x = mv_x$.

For simplicity, let us focus on one dimension, which I will call the x-axis. Suppose a ball is moving along this axis at constant momentum. Assume, also for simplicity (though this is not necessary in general), that the mass of the ball is constant. Then the velocity of the ball in the x-direction will be constant.

Suppose we film the ball's motion with a blank background, as we did previously for the pendulum. Watching the film, we can find no identifying feature that enables us to determine the position of the ball at any particular time. Just as we were unable to specify a special time for the pendulum (something we can't do here either), we also can't specify a special place. This is called space translation symmetry.

Suppose we next film a ball after as it has been tossed straight up in the air. We know from experience that the ball will gradually slow down and stop at the top of its trajectory, and then reverse it direction and drop, increasingly gaining speed as it approaches the ground. Evidently, the momentum of the ball is continually changing, in apparent violation of conservation of momentum. We infer that an invisible force called "gravity" is responsible for the change in momentum.

Note that space translation symmetry is also violated. Consequently, we now are able to locate the place where the ball stopped, the highest point of its trajectory. The place where the ball stops and turns around to go back down is a special point in space. As with the pendulum, we conclude that an agent external to the system must be in action. In this case, translational symmetry along the vertical axis of space is broken by gravity.

Just as conservation of energy is the natural effect of translational symmetry in time, conservation of momentum is the natural effect of translational symmetry in space. And just as the local violation of energy conservation in a nonisolated system implies an interaction with an outside system, so the local violation of momentum conservation implies a force applied from outside the system. Indeed, Newton's second law of motion defines force as the time rate of change of the momentum of a system.

In the four dimensional framework of Einstein's relativity, translation symmetry in space-time leads to conservation of a generalized four-dimensional momentum. The familiar three dimensional momentum is assigned to three of the components of the four-momentum, with the fourth component proportional to energy. Also note that the quantity that is conserved in each case is the "momentum" that is conjugate to the "coordinate" along whose axis the symmetry exists. In the case of the time "coordinate," the conjugate "momentum" is the energy.

ANGULAR MOMENTUM AND ROTATIONAL SYMMETRY

The third great conservation principle of physics is **conservation of angular momentum**. What I have previously called momentum is more precisely termed "linear momentum," since it relates to the rectilinear (straight-line) aspect of a body's motion. **Angular momentum** usually relates to a body's rotational (curvilinear) motion, although bodies moving in straight lines can also have angular momentum. In introductory physics, we are told that angular momentum conservation is responsible for the steady rotation of the earth and other planets, keeping a moving bicycle upright, and the speeding up of a figure skater as she pulls in her arms during a spin.

By arguments similar to those already used:

Space Rotation Symmetry ⇒ Angular Momentum Conservation

Like linear momentum, angular momentum has three components and these are individually conserved when the system is rotationally symmetrical about each of the three spatial axes x, y, z. The breaking of angular momentum conservation occurs when a torque is applied from the outside, where torque is the time rate of change of angular momentum, just as force is the time rate of change of linear momentum. The three components of angular momenta are the momenta conjugate to the "angular coordinates," the angles of rotation about the three axes.

When the system is symmetric under a particular transformation, such as the rotation of the coordinate system, it is said to be **invariant** to that transformation. This means that measurements made on the system do not depend on the particular space-time coordinate in question. For example, if the system is symmetric under translation along the x-axis, then our description of that system cannot depend on the x-coordinate. This is true both classically and quantum mechanically. In both cases, though the mathematical formalisms are somewhat different, a quantity called the **generator** of the transformation is conserved. Thus, energy, the "momentum" conjugate to the time "coordinate," is the generator of time translation. Linear momentum is the generator of space translation, and angular momentum is the generator of rotation.

In a profound relativistic generalization, rotational transformations in four-dimensional space-time not only encompass spatial rotation, but also the Lorentz transformations that describe the features of Einstein's special relativity. When these space-time rotational transformations are combined with space and time translations, we have a mathematical group of symmetry operations called the **Poincaré group**. This group contains all the transformations discussed so far, and the Lorentz transformations to boot!

THE SYMMETRIES OF THE UNIVERSE

And so, we have the following picture: Four-dimensional space-time obeys the symmetry of the Poincaré group. In this configuration, space-time has no special origin or direction. A universe with such unsurprising symmetry properties would then, automatically, possess the following physical laws:

The Natural Laws of Nature

1. Conservation of energy (or the first law of thermodynamics);
2. Conservation of momentum (including Newton's three laws of rectilinear motion);
3. Conservation of angular momentum (including the corresponding laws of rotational motion);
4. Einstein's special theory of relativity (including the invariance and limiting nature of speed of light, time dilation, space contraction, $E = mc^2$, and all its other features).

This already encompasses almost all of the physics that is taught in an introductory physics course. Except the students are told these are "laws of nature," not simple definitions that follow from a universe without law. But we see that the fact that the universe obeys these principles is not the profound mystery that students and lay people, as well as scientists and philosophers, are often led to believe. We all are told that the universe is so special, so unique, so improbable that only intelligent, supernatural design or some mystical, Platonic, logical necessity must be invoked to explain it. However, as we see, the real mystery would exist if our universe did not follow these laws.

CHARGE, BARYON, AND LEPTON CONSERVATION

Physicists are aware of other conservation principles besides energy and the two kinds of momentum. The most notable, also taught in introductory physics, is **conservation of electric charge**.

Charge conservation is a principle of electrical phenomena and of chemistry. (Chemistry is basically electrical in nature). A battery will move charge through an electrical circuit, but the amount of charge that leaves one terminal will be balanced by an equal amount entering the other. When various chemical ingredients are mixed together in an isolated test tube, charges may be exchanged among the constituents of the reaction, but the total charge will remain constant.

We now recognize that charge is just one of a number of observables that appear to be conserved in interactions among the basic constituents of matter. These were already described in chapter 9; however, I think that some repetition and review at this point will be helpful.

Baryon Number. Certain heavier particles called **baryons** do not immediately decay into lighter particles, like electrons and neutrinos, even though this would be allowed by energy, momentum, and charge conservation. For example, the positively charged proton might be expected to decay into three electrons, two positive (positrons) and one negative. This conserves charge, energy, and both kinds of momentum. The rest energy of three electrons is far less than that of a proton, so this reaction should happen in a fraction of a second. Of course, it doesn't.

If protons decayed with anything but the tiniest probability, the galaxies with their stars and planets made of protons, neutrons, and electrons would not be around today. Our individual bodies contain the order of 10^{28} protons, and none are seen to disintegrate over our lifetimes. Experiments (including one in which I have participated) have determined that the average lifetime of a proton is at least 10^{33} years, if it decays at all.

The observation that protons do not decay, or at least are very unlikely to do so, has led physicists to conclude that a "law" must be in operation to prevent this from happening. That law is **baryon conservation**. However, this may not be an infinitely precise law, as we will see.

Baryons form one class of basic particles. In the current standard model of elementary particles, most are composed of three quarks. That is, baryons such as the proton and neutron are not themselves elementary. They have a substructure of quarks and perhaps other stuff, such as gluons, as well. At the level of quarks, baryon number conservation is equivalent to the statement that the number of quarks minus the number of antiquarks is a constant.

Lepton Number. Unlike protons, electrons have so far remained elementary within the standard model. They join with neutrinos, and the heavier muons and tauons, in a class called **leptons**. Like baryons, the number of leptons minus the number of antileptons in a reaction is also observed to be constant. This is called lepton number conservation. Until recently, the data indicated that individual lepton numbers associated with the leptons of each generation—the electron, muon, tauon, and their associated neutrinos, were separately conserved. However, the observations in Super-Kamiokande (in which I participated), mentioned in chapter 9, are accounted for by a phenomenon called **neutrino oscillations** in which one type of neutrino, specifically the muon neutrino with muon lepton number $L_\mu = 1$ transmogrifies into, we experimenters think, a tauon neutrino with $L_\mu = 0$. The total lepton number remains constant, however, as the tauon lepton number L_τ changes from 0 to 1.

INNER SPACE

Having observed these and other conservation principles in nature, and having established in other cases a deep connection between conservation principles and symmetries, physicists have been naturally drawn to search for symmetries that connect to all conservation principles. However, except for space-time reflection symmetries, which I will discuss in a moment, we have largely exhausted the symmetry possibilities for four-dimensional space-time. The dimensional axes whose symmetry properties lead to charge, baryon and lepton conservation are not immediately obvious to eye or instrument.

In 1914, Gunnar Nordström proposed that a five-dimensional space be used to unify electromagnetism with gravity. In 1919, after Einstein had published his new theory of gravity within the framework of general relativity and four-dimensional space-time, Theodor Kaluza proposed that a fifth dimension be added to encompass electromagnetism. This theory was further extended in 1926 by Oskar Klein. Although Einstein was enthusiastic about it, the idea did not prove immediately fruitful.

In more recent times, **Kaluza-Klein theories** in five or more dimensions have acquired renewed interest. The extra dimensions might account for baryon, lepton, and other conservation principles in addition to charge. In fact, eleven dimensions will do the trick. The thought is that the universe formed with eleven or so dimensions of space and time, all curled up in a little ball of Planck size (10^{-35} meter). For an unknown reason, perhaps simply by chance, the exponential cosmic inflation acted to straighten out four of the eleven dimensions to give us our familiar axes of space-time. The seven axes that stayed curled up remain undetected by our most sensitive apparatus except indirectly through the conservation principles which result from symmetries to translation and rotation in this subspace.

Possibly, the inner space of charge, baryon, and lepton numbers has nothing to do with multidimensional space-time as such. This does not mean that the concept of inner dimensions must be discarded. Physicists have learned from mathematicians to think of space in more abstract terms, not necessarily visualized geometrically. In general, any graph in which the variables of a system are represented by axes and the state of the system is defined by a point on the plot, can be regarded as an abstract state-space or Hilbert space. For example, an economic state might be crudely defined as a point in a two-dimensional "supply-demand space," with supply one axis and demand the other. This is in fact the more sophisticated way to represent quantum states that was pioneered by Dirac. No wave functions nor waving continuous media are required.

In particle physics, the states of quarks and leptons are represented by vectors in Hilbert spaces that possess certain symmetry properties that are analogous to the symmetries of space-time discussed above. This picture allows us to form the following conception of the origin of conservation principles such as charge and baryon number that are not immediately evident from the symmetries of space and time.

When our universe came into existence, it may have possessed *inner* dimensions of some kind beyond the familiar four dimensions of space and time. It was, at least initially, symmetric under all possible transformations (rotations, translations) in that space. Mathematically, these transformations can be described in terms of generators, analogous to generators of transformations in space-time. Each generator would be conserved when the system is invariant under that operation. So, just as angular momentum is the generator of a rotation in normal space, electric charge, baryon number, lepton number, and other conserved or partially conserved quantities measured in the laboratory arise as the generators of symmetries in these inner dimensions.

As we saw in chapter 9, the theory of transformations called group theory has been utilized to systematically study the internal symmetries of elementary particles. Indeed the standard model is referred to technically as U(1)×SU(2)×SU(3), where U(1) is the one-dimensional group that gives charge conservation and electromagnetism, SU(2) is the two-dimensional group that gives weak isospin conservation and the weak interaction, and S(3) is the three-dimensional group that gives color charge conservation and the strong interaction.

BROKEN SYMMETRY

A profound series of discoveries, dating back to the 1950s, added further dimensions to our appreciation of the underlying role of symmetries in nature. Ironically, these were observations that not all symmetries are perfect; some are ever-so-slightly broken under certain, narrow circumstances.

Other possible symmetry operations exist besides translation along, and rotation about, the various axes of space-time or abstract inner axes. We can also think of reflecting a system in a plane perpendicular to a given spatial axis in an operation that is called **space reflection**. As has already been mentioned in earlier chapters, this is simply the familiar process of looking at the mirror image of an object, where the axis that is reflected is perpendicular to the mirror. If the reflected image is indistinguishable from the original, then the object possesses mirror symmetry. That is, if someone presents you with a picture of an object and you cannot tell if was taken in a mirror, then that object is mirror symmetric.

Many geometrical objects are mirror symmetric along various axes: spheres, cubes, cylinders, cones. Other more complex objects, such as common screws, are asymmetric under space reflection. Most screws are right-handed in real life, and left-handed as viewed in a mirror. A difference between left and right seems evident in everyday life.

Life violates mirror symmetry on every level. Most organisms are certainly not identical with their mirror images, and the basic building blocks of proteins, the amino acids, are not mirror symmetric. If your spouse happened to meet a person who was your exact mirror image, he would immediately recognize that she was not you. But your mirror twin would not be an impossible entity. The fact that complex structures of matter are asymmetric in various ways does not imply a violation of symmetries at the underlying level of the fundamental interactions between particles. Indeed, most of these asymmetries are very probably accidental.

The **parity** operator P, previously introduced, is the generator of space reflection—just as linear momentum is the generator of space translation. Until the 1950s, it was thought that space reflection symmetry is preserved in fundamental interactions of elementary particles. That is, these processes were assumed to be indistinguishable from their mirror images. However, weak nuclear reactions, such as the radioactive beta-decay of atomic nuclei, were found to occur at rates that were different than their mirror images. This implied that mirror symmetry is broken at the fundamental level.

The discovery that an elementary process like beta-decay prefers a particular handedness hit physics like an earthquake. It prompted physicists to rethink the whole role of symmetries in nature. Today, less than two generations later, we have a fully functional theory, the standard model, that rests heavily on broken symmetry.

Since the 1930s, experiments have indicated another kind of mirror symmetry, where the reflection is not from right to left but from matter to antimatter. As we saw in chapter 7, Dirac predicted the existence of a new particle that was identical to the known electron in every way except having positive charge. Antielectrons, or positrons, were discovered a few years later. The observation of other kinds of antimatter, such as antiprotons and antineutrons, followed after the completion of World War II.

Except for charge, the positron was otherwise identical to the electron. This suggested another reflectionlike symmetry operation, in inner space. Recall that C is the operator that changes a particle to its antiparticle. For example, $Ce^- = e^+$, where e^- refers to the ordinary negatively charged electron and e^+ is its antiparticle, the positron. Originally, C was called charge conjugation. However, a more accurate term is **particle-antiparticle reflection**. When a particle is electrically charged, the C operation causes that charge to be reversed in sign and we have charge reflection. But electrically neutral particles, such as neutrons and neutrinos, can also change from particle to antiparticle under C, so one should not visualize the operation specifically as a reflection of a "charge axis." Still, we can utilize the abstraction of inner dimensions and think of C as a reflection of some inner particle-antiparticle axis, the way P is a reflection of an outer, spatial axis.

The combination CP converts a process to its antiparticle mirror image: right-handed particles are replaced by left-handed antiparticles, left by right. For example, an electron spinning around its direction of motion like a right-handed screw is changed, by CP , to a positron spinning like a left-handed screw. The data, such as the beta decay reactions where space reflection symmetry is violated, seemed to indicate that reactions were indistinguishable from those in which the particles were all changed to antiparticles and viewed in a mirror.

However, when the decays of neutral kaons into pions were examined by Val Fitch, James Cronin, and their assistants in 1964, they found the decay rates were slightly different, by about one part in a thousand, for mirror-image reactions of antiparticles on neutral kaon decays. They concluded that the combined symmetry CP was violated in these decays. Their observations were quickly confirmed (I was involved in one of these confirming experiments), and CP violation joined P violation as two critical examples of broken symmetry in fundamental physics.

The violations of mirror-type symmetries was later used to help answer an important puzzle about the nature of matter in the universe. If matter were perfectly symmetric with antimatter, as was once thought, then why is the universe mostly matter? While some antimatter is present in cosmic rays, as far as we can tell the visible stars and galaxies are primarily normal matter. Only about one part in a billion of the observable universe is antimatter. Lucky thing. Since matter and antimatter annihilate, converting all their rest energy to photons, the universe would have no galaxies, stars, or planets if all its matter annihilated into photons. This would have happened early in the big bang, billions of years ago and the universe today would comprise the 2.7K microwave photon background, an equivalent background of neutrinos, and little more.

As first suggested by the famous dissident Russian physicist André Sakharov, the observation of CP violation, supplemented by a small violation of baryon number conservation, provides a possible answer to the antimatter puzzle. At some early time the universe may have contained exactly half matter and half antimatter. That is, at the extremely high temperatures of that epoch, CP symmetry was obeyed. The two types of matter annihilated into photons, but the energy of the produced photons was sufficiently high to allow the reverse reaction, the production of matter-antimatter pairs of particles, to take place. An equilibrium between the two reactions, forward and reverse, existed.

Luckily, before the temperature became so low that energy conservation prevented photons from making any more matter, CP symmetry was broken. The violation of CP symmetry led to an imbalance in the relative strengths of the various reactions involving particles and antiparticles, with the particle interactions accidentally stronger by a minuscule of one part in a billion. (If the "antimatter" had been stronger, we would be calling that matter and the other antimatter).

So when the photons cooled to the point where they no longer interacted with the rest of the universe except by simple elastic scattering, they expanded and cooled further on their own to become the 2.7K radiation of today. In doing so, they left behind a small residue of matter that eventually accumulated into the galaxies of the visible universe.

This is the first example of the scenario we will see repeated many times, as we go from the Planck era of perfect symmetry to the broken symmetric world of today. Structure is the result of broken symmetry. The first important element of structure in the universe is the dominating presence of matter itself, produced a billion times more frequently than antimatter.

TIME REFLECTION AGAIN

Let us move once more from inner space to outer space. Mirror symmetry, is a symmetry of space. It is strictly obeyed by certain fundamental particle interactions, such as those between quarks, and strictly disobeyed by other fundamental reactions, such as those between neutrinos and quarks. Indeed, the breaking of mirror symmetry is one of the structural factors that leads to the substantial difference between quarks and neutrinos.

Time is the fourth dimension of space-time. Just as we can reflect a spatial axis, $x \Rightarrow -x$, we can reflect the time axis, $t \Rightarrow -t$. This has the effect of reversing the direction of time, the mathematical equivalent of running a film of an event backward through the projector. When a system is indistinguishable under this operation, it possesses **time reflection symmetry**. In the film example, this means that a viewer has a way of knowing by just watching the film that it is running backward. Of course, we have already discussed time reflection symmetry in considerable detail. Now let us try to see where this symmetry fits into the general picture of elementary symmetries we have developed in this chapter.

The violation of CP reflection symmetry, described above, indirectly implies a violation of T reflection symmetry by virtue of the CPT theorem mentioned in chapter 8. The CPT theorem is derived in quantum field theory from assumptions that are thought to be so basic that physics would come crashing down should the theorem be violated. This theorem says that any combination of space reflection, time reflection, and particle-antiparticle conjugation is symmetric for all fundamental interactions. For example, the beta-decay of antinuclei filmed through a mirror will look just like normal beta decay when the film is run backward through the projector.

A corollary is: since CP is violated in kaon decays, then so, too, must T be violated in those reactions. Recent experiments have now directly observed T violation in kaon particle processes. So the evidence for T violation, at least at this level, seems secure.

It makes a certain relativistic sense to have time reflection symmetry broken, given that space reflection symmetry is known to be broken. As we have learned in previous chapters, thinking of time as "just another coordinate" of space-time can open up whole new ways of looking at the universe and demonstrate the fallacies in many of the old ways.

However, the tiny violation of time reflection symmetry in kaon interactions plays a negligible role in most material processes. As I have noted, most fundamental interactions do not seem to care about the direction of the time axis. They work just as well backward in time as forward. Chemical and strong nuclear reactions are all time reversible without having to go through the complications of changing to antimatter and viewing everything in a mirror.

Can the T violation observed in rare elementary particle processes have anything to do with the macroscopic arrow of time? It does not seem likely, for a number of reasons. The macroscopic arrow of time is certainly not limited to kaon processes, nor the other rare elementary particle interactions that are expected to also exhibit CP violation. All large scale phenomena manifest an arrow of time. What is more, kaon processes are still reversible. They just have very slightly different probabilities, on the order of one part in a thousand, for going one way or the other. It is hard to see how the breakdown of T symmetry at the microscale implies time irreversibility at the macroscale, although I am not prepared to rule it out.

Perhaps weak interaction T violation was more important in the early universe than it is now, and its effect was magnified at that time. After all, the partner of T violation is CP violation, and we have already seen how this may help account for the vast preponderance of matter over antimatter that we have in the current observable universe. Could this have also led to the cosmological arrow of time? Perhaps, although it is not clear how. Furthermore, the time when CP was violated in the early universe may have been already too late. As I indicated in chapter 4, entropy generating processes during the first instant of inflation, long before the breaking of CP symmetry, already specified the cosmological arrow.

Even if a connection can be made between T violation in the weak interaction and time's arrow, it is still a broken symmetry. At high enough energies, small enough distances, or early enough times, T symmetry would be present. The underlying basic processes would remain symmetric under time reflection, and so would "ultimate reality." Just as broken mirror symmetry is unsurprising when observed in complex material systems, so should broken time symmetry on the macroscopic scale be equally unsurprising. No one seems to be greatly bothered by the fact that the spatial asymmetries of common objects are not directly evident in the interactions between the fundamental particles that make up those objects. We would expect fundamental interactions to be simpler, and what is symmetry if not simplicity? There is no need to introduce a special law of physics to explain

Fig. 12.1. A sphere of water vapor has continuous rotational symmetry about all axes. The snowflake has discrete symmetry about one axis.

why most complex material systems are spatially or temporally asymmetric. In the objects of our experience, time reflection asymmetry can readily arise from the same source as spatial asymmetry, in which case it is no more profound than the mirror asymmetry of a human face.

The clear time symmetry of the quantum world is not in conflict with the fact that time asymmetry exists in the structure of the macroscopic universe, any more than the spatial symmetries implicit in the laws of conservation of linear and angular momentum deny the existence of spatially asymmetric structures. Structure, as I must continually emphasize, is not symmetry but broken symmetry.

STRUCTURE AND BROKEN SYMMETRY

Common intuition would say that chaos and symmetry are opposites, and since chaos and structure are also opposites, then structure must correlate to symmetry. But this is another instance where intuition leads us astray. As we saw in chapter 4, the highest level of symmetry is total chaos. In that case, no structure exists. The inside of a black hole, with the highest possible entropy, has perfect symmetry. A gas has more symmetry than a liquid, and a liquid more symmetry than a solid.

A snowflake has more structure but is less symmetric than the water vapor from which it formed (see figure 12.1). The symmetry of a snowflake is measured by the fact that it looks the same when rotated about a perpendicular axis in 60 degree steps, but only these angles and no other. If it were symmetric to a rotation about any axis, by any amount, it would be simply a boring sphere. A spherical cloud of water vapor can be rotated by any amount about any axis and still look the same. Gravity breaks the full spherical symmetry of a rain drop, but the rain drop is still more symmetric than a snowflake since it looks the same under any rotation about the vertical axis.

The structure of our universe is evident in snowflakes, trees, mountains, and animals—all less symmetric than the more disorderly systems of matter from which they formed. We have seen that the conservation laws of physics—energy, momentum, charge—are consequences of space-time and internal symmetries that need no explanation other than they exist as the most natural consequence of a totally chaotic, totally symmetric, totally unstructured beginning to things. Conservation laws and their underlying symmetry principles are not part of the structure of the universe; rather, they manifest the *lack* of complex structure everyplace but in a few incidental locations such as the earth where these symmetries are locally, and accidentally, broken. Thus, the structure of the universe results from off-hand broken symmetry, occurring by chance at specific places, but not everyplace, as a product of the natural fluctuations that exist in any stochastic system.

Consider handedness as one measure of structure. When we look at the mirror images of many simple geometric figures, such as spheres, cubes, and triangular pyramids, we cannot distinguish them from the original. An image may appear rotated, but a rotated object is still the same object, just viewed from a different angle. But no matter what angle you view your right hand, it will never look like your left hand. Simple geometrical shapes with mirror symmetry have less structure than complex shapes that look different in a mirror.

A normal right-handed screw has more structure than a simple cone of the same size that has an otherwise similar shape. It becomes left-handed when viewed in a mirror, while the image of the cone is indistinguishable from the original. The mirror image is distinguishable, and so we say it has a specific handedness or **chirality**. Chirality changes under space reflection, which produces an object that is empirically distinguishable from the original. Just try turning a left-handed screw in a right-handed thread. Note that if we apply a second mirror reflection, we return to the original.

Similarly, the normally left-handed single helical strand of an RNA molecule changes to a right-handed version upon mirror reflection. Only left-handed RNA is found naturally, although a right-handed version is physically possible and can be manufactured artificially. The double-stranded DNA that stores the information to manufacture a living cell is also left-handed. In fact, most of the complex molecules of life have a specific handedness. For example, when biochemists synthesize amino acids from inorganic materials in the laboratory, molecules of both types of handedness (called L-type and D-type) occur. On the other hand, if only biological materials are used as the ingredients, the naturally occurring types will be exclusively produced.

This apparent preference of the molecules of life for one particular handedness led Louis Pasteur to remark that this was the only difference he could find between living and dead matter. The matter of life is not mirror symmetric. However, Pasteur's observation should not be misinterpreted; no property has ever been found to distinguish the matter in biological systems and the matter found any where else. Inorganic matter, like neutrinos, can be equally mirror asymmetric.

Handedness is a structural property of matter. Objects with perfect mirror symmetry have no handedness and less diversity than objects possessing this property. Spheres and cubes are neither left-handed nor right-handed. Handedness happens when mirror symmetry is broken.

As I have indicated, the fact that some structures exhibit handedness does not imply the breakdown of parity conservation at the fundamental level, as long as both the structures and their mirror images are equally allowed by the laws of physics. Parity violation in nuclear beta-decay is probably not the source of the mirror asymmetry of organic molecules. More likely, that was an initial chance event that became embedded by evolution, as the progeny of chiral molecules retained their ancestors' chirality.

HAIRY PARTICLES

At the elementary particle level, one would think that parity should be conserved. By definition, particles appear to us as pointlike. If they did not, then we would be aware of some more elementary constituent inside the particle, and the particle would, by definition, not be elementary. Furthermore, a point is certainly mirror-symmetric. Despite this, some elementary particles still show a selection for one handedness or another by virtue of their spins. If the direction of a particle's motion is given by the thumb of your hand, the particle will have right-handed chirality if its spin is in the direction that the fingers of your right hand curl when you make a fist, and left-handed chirality when the spin direction is given by the fingers of your left hand.

Perhaps the most familiar example of chirality is circularly polarized light, discussed in chapters 6 and 8. The photons in a normal, unpolarized beam of light will divide into two separate beams when passed through a circularly polarizing beam splitter. In one outgoing beam we have left-handed photons; in the other, right-handed photons. This, however, does not mean that parity is violated in photon processes, since no special distinction is made between left and right in electromagnetic interactions. A mirror image photon interaction is almost indistinguishable from the original. A tiny asymmetry results from the fact that the photon also interacts by way of the weak interaction. This, incidentally, was another successful, and risky, prediction of electroweak unification in the standard model.

Processes involving neutrinos, on the other hand, make a strong distinction between left and right. The neutrinos associated with normal matter are left-handed, while antineutrinos are right-handed. Neutrinos and antineutrinos of the opposite handedness have not yet been observed, though they are included in some theories and may eventually be found. Now that we think particles have a small mass, these statements are only approximately true, but the approximation is a good one at the energies of neutrinos observed in current experiments.

So, point particles can still possess chirality by virtue of their spins, although many show no preference for left or right. All the confirmed elementary particles in the standard model have spin. The quarks and leptons have spin $1/2$; the gauge bosons, like the photon, are spin 1.

Spin is the intrinsic angular momentum of a body, and we have seen that it will be conserved in any system that is rotationally symmetric. Rotational symmetry means that no particular direction in space is preferred. Thus, a smooth sphere looks the same when rotated through any angle.

As we know, the earth is not a true sphere but an oblate spheroid flattened at the poles. It does not look the same when rotated about an arbitrary axis, but the earth still possesses a good deal of symmetry about its axis of rotation. This also disappears as you look closer and see the mountains and valleys which make the earth a more complex—and more interesting—object.

So the breaking of rotational symmetry is also a major source of structure on the macroscopic scale. The earth's rotational symmetry is broken because it is not an isolated system but interacts gravitationally with other solar systems objects such as the sun and moon. We can use broken rotational symmetry, generalized to abstract "inner dimensions," to provide a mechanism for producing most of the structure we observe in elementary particles.

The basic idea can be seen by analogy with the Stern-Gerlach experiment described in chapter 5 and utilized in several places. Recall that when you place a beam of particles with spin s (integer or half-integer) through a nonuniform magnetic field, 2s+1 beams come out (see figure 5.1) For a silver atom or an electron with s = $1/2$, you get two beams. An atom with spin 3 would give seven beams. Going into the apparatus, no particular orientation in space is singled out. The beam direction is arbitrary, depending on the observer's reference frame.

The Stern-Gerlach experiment provides us with a model for symmetry breaking. The particles coming in have a definite spin s, but no particular direction of that spin is singled out. This means that the particle has an indefinite component of spin along any axis you choose. The interaction with the external field of the magnet breaks this symmetry so that each beam coming out contains particles with a definite component along the direction of the axis of the magnet.

As with a multiple slit experiment, a particle's motion is separated into different, distinct paths. The direction of the field of the magnet is preselected by the experimenter, and the original symmetry is broken when the electrons emerge from the magnet. Let the z-axis be along the field. Then the electrons that emerge have a spin component $s_z = +\frac{1}{2}$ in the "up" beam and $s_z = -\frac{1}{2}$ in the "down" beam.

In the conventional interpretation of quantum mechanics that is found in most textbooks, and taught at most universities, the electron is said to not even possess the property s_z until it passed into the magnet. In the process of its interaction with the magnet, the electron obtains a new property s_z that has two possible values. It is as if the magnet attaches a little pink "stickie" label on half of the electrons and a blue stickie on the other half, thus giving each electron a new structural property it did not originally possess. Upon passing through the magnet, the electrons acquire a way to be distinguished. Someone who has only viewed the electrons from one beam might conclude that all electrons are pink. Someone who has viewed electrons from both beams, but not the electrons going into the magnet, might conclude that two different particles exist: "pinktrons" and "bluetrons."

Now stickie labels come off easily. Suppose we next pass the spin up beam, say the pinktrons, through a second magnet whose axis is rotated by 90 degrees with respect to the first so that it points along the x-axis. Two beams emerge, left and right, that contain electrons with $s_x = +\frac{1}{2}$ or $-\frac{1}{2}$. The second magnet has somehow removed the pink stickie from all the electrons passing through, replacing half of them with green stickies and half with yellow.

As this illustrates, the spin components of particles are not invariants. They depend on your choice of axes—the way you decide to orient the magnet. So spin components would not seem to be promising candidates for intrinsic properties of the primitive objects in our metaphysics. They are rather properties that are attached to the object upon the act of their measurement.

However, I must emphasize that not all properties of particles result from the act of measurement. This would lead us back into solipsism. As I have previously maintained, particles and composite objects possess invariant properties that do not depend on observers and arbitrary reference frames. These properties include proper mass (but not energy or momentum), spin magnitude (but not spin component), charge, and other quantities like baryon number and lepton number that seem to be conserved, although we may have to modify this somewhat, as we will see in a moment.

One of the famous theorems concerning black holes, derived from general relativity, is that "black holes have no hair" (for a good discussion, see Thorne 1994). Basically the theorem shows that black holes can have no structure. Any little protrusion ("hair") on the original object that collapsed to the black hole is smoothed out by the intense gravity produced by the mass being confined to a small region. However, black holes do possess properties other than hair, namely, those properties that are conserved invariants, like proper mass, charge, and spin. Black holes have no magnetic field because no invariant "magnetic charge" (magnetic monopole) apparently exists, at least as far as we now know. A black hole can have an electric field, but only the strictly radial one of a point charge (electric monopole). These properties are not "hair" because they all result from symmetries. Hair, or local structure, is the result of broken symmetry.

The sticky labels particles pick up when observed are the "hair" that makes them into the structured objects we perceive. Recall the weak isospin, which was introduced in chapter 9. In the standard model, the u and d quarks are regarded as up and down states of weak isospin for the same generic quark in a way that is completely analogous to the up and down spins of the electron. Similarly, the electron neutrino is the weak isospin up state of a generic lepton, while the electron itself is the weak isospin down state. The other two generations of quarks and leptons follow the same scheme. The charmed and strange quarks, and the top and bottom quarks, are up and down weak isospin states of a generic second and third generation quark, respectively. The muon and muon neutrino, and the tauon and tauon neutrino, are the up and down states of the second and third generation leptons.

The bosons in the standard model also fit into the weak isospin scheme, with two generic bosons—one with weak isospin zero and one with weak isospin one. The W+ and W- have weak isospin components +1 and –1. The photons and Z bosons are mixtures of the two weak generic bosons that give isospin zero states. The mixing parameter is the Weinberg "angle," described in chapter 9.

Going beyond the standard model, we find the theoretical foundation far less certain. Still, we can imagine that other properties might be cast in this same general framework. The generic quark might be considered the spin up state and the generic lepton the spin down state of an even more generic leptoquark, or quarkon. The three generations of quarkon are then pictured as the three states associated with an object having "generation spin" (more conventionally called "flavor") = 1.

Finally, supersymmetry treats bosons and fermion as two states of the same fundamental object. Perhaps they are the up and down states of an object with "supersymmetric spin" = $\frac{1}{2}$.

The following scenario suggests itself: The primitive object of reality is neither a boson nor a fermion, not a quark, nor a lepton. It possesses none

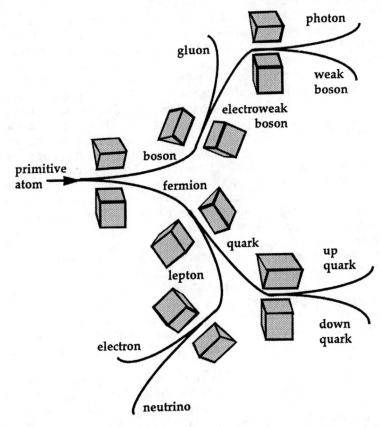

Fig. 12.2. How symmetry-breaking can be likened to the splitting of a spin $-1/2$ particle in the Stern-Gerlach experiment. The supersymmetric primitive atom is first split into a boson and a fermion. The boson is then further split into the gauge bosons. The fermion is split into leptons and quarks, and these are split into their weak isospin up and down components. Not all the details, such as generation splitting, are shown.

of the properties we associate with the particles we observe in experiments because it is perfectly symmetric and these properties are associated with broken symmetries. This primordial particle is as simple as it can possibly be, without structure or form. It is massless, chargeless, and spinless.

As this particle moves through space-time, it interacts with other particles—or possibly even itself curling back in time. At very small distances, requiring very high energies to observe in the laboratory, these interactions preserve the complete symmetry of the particle. All these particles form an extreme relativistic gas of massless, formless objects.

At larger distances, observed in the laboratory at low energies, individual symmetries are broken during interaction in which the particle picks up the corresponding properties, that is, hair. We can use the Stern-Gerlach experiment as a metaphor for this process. As we see in figure 12.2, our primitive particle first undergoes a breaking of supersymmetry that takes it along two separate paths, one as a fermion and another as a boson. The boson undergoes a further splitting into a gluon and an electroweak boson. The electroweak boson splits into a weak boson and photon. To keep things from getting too messy, I have left off generation splitting and details of the further splitting of the weak boson, but these would be included in the full picture.

The masses of particles are also picked up during this process, by the Higgs mechanism discussed in chapter 9. I have also not shown Higgs particles in the figure; we do not yet know where they occur in the symmetry breaking process, if at all. No doubt, the scheme illustrated in figure 12.2 should be regarded as a concept whose details may change as we learn more, especially for any parts that make inferences beyond the current standard model. For example, my placement of the breaking of **supersymmetry** (SUSY) to give bosons and fermions may turn out to be in the wrong place. We have not yet probed to that level in experiments and I have just put SUSY breaking where my own guess says it should be. That guess could turn out to be wrong.

That said, let us return to the fermion that is associated with the other beam of the first "magnet." It is split into a lepton and quark. The lepton splits into an electron (or muon or tauon) and the appropriate neutrino. The quark becomes an up and a down (or charmed and strange, or top and bottom). The antiparticles? These are, as usual, accommodated by the time-reversed paths. In fact, since the paths are reversible we can think of them all being traversed by a single particle. Perhaps that single particle traverses all space-time. But I will not pursue this crazy idea.

UNDERLYING REALITIES

This picture of certain particle properties resulting from a process analogous to magnetic beam splitting bears on the philosophical issue of the content of reality. In the conventional, Copenhagen interpretation of quantum mechanics, no physical property that we associate with a measurement exists intrinsically, as part of some "true reality" beyond the senses, but comes into existence as a result of the act of measurement. Thus, the underlying reality of particles, if any, cannot be directly observed. From the time of Einstein, this interpretation has been the subject of intense debate. However, all experiments, including several that were capable of falsifying conventional quantum mechanics, have proved consistent with that interpretation, while not ruling out other interpretations. The types of experiments used to test these ideas are precisely the beam-splitting type.

According to the Copenhagen interpretation the properties we associate with particles are human inventions, used to describe the outcome of specific experiments. The specific design of an experiment operationally defines the quantity being measured. Time is defined by the design of clocks. Charge is defined by the design of experiments to measure charge. Baryon number is defined by the design of experiments to measure baryon number.

Does this imply that no underlying reality exists and that everything is in our head? Not necessarily. Indeed, if that were true, experiments would have no predictable outcomes. Einstein recognized that predictability provides the best criterion for deciding whether or not an idea bears some relationship to reality. At least, this provides a definition of reality that is consistent with our common understanding of the term, one that stands in stark contrast to the unpredictability associated with dreams and fantasies.

The observations of particles have very many predictable aspects. If we collide a beam of electrons against a beam of protons, the particles in all the reactions produced will have a total lepton number $L = 1$ and a total baryon number $B = 1$. If I send an electron into a region of magnetic field, it will bend in a predictable way. The proton will bend predictably opposite. A photon or neutrino will not bend at all.

However, these are all "large distance" effects, meaning here not "macroscopic" but rather distances on the order of nuclear diameters. These distances can be explored with current accelerators, and when they go higher in energy they will be able to explore deeper. Unfortunately, the energies needed to explore down to the grand unification scale, or beyond to the Planck scale where all the symmetries are expected to hold exactly, is far beyond the horizon of current or projected technology.

Fortunately, another way to probe very small distances exists in the study of very rare processes such as proton decay. This enables us to probe the place in figure 12.2 prior to the beam of fermions being split into quarks and leptons. The observation of SUSY partners would also give us a clue where to place SUSY breaking.

Regardless of their ontological significance, we can continue to speak in terms of particle properties because these provide an accurate description of observations. After all, this is our only source of information about the universe. We do so with complete confidence that, while not necessarily describing primary reality, they give us the practical tools to make predictions and put our knowledge to use.

The invariant properties of particles, however, seem to be objective and absolute. They are the best candidates we can find for elements of reality.

THE STRUCTURE OF FORCES

Symmetry breaking can also be seen as the mechanism for the generation of the different forces in nature. Or, put another way, forces are part of the structure of things. We saw above that an external force is inferred when momentum conservation is not observed in a body's motion, and that this is connected to a breaking of space translation symmetry.

Consider magnetism. In the absence of an external magnetic field (speaking classically), a compass needle will move around randomly, pointing in an arbitrary direction at any given time. When placed in a magnetic field, such as that of the earth, the compass needle will align itself with the magnetic field lines and point toward a particular magnetic pole. Spherical symmetry is thus broken by the external magnetic field.

That is the textbook view. Alternatively, we can think of the structural property of magnetism as resulting from the breaking of spherical symmetry. When the symmetry is in effect, no magnetic force exists. When the symmetry is broken, the magnetic force appears.

Let us imagine a perfectly symmetrical sphere of iron. Above a critical temperature called the Curie temperature, this iron sphere has no preferred axis. When it cools below the Curie point, a specific but random, nonpredetermined direction is frozen in. This then becomes the direction of the magnetic field. Spherical symmetry, with its infinitude of equally possible axes, is broken so that only one axis of symmetry remains. The loss of symmetry is accompanied by an abrupt drop in entropy, as we move to a higher level of structure. As is the case for water vapor freezing to give a snowflake, a phase transition has taken place.

By the same means, electric fields, gravitational fields, and the other structural properties we classify as the forces of nature can be interpreted as resulting from symmetry breaking. At the very beginning of the universe with perfect symmetry, neither forces nor properties such as spin, charge, or baryon number existed. There was no structure; entropy was as large as it possibly could be.

If we abstract the idea of the magnetic field vector freezing in place, we can understand the spontaneous appearance of the structural properties of particles and forces. Each corresponds to the choosing of an axis among the infinite number of choices that are available prior to the phase transition. The axis chosen is spontaneous and arbitrary, and the underlying symmetry remains because no one axis—no one property—is preferred over any other. Just as a polarizer might select out left-handed photons, though right-handed ones exist, or evolutionary processes might select out left-handed DNA though the right-handed type is just as likely, so, too, the properties of matter, and the forces that exist to hold that matter together in our universe, could easily have been the result of random selection.

As we saw earlier, a falling object appears to violate space translation symmetry. The generator of space translation symmetry is the linear momentum, and momentum conservation is apparently violated as the object moves with increasing speed toward the ground. Since a visible force like a swift kick can also change a body's momentum, and we prefer to describe our observations in familiar ways, we introduce an invisible force of gravity as the external agent that adds momentum to a falling body.

We have learned that Einstein's general theory of relativity did away with gravitational fields and forces and describes the motion of bodies alternatively in terms of the non-Euclidean curvature of space. In the years since, general relativity has been subjected to a series of rigorous tests of ever increasing precision. So far it has survived every challenge and remains a complete description of all observed phenomena we associate with gravity.

Einstein hoped that he or others would eventually find a "unified field theory" that also encompassed electromagnetism within the same theoretical framework. However, no one has succeeded in achieving a similar geometrical explanation for electromagnetism. Later, when the strong and weak nuclear forces were discovered, the task became that much tougher.

Today little hope remains that the approach taken by Einstein in explaining gravity by explaining away the gravitational force can work for other known forces. Instead, a quantum field approach was taken that eventually led to the unification of electromagnetism with the weak nuclear force and the current standard model. The expectation is that this approach will eventually bring all the forces, including gravity, into the fold of one unified interaction between particles. This interaction would exist at extremely high temperatures, and the early stages of the big bang. At low temperatures, as in the current cold universe, the symmetry associated with the unified force is broken to give the forces we now see.

Actually, the gauge approach is not as far from the spirit of general relativity as you might be led to think. Einstein's theory is a geometrical one, based on the idea of non-Euclidean (curved) four dimensional space-time. As we have seen, the current standard model operates in a more abstract inner space and does not deal directly with the geometry of that space. Both theories rely on the idea that symmetry-breaking leads to forces, though Einstein did not state it in precisely those terms. String or m-brane theories, mentioned in chapter 9, may eventually show that these inner dimensions are indeed part of a multidimensional space-time, as Kaluza and Klein had suggested (Greene 1999).

We have seen that energy, linear momentum, and angular momentum conservation, along with all the features of Einstein's special theory, follow from the symmetry of the Poincaré group which includes all rotations and translations in four-dimensional space-time. Einstein's general theory provided the means for producing gravitational effects from the breaking of Poincaré symmetry. Let us now briefly reexamine our discussion on general relativity in chapter 3 from this new point of view.

Consider the bending of a particle as it passes by the sun. We can think of this bending as the result of the gravitational force of the sun, acting on the mass of the particle. Alternatively, Einstein proposed that the four-dimensional space-time in the vicinity of massive objects like the sun is greatly distorted. The particle then still travels its "natural" path, but that path appears curved to an outside observer.

In the latter perspective, the sun breaks the translational symmetry of space-time. Empty Euclidian space might be viewed as a three-dimensional flat grid of parallel, equally spaced lines, where the spacing can be made as small as you like (except, operationally, no smaller than the Planck length). That grid is three-dimensional, of course, but it helps to visualize just two dimensions. Every intersection point of the grid is as good as any other. Imagine rolling a ball bearing along a flat sheet. It will continue to travel in a straight line along its original path.

Suppose we represent the sun by a large metal sphere and place it at one of the intersection points. Now one point in space is different from all the others, and space translation symmetry is broken. The grid will become distorted as the sphere sinks down into the sheet. A ball bearing rolling in from the edge of the sheet will now dip down into the valley caused by the heavier sphere. If it is moving fast enough, the ball bearing will climb back out and continue off the sheet. At another speed, it might go into orbit. On the other hand, if it is moving at less than some critical speed, it will spiral down the value and eventually collide with the sphere representing the sun. This is analogous to what happens when bodies venture near the sun with insufficient speed to pull away or go into orbit.

The planets, asteroids, and comets of the solar system have been captured by the sun. They will eventually spiral in if left alone for many billions of years—though the sun is likely to burn out first. They have enough speed to climb partially out of the valley of the sun, but not enough to completely escape.

In any case, we see that Einstein's idea of curved space offers an appealing picture of the bodies following their natural paths. To make that picture quantitative, which means to place it in the realm of science rather than aesthetics, the geometry around massive bodies must be mathematically described and provide specific and risky predictions about the observed behavior of bodies. This Einstein succeeded in doing.

GAUGE THEORIES

In a manner quite different in mathematical detail, but rather alike in generic concept, the standard model breaks the symmetries of internal space to produce the electroweak and strong forces. It does so in a way consistent with quantum mechanics, whereas general relativity has not yet been reconciled with the quantum world, so major differences in approach remain. Still, the similarities should not be ignored.

Imagine that, after placing the metal sphere on the rubber sheet and letting it sink in, we apply a force from the bottom that restores the sheet to a plane. Then the grid lines on the sheet would once again be nice and straight, and symmetric to space translations. Thus, we restore Euclidean geometry, or more generally, Poincaré symmetry, but only at the expense of introducing a new external force—gravity. This process is similar to the way we introduce the fictitious "centrifugal force" to restore Newton's second law, $F = ma$, in a rotating reference frame.

This is basically the approach that is taken in what are called *gauge* theories, which were discussed in chapter 9. The electroweak and strong forces are introduced into the theory to restore what is believed to be underlying symmetries in inner space. The equations that describe interacting particles in terms of quantum fields are written down in the asymmetrical way implied by the details of the known interaction, and then a term is added to make them symmetric. This term corresponds to the field or force. In the case of quantum electrodynamics, the highly accurate theory of the electromagnetic force, the electromagnetic field is the gauge field and the underlying symmetry group is U(1). In electroweak unification, the symmetry is described by the product group SU(2)×U(1). In quantum chromodynamics (QCD), the gauge theory of the strong nuclear force, the symmetry is SU(3), as described earlier. While the actual procedure is to introduce the force into the theory to save the symmetry, we can ontologically view the mechanism of broken symmetry as the source of the force.

ECONOMY AND ELEGANCE

Now, it may not seem that it matters much whether we interpret a force as some external agent that breaks a symmetry, or we take the force to result from the broken symmetry, as long as both views give the same results. Einstein's non-Euclidean, forceless, gravity gives results, such as its famous prediction of the bending of light around the sun, that Newton's Euclidean, forceful gravity does not. But it is not fair to compare two theories developed centuries apart, and Einstein's theory can be reformulated in terms of a post-Newtonian gravitational force operating in Euclidean space and obtain agreement with experiment.

The issue is more of one of economy and elegance. Theories should always be formulated in the simplest way, with the fewest assumptions possible and a minimum number of parameters. Einstein's general relativity is so elegant and simple (once you learn the necessary mathematics—a task not beyond the capabilities of anyone with a college education), that it is difficult to see how it could be made more economical. Its existing flaw, that it is not quantum mechanical, presents no problem as long as quantum gravity effects remain unmeasurable, which may be a very long time. I see no reason to discard the beautiful paradigm of general relativity. But it's intrinsic nonquantum nature makes it less likely to be providing deep insight into the nature of "true reality" than the fully quantum and at least special-relativistic standard model.

The symmetry breaking in modern particle theory is more difficult to understand because of its operation in abstract inner space rather than concrete space-time. However, to those well versed in the arcania of particle theory, spontaneous symmetry breaking is also natural, elegant, and economical. In particular, it provides us with just the mechanism we need to understand the production of structure at all levels in the universe, from the elementary particles to the neural network in the human brain, thus giving us the ability to also understand ourselves.

13

THE
NATURAL
AND
THE
SUPERNATURAL

If God is a synonym for the deepest principles of physics, what word is left for a hypothetical being who answers prayers, intervenes to save cancer patients or helps evolution over difficult jumps, forgives sins or dies for them?

Richard Dawkins (1999)

READING THE TWO BOOKS OF GOD

In the last chapter, I argued that the primary laws of nature, the great conservation principles, are exactly what would be expected in a universe absent of any design. This contradicts the understanding of most people that the laws of nature are there by design, either set down by an external Creator or somehow built into the logical structure of the universe itself. That is, they exist either as the part of some supernatural plan, or they could not have been otherwise because only one set of laws are possible. In this chapter I will show why recent claims of evidence for "intelligent design" to the universe, whether natural or supernatural, are unfounded.

At least two-and-a-half millennia have passed since a handful of thinkers, notably Thales and Heraclitus in ancient Greece, had the idea that the world around us might be understood wholly in terms of familiar substances and forces such as water and fire. As we saw in chapter 1, they were the first to grasp the possibility that mysterious, undetectable agents need not be invoked in the explanation of phenomena. It was a revolutionary notion—and the world was far from ready to embrace it. At this stage, humanity still clutched the superstitions carried out of cave and forest. And so, with a few momentary appearances here and there, **naturalism** lay largely dormant for two millennia. In the meantime, human culture continued, and continues today outside of science, to be dominated by supernatural thinking.

In Christian Europe during the Middle Ages, the study of empirical phenomena did not necessarily exclude the supernatural. Indeed, most if not all of the scientists, or "natural philosophers," of the period were clerics or otherwise connected with the Church. Nevertheless, serious conflict between the fledgling science and religion broke out in the sixteenth century when the Church condemned Galileo for teaching that the earth circles the sun. Although Copernicus, a churchman, had published this notion posthumously fifty years earlier, his publisher Andreas Osiander (d. 1552) had carefully represented the result as a simply a *mathematical* description of the solar system. Galileo's offense was to propose that this was the way things really are. The earth moves.

However, religion and science soon reconciled. Newton interpreted his great mechanical discoveries, which followed on the earlier work of Descartes, Galileo, and others, as uncovering God's design for the physical universe. The success of Newtonian science was rapid and dramatic. People began to speak of the need to read two books authored by God: *Scripture* and *The Book of Nature*.

As science progressed, it offered natural explanations for phenomena previously attributed to supernatural agency. Static electricity, not Thor's hammer, produces thunder. Natural selection, not divine intervention, impels the development of life. The neural network of the brain, not some disembodied spirit, enables mental processes. Scientific explanations are frequently unpopular. Witness the continuing rejection of evolution by a many in the United States and the handful of other countries where organized religion maintains a dominant influence on public thinking.

It seems that science occupies a respected place in the United States, not because people find its ideas appealing, but for the simple reason that it works so well. Technological progress, fed by scientific discovery, testifies to the power of natural explanations for events. Technology has fueled the great economic growth of developed nations, especially in the United States. This has given science enormous stature and credibility. People listen to what science has to say, even if they do not always like what they hear—

particularly when science tells them that they are not the center of a universe designed with them in mind.

With the exception of the minority who insist on literal interpretation of scripture, religious scholars have largely deferred to science on those matters where the scientific consensus has spoken. Scholarly theologians are quite adept at reinterpreting the teachings of their faiths in the light of new knowledge. Nothing is wrong with this. Most academic scientists and theologians agree that both groups are in the business of learning, not preaching. Theologians argue, with some merit, that religion still has a role to play in moral matters and in the search to find the place of humanity in the scheme of things. Most scientists regard questions about the purpose of the universe to be beyond the scope of science and generally ignore such discussions. But when religious believers read the press reports about science and religion converging, many take heart that, when everything is said and done, the sacred purpose they yearn for will be confirmed by science.

THE SUPPOSED SIGNAL OF PURPOSE

For about a decade now, an ever-increasing number of scientists and theologians have been asserting, in popular articles and books, that they can detect a signal of cosmic purpose poking its head out of the noisy data of physics and cosmology. This claim has been widely reported in the media (see, for example, Begley 1998, Easterbrook 1998), misleading lay people into thinking that some kind of new scientific consensus is developing in support of supernatural beliefs. In fact, none of this reported evidence can be found on the pages of reputable scientific journals, which continue to successfully operate within an assumed framework in which all physical phenomena are natural.

The purported signal of cosmic purpose cannot be demonstrated from the data alone. Such observations require considerable interpretation to arrive at that conclusion. Those scientists who are generally not very familiar with recent deliberations in the philosophy of science might be inclined to scoff and say that the observations speak for themselves, with no interpretation necessary. Facts are facts, and neither God nor purpose are scientific facts.

However, as we will see in the following chapter, philosophers of science have been unable to define a clear line of demarcation between observation and theory. Most now agree that all scientific observations are "theory-laden." That is, empirical results cannot be cleanly separated from the theoretical framework used to classify and interpret them. This relatively recent development in philosophy, although based on ancient thinking (see chapter 1), has opened the door for a minority of religious scientists, theologians, and philosophers to reinterpret scientific data in terms of their preferred model of intelligent design and divine purpose to the universe. Some claim the data fit this model better than alternatives. Most say it is at least as good (Ross 1995; Swinburne 1990; Moreland 1998; Dembski 1998, 1999).

The data whose interpretation is being debated in the religion-science dialogue are not scraps of fading documents, nor the uncertain translations of ancient fables that over time have evolved into sacred texts. Rather, they consist of measurements made by sophisticated research teams using advanced scientific instruments. The new theistic argument is based on the fact that earthly life is so sensitive to the specific values of the fundamental physical constants and properties of its environment that even the tiniest changes to any of these would mean that life as we see it around us would not exist. This is said to reveal a universe that is exquisitely fine-tuned and delicately balanced for the production of life.

As the argument goes, the chance that any initially random set of constants would correspond to the set of values they happen to have in our universe is very small; thus, this precise balancing act is exceedingly unlikely to be the result of mindless chance. Rather, an intelligent, purposeful, and indeed personal Creator probably made things the way they are. The argument is well captured by a cartoon in Roger Penrose's best-seller, *The Emperor's New Mind*, which shows the Creator pointing a finger toward an "absurdly tiny volume in the phase space of possible universes" to produce the universe in which we live (Penrose 1989, 343).

Most who make the fine-tuning argument are content to say that intelligent, purposeful, supernatural design has become an equally viable alternative to the random, purposeless, natural evolution of the universe and humankind suggested by conventional science. However, a few theists have gone much further to insist that God is now *required* by the data. Moreover, this God must be the God of the Christian Bible. No way, they say, can the universe be the product of purely natural, impersonal processes.

A typical representation of this view can be found in *The Creator and the Cosmos: How the Greatest Scientific Discoveries of the Century Reveal God*, by physicist and astronomer Hugh Ross. Ross cannot imagine fine-tuning happening any other way than by a "personal Entity . . . at least a hundred trillion times more 'capable' than are we human beings with all our resources." He concludes that "the Entity who brought the universe into existence must be a Personal Being, for only a person can design with anywhere near this degree of precision" (Ross 1995, 118).

The apparent delicate connections among certain physical constants, and between those constants and life, are collectively called the **anthropic coincidences**. Before examining the merits of the interpretation of these coincidences as evidence for intelligent design, I will review how the notion first came about. For a detailed history and a wide-ranging discussion of all the issues, see *The Anthropic Cosmological Principle* by John D. Barrow and Frank J. Tipler (Barrow 1986). I also refer the reader there for the original references. But be forewarned that this exhaustive tome has many errors, especially in equations, some of which remain uncorrected in later editions, and many of the authors' conclusions are highly debatable.

THE ANTHROPIC COINCIDENCES

In 1919, Hermann Weyl expressed his puzzlement that the ratio of the electromagnetic force to the gravitational force between two electrons is such a huge number, $N_1 = 10^{39}$ (Weyl 1919). This means that the strength of the electromagnetic force is greater than the strength of the gravitational force by 39 orders of magnitude. Weyl wondered why this should be the case, expressing his intuition that "pure" numbers like π occurring in the description of physical properties should most naturally occur within a few orders of magnitude of unity. Unity, or zero, you can expect "naturally." But why 10^{39}? Why not 10^{57} or 10^{-123}? Some principle must select out 10^{39}.

In 1923, Arthur Eddington commented: "It is difficult to account for the occurrence of a pure number (of order greatly different from unity) in the scheme of things; but this difficulty would be removed if we could connect it to the number of particles in the world—a number presumably decided by accident (Eddington 1923, 167). He estimated that number, now called the "Eddington number," to be $N = 10^{79}$. Well, N is not too far from the square of N_1. This was the first of the anthropic coincidences.

These musings may bring to mind the measurements made on the Great Pyramid of Egypt in 1864 by Scotland's Astronomer-Royal, Piazzi Smyth. He found accurate estimates of π, the distance from the earth to the sun, and other strange "coincidences" buried in his measurements (Smyth 1978). However, we now know that these were simply the result of Smyth's selective toying with the numbers (Steibing 1994, 108–10; De Jager, 1992). Still, even today, some people believe that the pyramids hold secrets about the universe. Ideas like this never seem to die, no matter how deep in the Egyptian sand they may be buried.

Look around at enough numbers and you are bound to find some that appear connected. Most physicists, therefore, did not regard the large numbers puzzle seriously until one of their most brilliant members, Paul Dirac, took an interest. Few of this group ignored anything Dirac had to say.

Dirac observed that N_1 is the same order of magnitude as another pure number N_2 that gives the ratio of a typical stellar lifetime to the time for light to traverse the radius of a proton (Dirac 1937). If one natural number being large is unlikely, how much more unlikely is another to come along with about the same value?

In 1961, Robert Dicke pointed out that N_2 is necessarily large in order that the lifetime of typical stars is sufficient to generate heavy chemical elements such as carbon. Furthermore, he showed that N_1 must be of the same order of N_2 in any universe with elements heavier than lithium, the third element in the chemical periodic table (Dicke 1961).

The heavy elements did not get fabricated straightforwardly. According to the big bang theory (despite what you may hear, a strong consensus of cosmologists now regard the big bang as very well established), only hydrogen, deuterium (the isotope of hydrogen consisting of one proton and one neutron), helium, and lithium were formed in the early universe. Carbon, nitrogen, oxygen, iron, and the other elements of the periodic table were not produced until billions of years later. These billions of years were needed for stars to form and assemble these heavier elements out of neutrons and protons. When the more massive stars expended their hydrogen fuel, they exploded as supernovae, spraying the manufactured elements into space. Once in space, these elements cooled and accumulated into planets.

Billions of additional years were needed for our home star, the sun, to provide a stable output of energy so that the earth could develop life. Had the gravitational attraction between protons in stars not been many orders of magnitude weaker than the electric repulsion, as represented by the very large value of N_1, stars would have collapsed and burned out long before nuclear processes could build up the periodic table from the original hydrogen and helium. The formation of chemical complexity is only possible in a universe of great age—or at least in a universe with other parameters close to the values they have in this one.

Great age is not all. The element-synthesizing processes in stars depend sensitively on the properties and abundances of deuterium and helium produced in the early universe. Deuterium would not exist if the difference between the masses of the neutron and proton were just slightly displaced from its actual value. The relative abundances of hydrogen and helium also depend strongly on this parameter. They also require a delicate balance of the relative strengths of gravity and the weak interaction, the interaction responsible for nuclear beta decay. A slightly stronger weak force and the universe would be 100 percent hydrogen. In that case, all the neutrons in the early universe will have decayed, leaving none around to be saved in helium nuclei for later use in the element-building processes in stars. A slightly weaker weak force and few neutrons would have decayed, leaving about the same numbers of protons and neutrons. In that case, all the protons and neutrons would have been bound up in helium nuclei, with two protons and two neutrons in each. This would have lead to a universe that was 100 percent helium, with no hydrogen to fuel the fusion processes in stars. Neither of these extremes would have allowed for the existence of stars and life as we know it, which is based on carbon chemistry.

The electron also enters into the balancing act needed to produce the heavier elements. Because the mass of the electron is less than the neutron-proton mass difference, a free neutron can decay into a proton, electron, and antineutrino. If this were not the case, the neutron would be stable and most of the protons and electrons in the early universe would have combined to form neutrons, leaving little hydrogen to act as the main component and fuel of stars. It is also essential that the neutron be heavier than the proton, but not so much heavier that neutrons cannot be bound in nuclei, where conservation of energy prevents the neutrons from decaying.

CARBON TUNING

In 1952, astronomer Fred Hoyle used anthropic arguments to predict that the carbon nucleus has an excited energy level at around 7.7 MeV. I have already noted that a delicate balance of physical constants was necessary for carbon and other chemical elements beyond lithium in the periodic table to be cooked in stars. Hoyle looked closely at the nuclear mechanisms involved and found that they appeared to be inadequate.

The basic mechanism for the manufacture of carbon is the fusion of three helium nuclei into a single carbon nucleus:

$$3He^4 \rightarrow C^{12}$$

(The superscripts give the number of nucleons, that is, protons and neutrons in each nucleus, which is indicated by its chemical symbol; the total number of nucleons is conserved). However, the probability of three bodies coming together simultaneously is very low and some catalytic process in which only two bodies interact at a time must be assisting. An intermediate process had earlier been suggested in which two helium nuclei first fuse into a beryllium nucleus which then interacts with the third helium nucleus to give the desired carbon nucleus:

$$2He^4 \rightarrow Be^8$$
$$He^4 + Be^8 \rightarrow C^{12}$$

Hoyle showed that this still was not sufficient unless the carbon nucleus had an excited state at 7.7 MeV to provide for a high reaction probability. A laboratory experiment was undertaken, and, sure enough, a previously unknown excited state of carbon was found at 7.66 MeV (Dunbar 1953, Hoyle 1953).

Nothing can gain you more respect in science than the successful prediction of a new, unexpected phenomenon. Here, Hoyle used standard nuclear theory. But his reasoning contained another element whose significance is still hotly debated. Without the 7.7 MeV nuclear state of carbon, our form of life based on carbon would not have existed. Yet nothing in fundamental nuclear theory, as it is known even today, directly determines the existence of this state. It cannot be deduced from the axioms of the theory.

More recently, Tesla E. Jeltema (a physicist whose first name is Tesla!) and Marc Sher (1999) have shown that a key parameter of the standard model of elementary particles and forces, called the *Higgs vacuum expectation value*, cannot be less than 90 percent of its measured value for the 7.7 MeV carbon resonance to occur. V. Agrawal et al. (1998) earlier showed that this parameter cannot be more than five times its current measured value or else complex nuclei would be unstable. So, no doubt, the constants of nature do seem to be finely tuned for the production of carbon.

THE COSMOLOGICAL CONSTANT PROBLEM

Nowhere does the fine tuning of physics seem more remarkable than in the case of the cosmological constant of Einstein's general theory of relativity, which was introduced in chapter 3.

The cosmological constant has been taken to be zero for most of the century—for the simple and adequate reason that this value was consistent with the data. However, this parameter resurfaced around 1980 in the inflationary big bang model. Furthermore, recent observations indicate the universe may in fact be accelerating, which may indicate that the universe possesses some residual value of the cosmological constant.

A nonzero cosmological constant is equivalent to an energy density in a vacuum otherwise empty of matter or radiation.[1] Quantum fluctuations will also result in a vacuum energy density, and so the total energy density of the vacuum is the sum of the two. Weinberg pointed out, from simple dimensional arguments, that the standard model implies a quantum energy density of the order of 10^8 GeV^4 (expressed in units where $\hbar = c = 1$). Observations, on the other hand, indicate that the total vacuum energy density is of the order of 10^{-48} GeV^4 or less. To obtain this, the value of the cosmological constant had to be "fine-tuned" to some fifty-six orders of magnitude to cancel the quantum fluctuations! If this had not been the case, the universe would look vastly different than it does now and, no doubt, life as we know it would not exist.

However, just as I am making my own fine-tuned adjustments to this book prior to publication, a plausible solution to the cosmological constant problem is surfacing. Recent observations indicate that the expansion of the universe is accelerating. In addition to the yet-unidentified dark matter, the universe seems to be pervaded by another unidentified component called dark energy that has negative pressure and seems to be starting the universe on a new phase of inflation. This is much slower than the inflation that took place in the early universe, so it is nothing for humans to worry about.

The dark energy seems to currently dominate over all the other known energy and matter components of the universe, including the dark matter that is already known to exceed, by a factor of ten or so, the more familiar normal matter in galaxies and the cosmic microwave background.

At first, dark energy was identified with the cosmological constant. However, this interpretation still suffers from the fine-tuning problem described above. Additionally, another fine-tuning problem is implied. The various forms of matter and energy evolve very differently with time, so we need to explain why they just happen to be about the same order of magnitude in our epoch. The amount of fine tuning required here is even greater than the one above, some 120 orders of magnitude in the early universe.

However, the problem seems to have gotten better after briefly appearing worse. Models have now been developed in which the energy density usually represented by the cosmological constant is contained instead in a slowly evolving scalar field, something like the Higgs field, called **quintessence** (Aristotle's term for what came to be called aether, but not to be confused with this). These models do not require fine tuning in the early universe and are consistent with expectations from particle physics. In addition, symmetry arguments have been made that the cosmological constant is identically zero, in which case no fine tuning was ever necessary.

Quintessence, rather than a cosmological constant, could account for the original inflationary epoch thirteen billion years ago, when it comprised the dominant energy. Then as matter and radiation appeared, they came to dominate—first radiation, then matter. Structure formed during the matter-dominated epoch. Eventually, because of its different evolutionary behavior, quintessence may have once again taken over for the next round of inflation. I will explain inflation in more detail below.

I think a valuable lesson can be learned here. Theists need to stop using the "god of the gaps" argument, in which those phenomena that scientists cannot explain at a given point are taken as evidence for supernatural design. Science has a way of filling its gaps with purely natural explanations, and this seems to be occurring yet again.

THE ANTHROPIC PRINCIPLES

In 1974, Brandon Carter (1974) introduced the notion of the **anthropic principle**, hypothesizing that the anthropic coincidences are not the result of chance but somehow built into the structure of the universe. Barrow and

Tipler (1986) have identified three different forms of the anthropic principle, defined as follows:

Weak Anthropic Principle (WAP):

The observed values of all physical and cosmological quantities are not equally probable but take on values restricted by the requirement that there exist sites where carbon-based life can evolve and by the requirement that the Universe be old enough for it to have already done so. (16)

The WAP has not impressed too many people. All it seems to say is that if the universe was not the way it is, we would not be here talking about it. If the fine structure constant were not $1/137$, it would be something else and people would look different. If I did not live at 301 Avenue A, Bayonne, New Jersey, when I was a child, I would have lived at some other address and my life would have been different.

The WAP is sufficient to "explain" the Hoyle prediction. If the carbon nucleus did not have an excited state at 7.7 MeV, it would have been somewhere else and the abundance of chemical elements would have been different.

Most physicists leave it at that. However, others have stretched the imaginations mightily to read much greater significance into the anthropic coincidences. Barrow and Tipler define Carter's principle as follows:

Strong Anthropic Principle (SAP):

The Universe must have those properties which allow life to develop within it at some stage in its history. (21)

This suggests that the coincidences are not accidental but the result of a law of nature. But it is a strange law indeed, unlike any other in physics. It suggests that life exists as some Aristotelian "final cause."

Barrow and Tipler claim that this can have three interpretations (22):

(A) *There exists one possible Universe "designed" with the goal of generating and sustaining "observers."*

This is the interpretation adopted by theists as a new argument from design.

(B) *Observers are necessary to bring the Universe into being.*

This is traditional solipsism, but also is a part of today's New Age mysticism.

(C) *An ensemble of other different universes is necessary for the existence of our Universe.*

This speculation is part of contemporary cosmological thinking, as I will elaborate below. It represents the idea that the coincidences are accidental. We just happen to live in the particular universe that was suited for us.

The current dialogue has focussed on the choice between (A) and (C), with (B) not taken seriously in the scientific and theological communities. However, before discussing the relative merits of the three choices, let me complete the story on the various forms of the anthropic principle discussed by Barrow and Tipler. They identify two other versions:

Final Anthropic Principle (FAP):

Intelligent, information-processing must come into evidence in the Universe, and, once it comes into existence, it will never die out. (23)

This is sometimes also referred to as the *Completely Ridiculous Anthropic Principle*.

In their book, Barrow and Tipler speculated only briefly about the implications of the FAP. Tipler later propounded its consequences in a controversial book with the provocative title: *The Physics of Immortality: Modern Cosmology, God, and the Resurrection of the Dead* (Tipler 1994). Here Tipler carries the implications of the FAP about as far as one can imagine they could go. He adapts the fantasy of Teillard de Chardin, suggesting that we will all live again as emulations in the cyber mind of the "Omega Point God" who will ultimately evolve from today's computers. I have previously reviewed this book (Stenger 1995a, b) and so will not repeat that here, except to say that recent observations in cosmology make Tipler's scenario, which requires a closed universe, unlikely. Indeed, since Tipler made the prediction, from his theory, that the universe is closed, one might regard his theory as falsified by these developments.

INTERPRETING THE COINCIDENCES: (A) THEY ARE DESIGNED

Let us now review the first of the three possible explanations for the anthropic coincidences listed by Barrow and Tipler: (A) *"There exists one possible Universe 'designed' with the goal of generating and sustaining 'observers.' "*

Many theists see the anthropic coincidences as evidence for purposeful design to the universe. They ask: how can the universe possibly have obtained the unique set of physical constants it has, so exquisitely fine-tuned for life as they are, except by purposeful design—with life and perhaps humanity in mind? (See, for example, Swinburne 1990, Ellis 1993, Ross 1995).

Edward Harrison has stated it this way:

Here is the cosmological proof of the existence of God—the design argument of Paley—updated and refurbished. The fine tuning of the universe provides prima facie evidence of deistic design. Take your choice: blind chance that requires multitudes of universes or design that requires only one. (Harrison 1985, 252)

The fine tuning argument is a probabilistic one. The claim is that the probability for anything but external, intelligent design is vanishingly small. However, based on the data, the number of observed universes $N_o = 1$. The number of observed universes with life $N_L = 1$. Thus, the probability that any universe has life $= N_L / N_o = 1$, that is, 100 percent! Admittedly, the statistical error is large. The point is that data alone cannot be used to specify whether life is likely or unlikely. We just do not have a large sample of universes to study, and no probability argument can be made that rests on a single sample. It can only rest on theory, and, as we will see, physical and cosmological theories do not require design.

Let us examine the implicit assumptions here. First and foremost, and fatal to the design argument all by itself, we have the wholly unwarranted assumption that only *one type of life is possible*—the particular form of carbon-based life we have here on earth. Ross (1995, 133) typifies this narrow perspective on the nature of life:

As physicist Robert Dicke observed thirty-two years ago, if you want physicists (or any other life forms), you must have carbon. Boron and silicon are the only other elements on which complex molecules can be based, but boron is extremely rare, and silicon can hold together no more than about a hundred amino acids. Given the constraints of physics and chemistry, we can reasonable assume that life must be carbon based.

Carbon would seem to be the chemical element best suited to act as the building block for the type of complex molecular systems that develop life-like qualities. However, other possibilities than amino acid chemistry and DNA cannot be ruled out. Given the known laws of physics and chemistry, we can imagine life based on silicon or other elements chemically similar to carbon. Computer chips, after all, are made of silicon, and these operate a billion times faster than carbon-based biological systems. However, all elements heavier than lithium still require cooking in stars and thus a universe old enough for star evolution. The $N_1 = N_2$ coincidence would still hold in this case.

Only hydrogen, helium, and lithium were synthesized in the early big bang. These are probably chemically too simple to be assembled into diverse structures. So, it seems that any life based on chemistry would require an old universe, with long-lived stars producing the needed materials.

Still, we have no basis for ruling out other forms of matter than molecules in the universe as building blocks of complex systems. While atomic nuclei, for example, do not exhibit the diversity and complexity seen in the way atoms assemble into molecular structures, perhaps they might be able to do so in a universe with somewhat different properties. This is only speculation, and I am not claiming to have a theory of such systems. I merely point out that no known theory says that such life forms are impossible. The burden of proof is on those who claim it is.

Sufficient complexity and long life may be the only ingredients needed for a universe to have *some* form of life. Those who argue that life is highly improbable must accept the possibility that life might be likely with many different configurations of laws and constants of physics. We simply do not know enough to rule that out. Furthermore, nothing in anthropic reasoning indicates any special preference for human life, or indeed intelligent or sentient life of any sort. If anything, the current data indicate an inordinate preference for bacteria.

The development of intelligent life did not proceed smoothly and elegantly from the fundamental constants in the way that the term "fine tuning" seems to imply. Several billion years elapsed before the conditions for intelligent life came together, and the process of fashioning these conditions was accompanied by a staggering degree of waste (all that space, dust, and seemingly dead cosmic bodies). By human standards, it seems remarkably inefficient. Also, in the case of human life, it appears that (among other things) the earth would have suffered frequent catastrophic collisions with comets had it not been for the gravitational effect of Jupiter. This hardly seems consistent with divine creation. Setting in motion a myriad of threatening comets and then positioning a huge planet as a protection against the danger you have thus created seems like the work of a cosmic jerry-builder. (For more about the contingency of life on Earth, see Taylor, 1998.)

Even before we examine the other possibilities in detail, we can see another fatal fallacy in the fine-tuning argument. It is a probability argument that rests on a misconception of the concept of probability. Suppose we were to begin with an ensemble of universes in which the physical constants for each vary over a wide range of possible values. Then the probability that one universe selected randomly from that set would be our universe is admittedly very small. The fine-tuning argument then concludes that our specific universe was deliberately selected from the set by some external agent, namely, God.

However, a simple example shows that this conclusion does not logically follow. Suppose that a lottery is conducted in which each entrant is assigned a number from one to one million. Each has invested a dollar and the winner gets the whole pot of $1 million. The number is selected and you are the lucky winner! Now it is possible that the whole thing was fixed and your mother chose the winning number. But absent any evidence for this, no one has the right to make that accusation. Yet that's what the fine-tuning argument amounts to. Without any evidence, God is accused of fixing the lottery.

Somebody had to win the lottery, and you lucked out. Similarly, if a universe was going to happen, some set of physical constants was going to be selected. The physical constants, randomly selected, could have been the ones we have. And they led to the form of life we have.

In another example, estimate the probability that the particular sperm and egg that formed you would unite—that your parents, grandparents, and all your ancestors down to the primeval stew that formed the first living things would come together in the right combination. Would that infinitesimally small number be the probability that you exist? Of course not. You exist with 100 percent probability.

Michael Ikeda and Bill Jefferys (1997) have done a formal probability theory analysis that demonstrates these logical flaws and others in the fine tuning argument. They have also noted an amusing inconsistency: On the one hand you have the creationists and god-of-the-gaps evolutionists who argue that nature is too *uncongenial* for life to have developed totally naturally, and so therefore supernatural input must have occurred. Then you have the fine-tuners (often the same people) arguing that the constants and laws of nature are exquisitely *congenial* to life, and so therefore they must have been supernaturally created. (For further discussion of probability and the fine-tuning argument, see Le Poidevin 1996 and Parsons 1998.)

The fine-tuning argument rests on the assumption that *any* form of life is possible only for a very narrow, improbable range of physical parameters. This assumption not justified. None of this proves that option (A) cannot be the source of the anthropic coincidences. But it does show that the arguments used to support that option are very weak and certainly insufficient to rule out of hand all alternatives. If all those alternatives are to fall, making (A) the choice by default, then they will have to fall of their own weight.

INTERPRETING THE COINCIDENCES: (B) THEY ARE ALL IN THE HEAD

Let us look next at the second of the explanations for the anthropic coincidences listed by Barrow and Tipler: (B) *"Observers are necessary to bring the Universe into being."*

As the philosophers George Berkeley and David Hume realized, the possibility that reality is all in the mind cannot be disproved. However, any philosophy based on this notion is wrought with problems, not the least of which is: why, then, is the universe not the way each of us wants it to be? Furthermore, whose mind is the one that is doing the imagining? Berkeley decided it had to be the mind of God, which makes this interpretation of the anthropic coincidences indistinguishable from the previous one. However, another possibility that is more in tune with Eastern religion than Western is that we are all part of a single cosmic mind.

This idea has become very popular in the New Age movement. Triggered by the publication of *The Tao of Physics* by physicist Fritjof Capra (1975), a whole industry has developed in which the so-called mysteries and paradoxes of quantum mechanics are used to justify the notion that our thoughts control reality. Perhaps the most successful practitioner of this philosophy is Deepak Chopra, who has done very well promoting what he calls "quantum healing" (Chopra 1989, 1993).

Option (B) is certainly not taken seriously in the current science-religion dialogues. However, let me include a brief discussion for the sake of completeness (see Stenger 1995b for more details).

Basically, the new ideas on cosmic mind and the quantum begin with the confusing interpretive language used by some of the founders of quantum mechanics, most particularly Bohr. As we have seen previously, the Copenhagen interpretation seems to imply that a physical body does not obtain a property, such as position in space, until that property is observed. Although quantum mechanics has continued to agree with all measurements to very high precision, the Copenhagen interpretation has been further interpreted, indeed misinterpreted, to mean that reality is all in our heads.

Moreover, according to the idea of "quantum consciousness," our minds are all tuned in holistically to all the minds of the universe, with each individual forming part of the cosmic mind of God. As applied to the anthropic coincidences, the constants of physics are what they are because the cosmic mind wills them so.

Today, few quantum physicists take the notion of a cosmic quantum mind seriously. The success of quantum mechanics does not depend in any way on the Copenhagen interpretation or its more mystical spinoffs. As we saw in chapter 6, other interpretations exist, like Bohm's hidden variables (Bohm 1952, 1993), the many worlds interpretation (Everett 1957, Deutsch 1997), and the consistent histories interpretation (Griffiths 1984, Omnès 1994). Unfortunately, no consensus interpretation of quantum mechanics exists among physicists and philosophers.

I have argued in this book that time-symmetry provides a nonmystical alternative explanation for several of the so-called quantum paradoxes. However, the acceptance of this proposal is not necessary to eliminate option B. Suffice it to say that the admittedly strange behavior of the quantum world is mysterious only because it is unfamiliar, and can be interpreted without the introduction of any mystical ideas, including cosmic mind. Thus, quantum mechanics cannot be used to support the notion that human consciousness has created a universe suited for itself.

INTERPRETING THE COINCIDENCES: (C) THEY ARE NATURAL

Finally, let me move to the possibility that we can understand the anthropic coincidences naturally. I have deliberately discussed the other options first in order to make it clear that, by themselves, they are highly flawed and provide us little reason to accept their premises. I might stop here and claim the natural explanation wins by default. This can be somewhat justified on the principle of parsimony. Since all scientific explanations until now have been natural, then it would seem that the best bet is a natural explanation for the anthropic coincidences. Such an explanation would probably require the fewest in the way of extraordinary hypotheses—such as the existence of a spirit world either inside or outside the physical universe.

As we have seen, the standard model of elementary particles and fields has, for the first time in history, given us a theory that is consistent with all experiments conducted as of this writing. More than that, in developing the standard model physicists have gained significant new insights into the nature of the so-called laws of nature.

Prior to these recent developments, the physicist's conception of the laws of nature was pretty much that of most lay people: they were assumed to be rules for the behavior of matter and energy that are part of the very structure of the universe, laid out at the creation. However, in the past several decades we have gradually come to understand that what we call "laws of physics" are basically our own descriptions of certain symmetries observed in nature and how these symmetries, in some cases, happen to be broken. And, as we saw in chapter 12, the particular laws we have found do not require an agent to bring them into being. In fact, they are exactly what is implied by the absence of an agent. Let us look further at some cosmological issues, in particular the notion that multiple universes might exist.

IN THE BEGINNING

For almost two decades, the inflationary big bang has been the standard model of cosmology (Kazanas 1980, Guth 1981, 1997; Linde 1987, 1990, 1994). While the popular media occasionally report that the big bang is in trouble and the inflationary model is dead, no viable substitute has been proposed that has anywhere near the equivalent explanatory power of these complementary theories. Furthermore, new observations continue to provide support for the inflationary big bang.

The inflationary big bang offers a plausible, natural scenario for the uncaused origin and evolution of the universe, including the formation of order and structure—without the violation of any known laws of physics (see Stenger 1990b for a not-too-technical discussion).

Prior to the introduction of the inflationary model, Tryon (1973) had pointed out that the universe could have begun as a quantum fluctuation. The inflationary scenario that was later developed went through several versions, but the *chaotic inflation* of Andre Linde is the simplest and most plausible.

In chaotic inflation, tiny bubbles of energy spontaneously appear out of "nothing," that is, a void empty of matter or radiation. Physicists can still describe such a void in terms of general relativity. It is completely flat geometrically, with space and time axes that run from minus infinity to plus infinity. Anything else and matter, radiation, or space-time curvature would have to exist and this universe would no longer be a void.

In the absence of matter and radiation, Einstein's equations of general relativity yield the *de Sitter solution*, which simply expresses the curvature of space as proportional to the cosmological constant. When the universe is flat, this term is zero and the equation then reads: $0 = 0$. This denotes the void.

This is the way things would have stayed were it not for quantum mechanics, which we can also apply to an empty void. The uncertainty principle allows for the spontaneous, uncaused appearance of energy in a small region of space without violating energy conservation.

The fluctuation energy appears as a "bubble of **false vacuum**." This bubble still contains no matter or radiation, but is no longer a "**true vacuum**." The size of the bubble is of the order of the Planck length, 10^{-35} meter, and the time interval of the fluctuation is the order of the Planck time, 10^{-43} second. The energy is of the order of the Planck energy, 10^{28} eV, and the equivalent mass is the Planck mass, 2×10^{-5} gram. Indeed, this process can be said to define the fundamental scale of the quantities, since no other basis for setting a scale existed at that time.

The original "nothing" was like a rock poised precariously on the top of a steep mountain. The quantum fluctuation was then like a tiny, random breeze that nudged the rock over the side, in a random direction, where it fell into a deep valley of lower potential energy. At the bottom of the valley, the energy gained during the drop was eventually dissipated in the production of the quarks and leptons, or perhaps their still-unknown more elementary progenitors, that proceeded to form our universe. Other bubbles may have formed other universes.

While I am giving a nonmathematical explanation, and I will give more motivation below, I should at least mention at this point that the inflation scenario I am describing was in fact suggested by the standard model of elementary particles and forces. The unstable false vacuum corresponds to the underlying symmetric state of the standard model equations, while the lower energy true vacuum corresponds to the broken symmetric state that contains the Higgs field. Also, while I have been describing this process in terms of the usual field language, we can still maintain the ontological view that the Higgs field is discrete and composed of quanta, the Higgs particles of the standard model. Finally, while these remain to be observed at this writing, their failure to be observed will not necessarily kill the inflationary scenario. Something equivalent would then likely take their place.

The exponential increase in the size of the bubble can be understood in the language of general relativity. In those terms, the energy density of the false vacuum warps the space inside the bubble. This warped space can be represented by a cosmological constant.

According to the de Sitter solution of Einstein's equations, the bubble expanded exponentially and the universe grew by many orders of magnitude in a tiny fraction of a second. The energy density was constant for this brief interval and as the volume of the bubble increased exponentially during that time, the energy contained within also increased exponentially. Although the first law of thermodynamics may seem to have been violated (a miracle!), it was not. The pressure of the de Sitter vacuum is negative and the bubble did work on itself as it expanded. By the time it inflated to the size of a proton, 10^{-15} meter in perhaps 10^{-42} second, the bubble contained sufficient energy to produce all the matter in the visible universe today.[2]

As mentioned earlier in this chapter, the latest ideas from cosmology suggest that inflation may be generated by the negative pressure of a scalar field called quintessence. This field is very much like the Higgs field and helps avoid fine-tuning puzzles. Quintessence would be the energy in the original quantum fluctuation and evolve differently than the positive pressure matter and radiation that is produced at the end of inflation. At that point radiation (photons) briefly dominate, then matter. During matter domination, the inhomogeneities that became galaxies appeared, but quintessence now seem to have taken over again producing a new, much slower, round of inflation.

As the first expansion phase, in the early universe, continued, some of the curvature or quintessence energy was converted into matter and radiation and inflation stopped, leading to the more linear big bang expansion we now experience. The universe cooled, and its structure spontaneously froze out, just as formless water vapor freezes into snowflakes whose unique patterns arise from a combination of symmetry and randomness. This mechanism for structure formation is precisely the **spontaneous symmetry breaking** that we saw in chapter 9, which was a key notion in the development of the standard model.

In our universe, the first galaxies began to assemble after about a billion years, eventually evolving into stable systems where stars could live out their lives and populate the interstellar medium with the complex chemical elements such as carbon needed for the formation of life based on chemistry.

So how did our universe happen to be so "fine-tuned" as to produce these wonderful, self-important carbon structures? As I explained above, we have no reason to assume that ours is the only possible form of life. Perhaps life of some sort would have happened whatever form the universe took—however the crystals on the arm of the snowflake happened to be arranged by chance.

At some point, according to this scenario, the symmetries of the initial true vacuum began to be spontaneously broken. Those of the current standard model of elementary particles and forces were among the last broken, when the universe was about a trillionth of a second old and much colder than earlier. The distances and energies involved at this point have been probed in existing colliding beam accelerators, which represents about the deepest into big-bang physics we have so far been able to explore in detail. Higher energy colliders will be necessary to push farther, but we are still far from directly probing the earliest time scales where the ultimate symmetry breakdown can be explored. Nevertheless, it may surprise the reader that the physical principles in place since a trillionth of a second after the universe began are now very well understood.

After about a millionth of a second, the early universe had gone through all the symmetry breaking required to produce the fundamental laws and constants we still observe today, thirteen to fifteen billion years later. Nuclei and atoms still needed more time to get organized, but after 300,000 years the lighter atoms had assembled and ceased to interact with the photons that went off on their own to become the cosmic microwave background. The observations of this background, which in the past decade have become exquisitely precise, have provided our most important probe of that era in the history of the universe.

The first galaxies began to assemble after about a billion years, evolving eventually into stable systems where stars could live out their lives and populate the interstellar medium with the heavier elements like carbon needed for the formation of life.

Regardless of the fact that we cannot explore the origin of the universe by any direct means, the undoubted success of the theory of broken symmetry, as manifested in the standard model of particle physics, provides us with a mechanism that we can apply, at least in broad terms, to provide a plausible scenario for the development natural law and physical structure within the universe.

As we saw in chapter 12, the conservation laws correspond to universal or "global" symmetries of space and time. The state of an undesigned universe at the earliest definable time would have possessed space translation, time translation, rotational, and all the other symmetries that result when a system depends on none of the corresponding coordinates. As an unplanned consequence, that universe would have automatically possessed the "laws" of conservation of energy, momentum, and linear momentum. It would also obey the "laws" of special relativity, in particular, Lorentz invariance.

The force laws that exist in the standard model are represented as spontaneously broken symmetries, that is, symmetries that are broken randomly. This situation is likened to a pencil balanced on its eraser end. It possesses rotational symmetry about a vertical axis. But the balance is unstable—it is not the state of lower energy. When the pencil falls over to reach its lowest energy state, the direction it now points to breaks the original symmetry and selects out a particular axis. This direction is random.

In another analogy, which shows directly how a force can be spontaneously produced in the process, consider what happens when a ferromagnet cools below a certain critical temperature called the *Curie point*. The iron undergoes a change of phase and a magnetic field suddenly appears that points in a specific, though random, direction, breaking the original symmetry. No direction was singled out ahead of time; none was determined by any known preexisting law.

The forces of nature are akin to the magnetic field of a ferromagnet. The "direction" they point to after symmetry breaking was not determined ahead of time. The nature of the forces themselves was not prespecified. They just happened to freeze out the way they did. Just as no agent is implied by the global symmetries, none is implied by the broken symmetries, which in fact look very much like the opposite. The possibility of such an agent is not eliminated by this argument, just made a less economical alternative.

In the natural scenario I have provided, the values of the constants of nature in question are not the only ones that can occur. A huge range of values are in fact possible, as are all the possible laws that can result from symmetry breaking. The constants and forces that we have were selected by accident—as the pencil fell—when the expanding universe cooled and the structure we see at the fundamental level froze out. Just as the force laws did not exist before symmetry breaking, so, too, did the constants not exist. They came along with the forces, all by chance. In the current theoretical scheme, particles also appear, with the forces, as the carriers of the quantities like mass and charge and indeed the forces themselves. They provided the means by which the broken symmetries materialize and manifest their structure.

WHAT ABOUT LIFE?

Someday we may have the opportunity to study different forms of life that evolved on other planets. Given the vastness of the universe, and the common observation of supernovas in other galaxies, we have no reason to assume life only exists on earth. Although it seems hardly likely that the evolution of DNA and other details were exactly replicated elsewhere, carbon and the other elements of our lifeform are well distributed, as evidenced by the composition of cosmic rays and the spectral analysis of interstellar gas.

We also cannot assume that life in our universe would have been impossible had the symmetries broken differently. We cannot speak of such things in the normal scientific mode of discourse, in which direct observations are described by theory. But, at the same time, it is not illegitimate or unscientific to examine the logical consequences of existing theories that are well-confirmed by data from our own universe.

The extrapolation of theories beyond their normal domains can turn out to be wildly wrong. But it can also turn out to be spectacularly correct, as when physics learned in earthbound laboratories is applied to send spacecraft to other planets. The fundamental physics learned on earth has proved to be valid at great distances from earth and at times long before the earth and solar system had been formed. Those who argue that science cannot talk about the early universe or about life on the early earth because no humans were there to witness these events greatly underestimate the power of scientific theory.

I have made a modest attempt to obtain some feeling for what a universe with different constants would be like. It happens that the physical properties of matter, from the dimensions of atoms to the length of the day and year, can be estimated from the values of just four fundamental constants. Two of these constants are the strengths of the electromagnetic and strong nuclear interactions. The other two are the masses of the electron and proton.

Of course, many more constants are needed to fill in the details of our universe. And our universe, as we have seen, might have had different physical laws. We have little idea what those laws might be; all we know are the laws we have. Still, varying the constants that go into our familiar equations will give many universes that do not look a bit like ours. The gross properties of our universe are determined by these four constants, and we can vary them to see what a universe might grossly look like with different values of these constants.

As an example, I have analyzed 100 universes in which the values of the four parameters were generated randomly from a range five orders of magnitude above to five orders of magnitude below their values in our universe, that is, over a total range of ten orders of magnitude. Over this range of parameter variation, N_1 is at least 10^{33} and N_2 at least 10^{20} in all cases. That is, both are still very large numbers. Although many pairs do not have $N_1 = N_2$, an approximate coincidence between these two quantities is not very rare (for more details, see Stenger 1995b, 1999c).

The distribution of stellar lifetimes for these same 100 universes has also been examined. While a few are low, most are probably high enough to allow time for stellar evolution and heavy element nucleosynthesis. Over half the universes have stars that live at least a billion years. Long life is not the only requirement for life, but it certainly is not an unusual property of universes.

Recall Barrow and Tipler's option (C), which held that an ensemble of other, different universes is necessary in any natural explanation for the existence of our universe. Another assertion that has appeared frequently in the literature (see, for example, Swinburne 1990) holds that only a multiple-universe scenario can explain the coincidences without a supernatural creator. No doubt this can do it, as we will see below. If many universes beside our own exist, then the anthropic coincidences are trivial. But even if there is only one universe, the likelihood of *some* form of life in that single universe is not provably small.

AN INFINITY OF UNIVERSES

Within the framework of established knowledge of physics and cosmology, our universe could be one of many in an infinite super universe or "multiverse." Linde's chaotic inflation, described above, provides a plausible scenario in which an endless number of universes form by quantum fluctuations in an unstable background vacuum that contains nothing, not even limits.

Each universe within the multiverse can have a different set of constants and physical laws. Some might have life of a different form than us, while others might have no life at all or something even more complex or so different that we cannot even imagine it. Obviously we are in one of those universes with life.

Several commentators have argued that a multiverse cosmology violates Occam's razor (see, typically, Ellis 1993, 97). This is disputable. The entities that Occam's rule of parsimony forbids us from "multiplying beyond necessity" are theoretical hypotheses, not universes. For example, although the atomic theory of matter greatly multiplied the number of bodies we must consider in solving a thermodynamic problem, by 10^{24} or so per gram, it did not violate Occam's razor. Instead, it provided for a simpler, more powerful, more economic exposition of the rules that were obeyed by thermodynamic systems, with fewer hypotheses.

As Max Tegmark (1997) has argued, a theory in which all possible universes exist is actually more parsimonious than one in which only one exists. That is, a single universe requires more explanation—additional hypotheses. Let me give a simple example that illustrates his point. Consider the two statements: (a) $y = x^2$ and (b) $4 = 2^2$. Which is simpler? the answer is (a), because it carries far more information with the same number of characters than the special case (b). Applied to multiple universes, a multiverse in which all possible universes exist is analogous to (a), while a single universe is analogous to (b).

The existence of many universes is in fact consistent with all we know about physics and cosmology. No new hypotheses are needed to introduce them. Indeed, it takes an added hypothesis to rule them out—a super law of nature that says only one universe can exist. That would be an uneconomical hypothesis! Another way to express this is with lines from T. H. White's *The Once and Future King*: "Everything not forbidden is compulsory."

An infinity of random universes is suggested by the inflationary big bang model of the early universe. As we have seen, a quantum fluctuation can produce a tiny, empty region of curved space that will exponentially expand, increasing its energy sufficiently in the process to produce energy equivalent to all the mass of a universe in a tiny fraction of second. Linde proposed that a background space-time "foam" empty of matter and radiation will experience local quantum fluctuations in curvature, forming many bubbles of false vacuum that individually inflate into miniuniverses with random characteristics (Linde 1987, 1990, 1994; Guth 1997). In this view, our universe is one of those expanding bubbles.

THE EVOLUTION OF UNIVERSES BY NATURAL SELECTION

Philosopher Quentin Smith (1990) and physicist Lee Smolin (1992, 1997) have independently suggested a mechanism for the evolution of universes by natural selection. They propose a multiuniverse scenario in which each universe is the residue of an exploding black hole that was previously formed in another universe. Since I discussed this in *The Unconscious Quantum* (Stenger 1995b, chapter 8), I will only summarize.

An individual universe is born with a certain set of physical parameters—its "genes." As it expands, new black holes are formed within. When these black holes eventually collapse, the genes of the parent universe get slightly scrambled by fluctuations that are expected in the state of high entropy inside a black hole. So when the descendant black hole explodes, it produces a new universe with a different set of physical parameters—similar but not exactly the same as its parent universe.

The black hole mechanism provides for both mutations and progeny. The rest is left to survival of the survivor. Universes with parameters near their "natural" values can easily be shown to produce a small number of black holes and so have few progeny to which to pass their genes. Many will not even inflate into material universes, but quickly collapse back on themselves. Others will continue to inflate, producing nothing. However, by chance some small fraction of universes will have parameters optimized for greater black hole production. These will quickly predominate as their genes get passed from generation to generation.

The evolution of universes by natural selection provides a mechanism for explaining the anthropic coincidences that may appear extreme, but Smolin suggests several tests. In one, he predicts that the fluctuations in the cosmic microwave background should be near the value expected if the energy fluctuation responsible for inflation in the early universe is just below the critical value for inflation to occur.

The idea of the evolution of universes is clearly akin to Darwin's theory of biological evolution. In both cases we are faced with explaining how unlikely, complex, nonequilibrium structures can form without invoking even less likely supernatural forces.

TEGMARK'S ENSEMBLES

Tegmark has recently proposed what he calls "the ultimate ensemble theory" in which all universes that mathematically exist also physically exist (Tegmark 1997). By "mathematical existence," Tegmark means "freedom from contradiction." So, universes cannot contain square circles, but anything that does not break a rule of logic exists in some universe. Note that "logic" is merely a human-invented process designed to guarantee that words are used as they have been defined and statements are not self-contradictory. So "freedom from contradiction" it should not be viewed as some property that must have been designed into the universe.

Tegmark claims his theory is scientifically legitimate since it is falsifiable, makes testable predictions, and is economical in the sense that I have already mentioned above—a theory of many universes contains fewer hypotheses than a theory of one. He finds that many mathematically possible universes will not be suitable for the development of what he calls "self-aware structures," his euphemism for intelligent life. For example, he argues that only a universe with three spatial and one time dimension can contain self-aware structures because other combinations are too simple, too unstable, or too unpredictable. Specifically, in order that the universe be predictable to its self-aware structures, only a single time dimension is deemed possible. In this case, one or two space dimensions is regarded as too simple, and four or more space dimensions is reckoned as too unstable. However, Tegmark admits that we may simply lack the imagination to consider universes too radically different from our own.

Tegmark examines the types of universes that would occur for different values of key parameters and concludes, as have others, that many combinations will lead to unlivable universes. However, the region of the parameter space where ordered structures can form is not the infinitesimal point only reachable by a skilled artisan, as asserted by proponents of the designer universe.

THE OTHER SIDE OF TIME

In *The Creator and the Cosmos*, Hugh Ross gives the following "proof of creation":

> The universe and everything in it is confined to a single, finite dimension of time. Time in that dimension proceeds only and always forward. The flow of time can never be reversed. Nor can it be stopped. Because it has a beginning and can move in only one direction, time is really just half a dimension. The proof of creation lies in the mathematical observation that any entity confined to such a half dimension of time must have a starting point of origination. That is, that entity must be created. This necessity for creation applies to the whole universe and ultimately everything in it. (Ross 1995, 80)

This "proof" is based on the Islamic *kalām cosmological argument*, which has been used in recent years by theistic philosopher William Lane Craig during his frequent debates on the existence of God and by many other theistic debaters who follow his lead. Craig states the argument as a simple syllogism (Craig 1979):

1. Whatever begins has a cause.
2. The universe began to exist.
3. Therefore, the universe has a cause.

Ross interprets that cause as the creation.

Craig gives the following justification for (1): "The first premiss [his spelling] is so intuitively obvious, especially when applied to the universe, that probably no one in his right mind *really* believes it to be false" (1979, 141). Somehow I do not find this very convincing. His debate opponent might reply: "The first premiss is so intuitively obviously wrong, especially when applied to the universe, that probably no one in his right mind *really* believes it to be true."

Note that Craig is not saying that everything must have a cause, which is a frequent misinterpretation. Only something with a *beginning* is supposed to require a cause. Craig uses the empirical evidence for the big bang to justify the second premise, and also an elaborate philosophical argument that in essence says that an infinite regress into the past cannot occur and so time must necessarily have a beginning.

The first premise has been disputed on the basis of the noncausal nature of quantum phenomena. As we have seen in earlier chapters, the spontaneous appearance of electron-positron pairs for brief periods of time, literally out of "nothing" is basic to the highly successful theory of quantum electrodynamics. Similar processes are built into the standard model. Thus, we have a counter example to statement (1), something that begins without cause. This and other arguments can be found, along with Craig's updated case in the 1993 Oxford book he coauthored with philosopher Quentin Smith, *Theism, Atheism, and Big Bang Cosmology* (Craig 1993). Smith also presents a good case against the second premise, and the two respond to each other in several go-rounds

Rather than elaborate further on this discussion, I would like to propose an objection to the second premise that I have not seen elsewhere, and so perhaps represents an original contribution to the dialogue. Previous responses to Craig, by Smith and others, have not disagreed with (2) per se, but questioned whether it even made any sense to talk about a cause before the existence of time, which is assumed to start with the beginning of the big bang. I propose that the beginning of the big bang was not the beginning of the universe, nor the beginning of time.

Ross also uses the kalām argument to counter the common atheist taunt: "Who created God?" He claims that God is not confined to a "half dimension" of time, and so need not have been created. I take this to mean that if I can demonstrate the universe is not necessarily confined to a half dimension, then Ross, Craig, and other theists who use the kalām argument will be forced to admit that the universe was not necessarily created. (Of course they won't.)

In chapter 4, I described a scenario in which the universe inflates in both time directions. I labeled as $t = 0$ the time at which the quantum fluctuation in the chaotic inflation model takes place. This is a random point on the time axis. The expansion then proceeds on the positive side of the t-axis, as defined by the increasing entropy on that side. The direction of time is by definition the direction in which the entropy of the universe increases.

As we saw, completely time-symmetric solution of Einstein's equations for the vacuum will give exponential inflation on both sides of the time axis, proceeding away from $t = 0$ where the initial quantum fluctuation was located. This implies the existence of another part of our universe (not a different universe), separated from our present part along the time axis. From our point of view, that part is in our deep past, exponentially *deflating* to the void prior to the quantum fluctuation that then grew to our current universe. However, from the point of view of an observer in the universe at that time, their future is into our past—the direction of increasing entropy on that side of the axis. They would experience a universe expanding into their future, just as we experience one expanding into our future.

Furthermore, as we have seen in this book, no absolute point in time exists in the equations of physics. Any point can be arbitrarily labelled t = 0. In fact, the most important law of physics of them all, conservation of energy, demands that there be no distinguishably special moment in time. Energy is the generator of time translation symmetry; when a symmetry is obeyed, its generator is conserved. This is why it so important to theologians that there be a unique t = 0. The existence of such a special point would imply a miracle, the violation of energy conservation, thus leaving room for God.

Fundamentally, then, the universe as a whole could very well be time-symmetric, running all the way from minus eternity to plus eternity with no preferred direction, no arrow of time. Indeed, the whole notion of beginning is meaningless in a time-symmetric universe. And, without a beginning, the kaläm cosmological argument for a creator fails because of the failure of step (2) in Craig's syllogism.

I have described a scenario for an infinite, eternal, and symmetric universe that had no beginning. The quantum fluctuation occurs at one particular spatial point in an infinite void. Obviously it could have happened elsewhere as well, which gives us the multiple universe scenario discussed above. While multiple universes are not required to deflate the kaläm argument, we have seen that they provide a scenario by which the so-called anthropic coincidences may have arisen naturally.

By showing that the universe did not necessarily have a beginning, we can counter another common theist line of argument used whenever the claim is made that a spontaneous "creation" violates no known physics. The theist will say, "Where did physics come from?" If their imagined God did not have to come from something, because she had no beginning, then neither did physics.

NOTES

1. Cosmologists make a somewhat artificial distinction between matter and radiation. Both are matter. Radiation is just matter, like photons, that moves at or near the speed of light. The energy contained by matter is mostly rest energy, while the energy contained by radiation is mostly kinetic energy.

2. The energy in the original fluctuation as was of the order of the Planck energy, 10^{28} eV. The energy density is this divided by the volume of a Planck sphere, about 10^{127} eV/cm^3. When the sphere expands to the volume of a proton, the energy contained is about 10^{88} eV. This is the equivalent to the rest mass of 10^{79} hydrogen atoms, the estimated number in the visible universe.

14 TRUTH OR CONSEQUENCES

I link, therefore I am.

S. J. Singer (as quoted in Wilson 1998)

SAMURAI SCIENCE

The remarkable success of the scientific enterprise, along with a general dissatisfaction with the absurd extremes toward which idealistic metaphysics was moving, led, in the early twentieth century, to the development of a philosophical school of thought called *logical positivism*. As had Comte and Mach before them (see chapter 1), the logical positivists viewed metaphysics as a useless exercise. They asserted that a proposition that was not empirically verifiable was meaningless. And propositions that are empirically verifiable are physics, not metaphysics.

The new positivists sought to save philosophy from further embarrassment and restore it to respectability as the discipline that clarifies and universalizes the results of others. They proposed to apply recently developed techniques in symbolic logic to the analysis of language and mathematics. They hoped to place empirical knowledge and scientific theory on a firm footing, and to establish rules for the validity of statements made in the sciences an elsewhere.

In *Tractatus Logico-Philosophicus*, Ludwig Wittgenstein attempted to provide a systematic theory of language that specified what can and what cannot be said in a proposition (Wittgenstein 1922). He noted that mathematics and logic are self-consistent but express nothing but tautologies. On the other hand, he thought that a set of discrete propositions could be established that truthfully describe the world. As we will see, he eventually changed his mind.

A variation on logical positivism, which is sometimes distinguished as *logical empiricism*, was developed by Karl Popper in *The Logic of Scientific Discovery* (Popper 1934). Popper believed in the superiority of scientific method and toiled mightily to establish exactly what it was. He agreed with Hume that induction is not a source of knowledge. Instead he emphasized the hypothetico-deductive method in which the scientist makes a set of hypotheses and then empirically tests statements that are logically deduced from these hypotheses.

Although he later divorced himself significantly from certain features of earlier positivism, in *Language, Truth, and Logic* (1936) Alfred Jules Ayer spelled out the basic logical positivist position: philosophers should cease concerning themselves with speculations about ultimate reality that can never be verified, but instead act as critics and analysts who clarify the language and symbols used in science, history, ethics, and theology. Synthetic (nontautological) propositions were deemed meaningful only if verifiable from experience or logically deduced as a consequence of other experiences. Metaphysics was neither analytic (tautological) nor verifiable.

The positivists viewed scientific theories as languages. Their program was intended to provide the rules for translation between the languages of rival theories to a generic observational language that could be used to test the theories against experiment. Rudolf Carnap tried to reduce all scientific disciplines to physics so their claims could be verified, as are physics claims, by direct empirical testing (Carnap 1936). He understood that a scientific theory can never be proven with complete certainty, even after repeated success of its predictions. Carnap agreed with Popper (1934), that theories can only be falsified by the failure of a prediction—never completely verified by any empirical or theoretical test.

Since we can never be sure we have all the data, we cannot conclusively demonstrate the truth of any hypothesis. Thus, according to Popper and Carnap, falsification is the only logical handle we have on the validity of a theory. Popper argued that a theory that is not falsifiable is not science, since it can never be tested. Like the samurai who must always have his sword handy in case he is called upon to disembowel himself, a theory must provide the seeds of its own destruction by telling you what observations would show itself to be wrong.

In this view, a wrong theory can still be legitimate science by virtue of the very fact that it is demonstrably wrong. On the other hand, a theory that explains everything is not science but pseudoscience. In the words of the physicist Wolfgang Pauli, such a theory "is not even wrong." Popper used the Marxist theory of history and Freudian psychology as examples of theories that are not scientific because they are not even wrong. That is, no observation could ever prove either false.

Nevertheless, falsifiability was at the same time insufficient and too general to act as the primary criterion for the determination of the validity of a scientific theory. While no doubt a powerful tool in eliminating theories that disagree with the data, falsifiability does not succeed as the primary criterion for deciding when a theory is "scientific." The prediction that the sun will come up tomorrow is falsifiable, but can be made without any knowledge of astronomy. Generally regarded nonscientific or pseudoscientific theories such as astrology, and most "alternative medicine" such as homeopathy and therapeutic touch, are quite falsifiable, and indeed have been falsified. The fact that something is falsifiable does not make it science.

On the other hand, many indisputably scientific theories are weakly falsifiable at best. Indeed, almost any theory you can think of is based on so many other prior theories and assumptions that the culprit responsible for a false prediction is often impossible to identify. Modifications can usually be made that bring a theory back into agreement with the data. When scientific theory is discarded after falsification, it is usually because a better, often simpler one has arisen to take its place.

PARADIGMS AND POSTPOSITIVISM

The logical positivist/empiricist program gradually grew out of fashion among philosophers, although most scientists today would still maintain that observation is the final arbiter of truth. Wittgenstein rejected his own earlier doctrines, which he originally thought provided a solid foundation for the logical positivist program. After years of reflection, he finally came to the conclusion that no precise "metalanguage" could be developed that allows for a completely self-consistent set of propositions about the external world (Wittgenstein 1953). Wittgenstein is often said to have decided that languages are games played according to different rules by diverse cultures and no common agreement on universal meanings can ever be extracted from them, but whether this was his intention remains debatable.

Williard Van Orman Quine, although a student of Carnap and always respectful of his mentor, agreed with the later Wittgenstein that the logical positivist program was doomed to failure. He disputed the notions that analytic and synthetic truths were distinguishable and that knowledge can be reduced to sense experience, calling these "the two dogmas of empiricism" (Quine 1951). He insisted that linguistic analysis will never provide for the unequivocal interpretation of the meaning of observational statements (Quine 1969).

Ayer thought Wittgenstein, Quine, and other critics of positivism overreacted. He maintained that the failure to formalize a procedure for confirming scientific statements should not cause us to discard the "scientific touchstone" of empirical confirmation (Ayer 1992, 302).

No one in recent times has been more influential in changing attitudes about science than Thomas Kuhn, who was mentioned in chapter 2. Kuhn's 1962 academic best-seller, *The Structure of Scientific Revolutions*, is perhaps most famous for introducing the term *paradigm* into the popular lexicon.

From an examination of the historical data, Kuhn concluded that science is basically a puzzle-solving activity. Scientific theories are not descriptions of the objective world so much as sets of shared beliefs, "paradigms," that serve as models or patterns for the solution of problems within a community of practitioners.

Normal science proceeds within the framework of the accepted paradigms of each field, Kuhn says, employing them to solve problems and further articulating their rules and procedures. As long as the paradigms work, scientists strongly resist any changes. However, historical moments occur when a paradigm fails and a new one must be invented to solve an outstanding problem. Change is then instigated by a small number of usually very young scientists who meet with resistance but eventually win over the consensus when the new paradigm proves successful. These are the scientific revolutions that lead to dramatic changes in thinking and whole new models and frameworks for problem solving that previously did not exist. Scientific development is thus portrayed as "a succession of tradition-bound periods punctuated by noncumulative breaks." Examples include the Copernican, Newtonian, Darwinian, and Einsteinian revolutions, but also lesser revolutions that may only be noticed within a discipline (Kuhn 1970, 208).

As I mentioned in chapter 2, Weinberg has recently disputed Kuhn's notion of scientific revolutions, arguing that the last great "mega-paradigm shift" occurred when Newton's physics replaced that of Aristotle (Weinberg 1998). Kuhn later allowed that *Structure* was more of an unstructured essay than a philosophical treatise and lamented that it was misunderstood by many of its critics. In a postscript to the second edition (Kuhn 1970, Horgan 1996), he admits that he used the word "paradigm" in at least two distinct ways and mentions that one sympathetic reader had counted twenty-two different usages. Kuhn suggested "disciplinary matrix" or "exemplar" as alternate expressions of his intent, a set of ordered elements that function together within a particular discipline.

No doubt science is to a great extent a social activity. Indeed, the classification of scientific knowledge has always been based on consensus. Kuhn claimed he was misinterpreted by those who drew the inference that the paradigms of science, being socially constructed, are thereby arbitrary. He denied he was a relativist, arguing that later scientific theories are generally better than older ones at solving puzzles and that this constitutes scientific progress.

However, Kuhn did indicate rather explicitly in *Structure of Scientific Revolutions* that, in his opinion at least at that time and not subsequently modified in print, scientific progress is primarily in the direction of better problem-solving ability rather then deeper ontological understanding. He agreed with developing trends in the philosophy of science that no theory-independent way can be found to determine, with any confidence, what is "really there." While not questioning the progress of physics from Aristotle to the present day, he still saw no evidence in this succession for progress in understanding reality, "no coherent direction for ontological development" (Kuhn 1970, 206).

According to Paul Hoyningen-Huene, Kuhn gradually refined his views on the ontological significance of science: "One theoretical consequence which now seems unavoidable stands in sharp contrast to Kuhn's original intentions. For we have moved substantially nearer to the form of realism generally taken for granted in empirical science and characteristic of Popperian philosophy. I will call this 'Peircean realism.' According to this form of realism, science captures reality, albeit neither absolutely nor incorrigibly, still to a fair and, over the years, ever-improving approximation" (Hoyningen-Huene 1993, 53).

But, whether or not Kuhn personally recanted, the damage caused by *Structures* was already done. Largely as the result of Kuhn's work, many social scientists and nonscientist academics have parted company (if they ever shared their company) with the majority of scientists in promoting a revolutionary idea: after thousands of years of philosophy and science, humanity knows no more about the "real" universe today than it did before Thales.

Kuhn's "socialization" of science, and the general inability of philosophers of science to agree among themselves on what constitutes science and scientific method, prompted philosopher Paul Feyerabend to pronounce his now-famous doctrine that "anything goes" (Feyerabend 1975, 1978). Feyerabend insisted that this is the only principle that applies in all circumstances. We should feel free to use any and all methods in coming to grips with the problems in all areas of human activity.

To Feyerabend, science is as dogmatic as any religion and we should demand separation of science and state just as fervently as we, in secular nations, demand separation of church and state, if humanity is to achieve its goals. According to Feyerabend, "We need a dream-world in order to discover the features of the real world we think we inhabit (and which may be another dream world)" (Feyerabend 1975, 1988, reprinted in Anderson 1995, 202). Unlike Kuhn, who still thought science had some value, Feyerabend espoused a methodological relativity where not only does scientific knowledge have no special claim to the truth, the methods of science are no better than any others.

Kuhn and Feyerabend agreed that no description of scientific observations is possible without making some theoretical assumptions. As philosophers now put it, all observations are "theory-laden." According to this doctrine, a sharp distinction between theory and experiment does not exist. This conclusion has been extrapolated by some to mean that since theory is language, and language is cultural with inter-translation not always possible, then all languages and all cultures have an equal claim on "scientific" truth.

As philosopher Larry Laudan puts it, Kuhn, Feyerabend and their followers have "turned scientific theory comparison into cultural anthropology." In their hands, rival theories have become "not merely different worldviews, but different worlds, different realities" (Laudan 1996, 9).

Laudan points out that the postpositivists and the positivists before them make the same over-arching assumption: that a choice between theories can only be made by translating between them. Thus, when it was shown that no such translation provided for a rational choice between theories because no metalanguage existed for making this choice, the postpositivists concluded that no choice was possible. This contributed to the strong epistemic relativism that we find today in what is called *postmodernism*. But, Laudan argues, rational choices can still be made in the absence of a metalanguage. Basically, this was also Ayer's position.

Rational criteria that need not be cast in terms of a rigorously self-consistent language such as symbolic logic or mathematics can still be applied to make choices. One obvious criterion is to simply choose what seems to work the best, arguing pragmatically about what is best rather than seeking out some flawless algorithm. Other criteria might include economy and elegance, and consistency with common sense. None can be "proved." None can be written out in a cookbook or programmed on a serial computer (I leave the door open here for massively parallel, or neural network, computers). But this does not mean we are being irrational in applying such criteria. Indeed, scientists do not follow a cookbook in doing science. They seldom apply precise algorithms in carrying out their tasks. They solve problems and build technologies by "rational" methods that are justified by their utility in solving the problems at hand.

The failure of linguistic analysis to define science or provide for a consistent set of rules for scientific decision making has had zero affect on the conduct of science, except indirectly as it has influenced lay perception, political support, and consequent funding. Natural scientists, at least, have carried on their work oblivious to the disputes among philosophers and sociologists.

Outside of the natural sciences, however, the message of the postpositivists has received a warm welcome. Part of the reason is the postpositivist assurance that lack of consensus is a strength, rather than a weakness. Social scientists no longer need feel inferior to natural scientists, nor compelled to emulate the methods of physics, chemistry, or biology. The consensus-building that characterizes the natural sciences is often regarded as dogmatism, especially in the social sciences where significant consensus is usually absent, or in literary criticism where consensus is despised.

The new ontological relativity resonates with recent multicultural trends. Radical multiculturalists find support for their convictions in the philosophy and sociology of science. Unfortunately, instead of welcoming an openness to all views, some elements of postmodernism seem to be characterized by intolerance to any idea that is not "politically correct" (Gross and Levitt 1994, Ortiz de Montellano 1997, Sokal and Bricmont 1998, Levitt 1999). Since ultimate reality cannot have much to do with the petty food fights of earthlings, I can safely ignore this aspect of postmodernism.

Not all of postmodernism, however, can be summarily dismissed. The basic thesis of this book is that physics has painted for us a simple picture of a material reality that is well within our general understanding. Since this flies directly in the face of current postmodern doctrine, I should respond to this. So, let me take a moment to expand on this currently fashionable philosophy, despite the fact that I fully expect it will, like all fashions, soon fade away—perhaps even by the time this book is published.

POSTMODERNISM AND PRE-POSTMODERNISM

According to Walter Truett Anderson, four distinguishable worldviews characterize contemporary Western society:

- the *social-traditional*, in which truth is found in the heritage of American and Western civilization;
- the *scientific-rational*, in which truth is found through methodological, disciplined inquiry;
- the *neo-romantic*, in which truth is found either through attaining harmony with nature or spiritual exploration of the inner self;
- the *postmodern-ironist*, who sees truth as socially constructed.

(Anderson 1995, 111)

Social-traditionalists include most nationalist and religious thinkers, but also literary figures such as Allan Bloom whose *Closing of the American*

Mind (1987) called for a return to Shakespeare, the Founding Fathers, and the wise men of ancient Athens as role models. Anderson unites the social-traditionalists with scientific-rationalists, labeling them both as *modernists*. Thus, Pat Robertson and Stephen Hawking find themselves in the same lineup, no doubt to the great surprise and amusement of both.

The neo-romantics, according to Anderson, are even more strongly oriented toward the past, longing for the days before the Industrial Revolution and the Enlightenment. These *premodernists* include New Agers, many feminist and environmental activists, and those who have revived tribal rituals, values, and beliefs.

Anderson thinks that the postmodern-ironist has provided the deepest insight into the nature of truth. He further divides them into three categories: *players* who surf the waves of cultural change, dabbling in whatever strikes their fancy without making any commitments; *nihilists* who have decided that everything is phony; and *constructivists* who are actively engaged in thinking through a new, postmodern worldview (Anderson 1995, 112).

I am not sure where the so-called *deconstructionists* in literary criticism, who also question traditional assumptions about truth, fit into Anderson's scheme. Literary deconstructionists assert that words can only refer to other words, and argue that statements about any text undermine its own meaning. I suppose they are deconstructionist constructionists, a locution that presents no problem in a discourse in which the need for consistency is not heavily emphasized, or even derided as the "hobgoblin of little minds."

Postmodernism has received intellectual support from the postpositivist philosophy of science. In a study on the state of Western knowledge commissioned by the Council of the Universities of Quebec in the late 1970s, French philosopher Jean-Francois Lyotard reported that modern systems of knowledge, including religion and science, were basically "narratives" (Lyotard 1984). These narratives assume mythic proportions and are extended out of their domains and assumed to apply universally by their proponents. The postmodern thinker, according to Lyotard, recognizes these stories for what they are, and that they do not really work so well when taken to universal extremes. He or she views any "grand narrative" with suspicion, preferring instead more localized language games that do not claim to provide all the answers.

Influential postmodern philosopher Richard Rorty agrees that all narratives are subject to change and reinterpretation. Attempts to use common sense or logical language games result only in the creation of platitudes. The biggest platitude is that a single permanent reality exists behind temporary appearances (Rorty in Anderson 1995, 102).

Rorty defines realism as "the idea that inquiry is a matter of finding out the nature of something that lies outside our web of beliefs and desires." The goal of such inquiry is to obtain knowledge and not merely satisfy our

individual needs. The *realist* believes that the object of the inquiry has a privileged context of its own that lies beyond that of the inquirer. Rorty contrasts this view with that of *pragmatists* who recognize that "there is no such
thing as an intrinsically privileged context" (Rorty 1991, 96).

Rorty disagrees with those who call this view "relativist," alluding to
the doctrine that truth depends on your particular cultural reference frame.
He scoffs at their astonishment that anyone could deny the existence of
intrinsic truth. Even formulating the question in these terms assumes a
Eurocentric perspective, he insists, appealing to some Platonic Form of
Reason that gives humans a transcultural ability to perceive reality.

The philosophy of pragmatism is usually associated with Charles
Sanders Peirce and William James. Each recognized that for a theory to be
meaningful it must be testable in experience. James famous definition says
that "truth is what works." As he explains, "The true . . . is only the expedient in the way of our thinking, just as 'the right' is only the expedient in
the way of our behaving" (James 1907).

Rorty is a pragmatist with a difference, however, insisting that you can
find no road to objective truth, even a pragmatic one. The Rorty pragmatist,
unlike the James pragmatist, has no theory of truth, much less a relativistic
one. His account of inquiry contains neither an epistemological nor ontological base, only an ethical one (Rorty 1991, 24).

Rorty suggests that Western philosophy's obsession with objectivity,
going back to Plato, is a disguised form of the fear of death, an attempt to
escape from time and chance (Rorty 1991, 32). Thus, agreeing with Feyerabend, Rorty views science is a form of secular religion.

In Western culture, the humanities and the "soft" social sciences have
felt compelled to follow "hard" natural science as an exemplar of rationality,
objectivity, and truth by attempting to emulate natural science's commit-
ment to strict methodology. Their inability to match the success of the hard
sciences and develop a rational system of human and social values has left
them with the inability to compete for people's minds, while natural science
is seen as the activity that keeps us in touch with reality.

Rorty wants us to end distinctions between hard facts and soft values,
objectivity and subjectivity, and start afresh with a new vocabulary. He
insists he does not wish to debunk or downgrade natural science, just stop
treating it as a religion.

He begins by redefining the meaning of the term "rational." Instead of
denoting methodological inquiry, he suggests that the word should refer to
acts that are "sane" and "reasonable." It should connote the application of
moral virtues like tolerance and the respect for the opinions of others, in
other words, civilized behavior. Rorty asserts: "To be rational is to discuss
any topic—religious, literary, or scientific—in a way which eschews dogmatism, defensiveness, and righteous indignation" (Rorty 1991, 37).

Rorty suggests that the goal of objectivity be replaced by "solidarity." Using the new standard of rationality given above, we attempt to reach unforced agreements by free and open encounter (dialectic?). He admits that science is exemplary in one respect: it is a model of human solidarity. Its institutions and practices provide concrete suggestions on how the rest of culture might effectively organize itself to obtain agreement. Solidarity, which apparently is to be distinguished from consensus, is to be sought within and across different disciplines and cultures, so we can agree on common ends and the truth, whatever it may be, will rise to the surface.

The theme that science is just another form of religion can be found in much current literature. Like most scientists, I can be expected to object. The argument usually is based on the observation that science and religion share certain goals, such as seeking knowledge about ultimate reality or improving human life. However, those who use this type of argument are committing a fundamental logical fallacy. The fact that religion and science have some goals in common does not make them equivalent. They have other characteristics that are sufficiently different to distinguish the two.

One can think of many examples where science differs markedly from religion. Indeed, anyone looking for a simple rule to demarcate science from nonscience need look no further than this: despite what Feyerabend and Rorty have said, *science is never dogmatic*. Or, at least it pledges not to be. What is often mistaken for dogmatism in science is conservatism, an unwillingness to change unless change is required by the data. A good scientist is always skeptical of any new claim. It was not dogmatism that prevented cold fusion from being accepted by the physics and chemistry community, it was lack of a theoretical basis, insufficient evidence, and failure to be consistently replicated. If evidence had been found in many independent experiments, cold fusion would have been readily accepted into the ranks of scientific knowledge, with the Nobel prize assured for its discoverers. Similarly, the absence of credible evidence, not dogmatism, keeps most scientists from accepting the existence of psychic phenomena. Find the evidence, and scientists will believe.

Rorty and his followers may disagree and still insist science is dogmatic, although none has presented any convincing examples where what has been decried as scientific dogmatism has in fact been true dogmatism or harmful in any way to society. But, rather than argue over the past conduct of science, let me accept Rorty's suggestion that we all adopt the same pledge: *we will never be dogmatic*. Let all activities, in natural sciences, social sciences, humanities, arts, and religion make the same pledge: *we will never be dogmatic*. We should all stand ready to talk and argue, with respect for opposing views. Let all cultures and the other human institutions that academics left and right hold so dear also make the pledge: *we will never be dogmatic*. We should all stand ready to listen to the concerns of others and work out solutions to common problems.

FINDING A PATH THAT WORKS

Under the pledge to be nondogmatic, we must guard against making any claim that we possess some special revelationary equipment that guarantees that our insights are to be preferred over others. However, nothing prevents us from pressing our own proposals with vigor, and supporting them with rational and reasonable criteria. Those of us who value consistency can still consistently argue that our picture of reality is preferable to others, based on rational criteria that we may propose but are also themselves arguable. We only need to avoid claiming that our view represents intrinsic truth just because we said so. In other words, we can be nondogmatic and still turn out to be correct.

Unlike a contestant on the 1950s quiz show *Truth or Consequences*, we cannot provide the Truth. So, we have no choice but to accept the Consequences. Our primary criterion will be one of usefulness: Do our proposed ideas serve our needs? How might they be superior to other choices in achieving these needs? Note that this criterion does not reject nonscientific paradigms, as long as they can be demonstrated to serve rationally defined needs.

Our prime need is to function as human beings, solving the problems of life. These problems are of both a personal and professional nature. We apply tools learned from mentors and by example in solving these problems.

Now, we might imagine a Darwinian process whereby paradigms are selected from a random set and only the most effective survive. A trivial example of this is deciding on the best path to follow in walking from one place to another. If possible, you would normally follow a straight line rather than stagger randomly (assuming you are sober). If an obstruction such as a wall blocks your path, you will detour around it. Or, if it is raining, you might take a longer path that offers more shelter.

In principle, no model of reality is required to select the optimum path. We could just toss the dice and keeping trying out random paths until we find one that does the job satisfactorily. Note that this does not necessarily have to be the best one, as long as it meets our needs. However, this is an inefficient process. The great majority of paths merely pulled out of the hat will not work very well. The random walk between two points might eventually get you to your destination, but you can do better.

To do better, you need some basis for deciding on a few plausible methods to apply to the problem at hand, without trying them all out in the finite time you have to solve the problem. This is where concepts about objective reality come in, helping you select from the infinite sample of possible paths, at least as the first cut before natural selection can get to work in further refining the sample. In the example above, you assume the reality of the wall blocking your way and devise a path around it. You do not attempt to will the wall away, or try to walk through or fly over it by force of mind-over-matter. That is, you adopt an ontology in which the wall is objectively real, not just a creation of your mind. That reality is manifested by placing a constraint on your behavior.

This is precisely the nature of the process of scientific paradigm-making. We adopt an ontological picture of reality and draw from it sets of rules that constrain the behavior of the objects of our domain of interest. In physics, we have the conservation principles of energy, momentum, angular momentum, electric charge, and others that represent the constraints that reality places on the motion of bodies. These and the other principles formulated in our theories developed over the centuries from increasingly sensitive observations about the world. Many proposed paradigms were discarded because they did not work. Those that survived original cuts and worked satisfactorily have become part of scientific lore. Thus, by natural means, paradigms evolve and along with them, our knowledge of reality.

15

EINSTEIN'S BUILDING BLOCKS

The most incomprehensible thing about the world is that it is comprehensible.

Albert Einstein

s I sit down to start this final, summary chapter, a few hours remain in the twentieth century and second millennium of the common era (counting from 0 C.E., as astronomers do). *Time* magazine has just named Albert Einstein as its Person of the Century. No individual's name appears more frequently in books written about physics or cosmology for a general audience, including this one.

Stories about Einstein tend to emphasize how the nuclear bomb grew out of relativity and $E = mc^2$. However, sustained nuclear energy was in fact an accidental property that he never imagined. The chain reaction is another one of those coincidences that seems to imply a "fine tuning" of the constants of physics. Those who see supernatural purpose in such coincidences might ponder if nuclear annihilation is the cosmic plan for humanity. Is Apocalypse the purpose they are seeking? Personally, I find a lack of purpose more comforting.

Einstein was not involved in the development of the nuclear bomb, and was quite surprised when told secretly about its feasibility in order that he might lend his authority in convincing President Roosevelt to approve the Manhattan Project. Nevertheless, the Person of the Century provided the basic theories from which nuclear power and many of the applications of modern physics derive.

Quantum mechanics has had far wider technological ramifications than relativity, and there Einstein was not the central figure. Planck started it all, and people like Bohr, Pauli, Heisenberg, Schrödinger, and Dirac contributed more than Einstein to its development. Still, Einstein's photon theory was a crucial step beyond Planck and the others took it from there.[1]

Time's selection of Einstein as the most influential person of the century testifies to the powerful role that physics has played in our society and culture. This started with Thales in the sixth century B.C.E. and has continued, through Democritus, Aristotle, Galileo, Newton, Maxwell, and twentieth-century physicists until today. Some say that physics (and, perhaps, all of science) is coming to an end (Horgan 1996). When I see the handful of students sitting despondently in advanced physics classes today, I worry that the doomsayers may be correct. Still, I cannot imagine how humanity will continue to progress without at least a handful of its members, ignored as they may be by the rest of society, striving to understand the basic nature of reality. At least now I can tell my students: "Keep heart. Perhaps you will be the person of the twenty-first century!"

Einstein made a number of contributions that were central to the model of timeless reality that I have proffered in the preceding pages. We begin, of course, with time itself. Einstein showed that time is not simply read off some cosmic clock that marks the intervals between the changes that take place as bodies move from place to place. Rather, each body has a clock that marks off its own, personal time. That clock will run in synchronization only with the clocks of other bodies sharing the same rest frame of reference, that is, with zero relative velocities.

Milne demonstrated how Einstein's relativity can be derived from the notion that spatial separations are defined in terms of measurements we make with our personal clocks, as we register the time differences between signals sent to other observers and returned by them, or reflected from them like radar signals. In this view, space is a contrivance used to picture the various observations we make with our clocks. Although generally unrecognized and unacknowledged, the Milne interpretation is now deeply encoded into science by the international convention for the definitions of time and space. After defining time in terms of a standard atomic clock, the convention then specifies spatial distance as the time it takes for light to go from one point or another.

In the Milne formulation, the distance between two objects (which, we assume, are "really" out there) is defined as half the time interval between sending a signal and observing its return. The speed of light is thus not a constant by observation, but a constant by definition![2] Those who suggest new theories in which the speed of light is variable must first redefine distance and then get the international community to agree to the new definition. This is not likely to happen until something is found to be terribly amiss with the current scheme in which space is defined in terms of what is read on a clock. Likewise, if anyone should claim to have measured a non-constant speed of light within the framework of the current convention, you can be assured it is a mistake. It would be certainly wrong, because it would be nonsense, like claiming to have discovered the West Pole.[3]

So far, clocks seem to be all we need, at least conceptually, to define all units of scientific measurements that seem so varied and complicated to the student—unnecessarily so, as we will see. While we still use meters for distance, an object is certified to be one meter away when a light signal sent out to that object is returned in six and two-thirds nanoseconds (three and one-third out, three and one-third back). This definition corresponds to the familiar convention that the speed of light $c = 0.3$ meters per nanosecond. Note this is a definition, not a measurement. A much better system, the one used by most particle physicists today, is to simply take $c = 1$. This is also the system used by astronomers when they measure distances in light-years (although, unfortunately, most still use the parsec, which is about 3 light-years).

All measurements in physics can then be reduced, operationally, to clock measurements. Most such measurements, such as temperature or magnetic field, are made by reading scales or dials; but these are still calibrated with distance or angle markings that in principle reduce to clock measurements. Of course, few distance measurements are actually performed with atomic clocks, but this is their fundamental basis.

One additional definition further simplifies our system of measurements and the units we use in representing these measurements, and helps also to elucidate the meaning of those measurements. We take Planck's constant to be dimensionless, or, more specifically, the quantum of action $\hbar = 1$. While this may seem arbitrary, we now recognize that quantities such a momentum and energy are also defined in terms of space and time (and thus, again, basically just time). Mathematically, momentum is the generator of translations in space, and energy the generator of translations in time.

We can determine the energy E and momentum p of an electron by measuring the frequency f and wavelength λ of the corresponding wave, since the wave and particle descriptions are equivalent.[4] That is, $E = hf$ and $p = h/\lambda$. If $\hbar = 1$, then $h = 2\pi \hbar$ is dimensionless and the dimensions of E and p are units of reciprocal time.

As a result of these simplifications, all physical units can be expressed as a power of the basic time unit, say the second. That is, they can always be expressed as s^n, where s stands for "second" and n is a positive or negative integer. Some examples are given in the table below, where "s^0" means the quantity is dimensionless ($s^0 = 1$). For convenience, particle physicists actually use "energy units" where the basic unit is usually the GeV, so I have added a third column indicating these.

time, distance	s^1	GeV^{-1}
velocity, angular momentum, electric charge	s^0	GeV^0
acceleration, mass, energy, momentum, temperature	s^{-1}	GeV^1
force, electric and magnetic fields	s^{-2}	GeV^2
Newton's constant G	s^2	GeV^{-2}
pressure, mass density	s^{-4}	GeV^4

This also serves to illustrate how the unfamiliar in physics can sometimes be much simpler than the familiar, and the simpler more profound than the more complicated. Simplicity and profundity both point in the direction of progress.

And so, time is the basic operational quantity of physics, and consequently all of quantitative natural science. The clock is our fundamental measuring instrument. Time intervals measured between events will be different for different observers in general, but the **proper time** measured by a clock at rest is an *invariant*. That is, it can be determined from what we call coordinate space and coordinate time measurements, made in any reference frame.[5]

Similarly, the mass of a body is invariant; that is, its rest mass or proper mass can be determined from energy and momentum measurements made in any reference frame. Other invariants can be formed from the other quantities of physics. The quantities that are measured in any given reference frame are seen as components of a four-dimensional vector. Just as familiar three-vectors do not depend on your choice of coordinate system, although their coordinates do, four-vectors are independent of reference frame and can be objectively determined. They may reasonably be taken as the most promising candidates for the elements of objective reality.

While Einstein made the huge intellectual leap required to go from absolute to relative (coordinate) time, he was not prepared to do the same for the related notion of causality. In his mind, as in the minds of most people, cause and effect are distinctive labels we place on connected events. If event A occurs before event B in some reference frame, then the same relationship must be maintained in all other reference frames. Einstein discovered, however, that the roles of cause and effect are reversed for two reference frames moving at a relative velocity faster than light. He found this unacceptable, and so made an additional hypothesis beyond those needed for relativity, the **Einstein principle of causality**, which demanded that cause must always precede effect.

You will often hear that Einstein's special relativity forbids superluminal motion. However, this is not technically true. What is precisely forbidden is the *acceleration* of a body from a speed slower than light to one faster (as measured in any given reference frame). Objects called **tachyons**, which always move faster than light, are allowed. They cannot be *decelerated* below the speed of light. However, none have been seen. (But, see below.)

Einstein was deeply disturbed by the philosophical implications of quantum mechanics, especially its break with the determinism inherent in classical physics: "God does not play dice," Einstein said. He proposed many objections that were answered by Bohr and others to the satisfaction of most physicists. Quantum mechanics agreed with all the data, and still does. However, although the twentieth century began with Planck's observation of quanta, arguments about what quantum phenomena imply about the nature of reality have continued into the twenty-first century, still without a consensus on the resolution of the dispute in sight.

One of Einstein's earliest objections, which later became elaborated in the EPR paradox, was the apparent superluminality implied by the conventional interpretation of quantum mechanics promoted by Bohr, Heisenberg, and others, that had rapidly became conventional. Once Bell showed how the EPR paradox could be tested experimentally, and the experiments were actually performed confirming the predictions of quantum mechanics exactly, the common belief developed that quantum phenomena imply the existence of some kind of superluminality. However, both superluminal motion and superluminal signalling are forbidden in any theory that is consistent with the axioms of relativistic quantum field theory. So, what is it that's superluminal, if not motion or signals?

The easiest answer is that superluminality resides in theoretical objects, like wave functions and quantum fields, rather than real objects like particles. That is, we can talk about some abstract field having a simultaneous value at each point in space and still not have any information moving around faster than the speed of light. However, arguments persist on the reality of fields, and most theoretical physicists think they are more real than particles. No doubt they represent properties of reality. But does that make them real? When you count 52 playing cards in a deck, does that make the number 52 real? Pythagoreans and Platonists think so—indeed more real than the cards themselves.

As was later to become clear, superluminality is not required when a theory is indeterministic. Thus Einstein could not win. Either superluminality existed, in which case cause and effect are relative, or superluminality need not exist and the universe is indeterministic. One way or the other, the notion of a causal universe in which cause always precedes effect seems to be ruled out.

An explanation for at least some of the puzzling features of quantum mechanics has been lurking all along in the fundamental equations of physics. These equations, both classical and quantum, make no distinction between the two directions of time. Furthermore, they describe the data. So, it is not just theory but real world data that seem to be telling us that time is reversible. Nevertheless, this notion has been generally rejected because of the deep-seated belief, based on common experience, that time can only change in one direction. If time is reversible, then cause and effect can also be reversed, and Einstein is not the only one who has found this hard to accept.

A century ago, Boltzmann showed how the arrow of time we follow in our everyday lives can be simply understood as the statistical consequence of the fact that we, and the objects around us, are made of many particles that move around to a great extent randomly. Certain observations, like a broken glass reassembling when all its molecules happen, by chance, to move in right direction, are possible but so unlikely as to not be observed by anyone in the lifetime of the planet.

At the quantum level, however, backward causality is implied in the observed interactions of small numbers of particles. Experimental results are found to depend on the future as well as the past. Most physicists as well as laypeople find it hard to believe that what these experiments indicate is, in fact, time reversal. Instead, the observations are explained away by other means that require, at least from my perspective, more extreme deviations from rational if not common sense. We need imaginary mass particles, superluminal connections, subquantum forces, cosmic consciousness, or parallel worlds. Or, we need reversible time at the quantum level. One or more of the schemes that retain directional time may be true, but none are provably necessary to describe observations at the quantum scale.

With particles going backward in conventional time, we can account for ostensible superluminal correlations without actual superluminal motion or signalling. The need to introduce nonlocal subquantum forces or mystical wave functions controlled by cosmic consciousness can readily be eliminated. The many worlds formalism, which has advantages over the conventional methods of doing quantum mechanics, can still be used without carrying along the ontological baggage of each world existing in a separate universe.

While multiple universes may exist, our universe, the one reachable by our scientific instruments, can be understood as one in which time progresses on the large scale pretty much as we normally think of it doing. On the other hand, at very small scales particles are zigzagging back and forth in time as well as space. Thus, an interaction between two electrons inside an atom takes place by the exchange of a photon that goes back and forth in space and time, covering all possible paths between the electrons and appearing multiple places at once. For each path in one time direction, a corresponding path in the other time direction exists, and the photon itself can be thought of as timeless or tenseless. This is timeless reality.

No doubt, timeless reality is a difficult world to visualize or describe in terms of human concepts and words that are so heavily laden with notions of tense and time order. Perhaps someday someone will invent new words to describe what is already contained in the mathematics by which physicists make calculations on these fundamental processes.[6]

Without greatly reducing my audience by switching to the language of mathematics, all I can do with the limited and imprecise English lexicon is try to indicate that these concepts are intrinsically simple—just unfamiliar. And, simplicity is my justification for claiming that time reversibility, or if you prefer, timelessness or tenselessness, provides a more rational basis for a model of objective reality.

Remember, this is what the equations seem to be saying, and since these equations describe the data, this is what the data are saying. For the umpteenth time, I am proposing no new physics. It strikes me as highly uneconomical, indeed highly irrational, to introduce new equations, new concepts, or new hypotheses, just to maintain common notions that are absent in both the data and the theory.

With time reversibility we can then proceed to build a simple model of reality. Once again we must take a bow to Einstein for telling us which building materials are needed and which are not. In particular, we need not purchase a tank of aether when we go shopping for the necessary ingredients. We do not have to fill the universe with some smooth, continuous medium to carry light waves. In fact, such a medium would imply an absolute reference frame, in violation of Galilean relativity. Einstein gave us the photon theory of light as an alternative and preserved the fundamental relativity of motion.

While Planck had earlier demonstrated the quantized nature of electromagnetic energy, Einstein showed that actual particles carry that energy from place to place. While this, in a sense, restored Newton's corpuscular theory of light, special relativistic kinematics was needed to describe particles moving at the speed of light.

Once special relativity was in place, physicists no longer had to make a distinction between "radiation" fields and "matter" particles (though this is still pointlessly done in many disciplines). Radiation, like light, was once assumed to be a substance separate and distinct from matter, a kind of "pure energy." With Einstein's demonstration that mass and rest energy are equivalent (indeed, equal when we work in units where $c = 1$), the arbitrary distinction between radiation and energy disappears.[7] Once again, we see how increasing scientific sophistication results in simplification along with deeper understanding.

The quantum mechanics that appeared around 1925, replacing the "old quantum theory" of the previous quarter century, was originally nonrelativistic. Thus, it retained the older distinction between particle and field, carrying along as useless baggage the mysterious "wave-particle duality." Unfortunately, these anachronisms remain in service today, where applications in chemistry and condensed matter physics generally avoid relativity. This is very limiting, and indeed incomplete. The photon is the carrier of electromagnetic energy, which is central in these disciplines. It cannot be described, even approximately, by nonrelativistic quantum mechanics.

By the late 1940s, quantum mechanics had become fully relativistic with the successful theory of quantum electrodynamics. This utilized the concept of the relativistic quantum field developed earlier. In relativistic quantum field theory, each particle is the quantum of a corresponding field. The field and its quantum are not separate entities as they are in classical physics. Rather, they are alternative ways of mathematically describing the same basic entity. The quantum behaves pretty much like a relativistic particle, obeying traditional particle equations of motion carried over from classical physics and "quantized" by the new rules of quantum mechanics. The field follows field equations of motion, but, unlike its quantum, it is nothing like the traditional smooth, continuous medium of the matter field in classical physics. While the mathematics is similar, the quantum field has many more components (dimensions) than classical fields.

Only the interactions of electrons and photons were considered in QED. By the 1970s, the current standard model of elementary particles and forces had been successfully developed along the same lines as quantum electrodynamics. Now, all known elementary phenomena were covered by a well-established theory. The particles are still field quanta in this theory, but the fields themselves have become even further dissociated from the notion of a material plenum. We simply do not picture waves in quantum fields as vibrations of an aetheric medium in three-dimensional space. That medium, if we are to consider it as real, must exist in an abstract, many-dimensional, Platonic space.

In classical physics, the interactions between particles are described in terms of fields. Each particle is the source of a field that spreads out into space where it interacts with other particles. This picture is to a great extent retained in the formal mathematics of modern quantum field theory. However, interactions can equivalently be viewed in a purely particle model, where the quanta of the fields are exchanged back and forth between the interacting bodies. These processes are illustrated in Feynman diagrams, and most calculations are more conveniently made by reference to these diagrams rather than by direct, formal field theory. That formal theory was used to derive Feynman's rules *after* he had inferred them by more intuitive means.

While the particles exchanged in Feynman diagrams have most of the properties of the same objects when observed in a free state, they have in most cases imaginary mass. Thus, they are referred to as **virtual particles**, implying they are not really real. Hypothetical tachyons also have imaginary mass, but they are supposed to travel faster than light. By contrast, exchanged particles must move at the speed of light or less. If tachyons were in fact being exchanged between objects, we would have faster-than-light, even instantaneous, connections over vast distances. We simply do not see this. I conclude that virtual particles are not tachyons, even though Feynman himself had said so.

The mysterious nature of virtual particles has been at least one reason why physicists have shied away from adopting a fully particulate ontological model of reality. With reversible time, however, we can make the masses of virtual particles real numbers. Those particles then become as real as any other particles that move through space along definite paths.

Perhaps M-theory, or one of its descendants, will provide us with a successful model of a multidimensional space and we will be able to visualize quantum fields as the vibrations of an aetheric medium contained in that space. For now, however, if we force ourselves to remain solely within the framework of today's working theories, we cannot produce a realistic model of a material continuum to describe fundamental processes. Any model of reality that treats fields as the basic entities must be regarded as a Platonic model, where abstract mathematical entities are more real than material objects like electrons and planets. We cannot prove it one way or the other, but somehow particle reality seems the simpler and more rational choice. We only have to accept time symmetry at the quantum scale to make it consistent with all observations.

Physicists have long recognized that time reversibility can solve the EPR paradox and many other conceptual problems of physics. But, almost uniformly, they have rejected this solution because of the causal paradoxes implied by time travel. However, these paradoxes appear to be absent at the quantum level. Backward causality in quantum mechanics is completely equivalent to forward causality in every instance, provided you change particles into antiparticles and exchange left and right handedness (CPT symmetry). If a causal process is logically possible in one time direction, it is equally possible, with the exact same probability under the conditions specified, in the opposite time direction.

What is more, you cannot send quantum particles back in time to kill themselves or their grandfathers, which means that we have no causal paradox at that level. A backward-travelling quantum particle does carry any special information that allows it to select itself from the coherent superposition of states of which it "was" part in the "past." Thus, it can do no better than a random shot, which requires no information from the "future."

Macroscopic time travel, on the other hand, seems unable to avoid paradox, unless that travel ends up in a parallel universe whose future need not be the same as the original universe.

Interestingly, parallel universes, or parallel worlds, also provide a solution to several of the other problems we have discussed. However, the parallel worlds must be separate if we are to maintain the conventional direction of time. On the other hand, with time symmetry it appears possible to keep them all in one world. The multiple paths that quantum particles simultaneously take can be fitted into a single world when those particles are allowed to go back and then forward again along a new path.

On the cosmological scale, parallel universes may be unavoidable if we are forced to provide an economical explanation for the anthropic coincidences. Again, science fiction can take us from one universe to the next by way of wormholes in space-time, but any such possibility will have to await the undreamed-of technologies that may or may not lie ahead in the conventional future. At the moment, we must take these parallel universes as being beyond reach. Any discussion that includes them must necessarily be speculative.

Now, speculation is not forbidden in science. Indeed, speculation often precedes major discoveries. So, we are not being unscientific to speculate, as long as those speculations are consistent with all known facts and well-established theories. Speculations about other universes meet those criteria. First, no known principle of physics or cosmology rules them out. Second, the best current cosmological theories suggest multiple universes. Third, they offer a simple, natural explanation for the claimed "fine tuning" of the constants of physics that seem to be required for the existence of carbon-based life.

Multiple universes provide a more parsimonious account of the observed universe than a theory that assumes the possibility of only one universe. We could just happen to be in the one that was suitable for our existence. Many other universes might exist, with other properties that would not lead to life as we know it on earth. But that does not rule out life of other forms in these different universes. No doubt, many of these universes would contain no life of any form that could reasonably be called "life." Others may contain something even more wonderful than life. Again, we can only speculate. But, based on what we now know, to assume our universe is all that is possible, and that life as we have it here on earth represents some pinnacle of existence, is the height of arrogance.

Today many people believe that all they need do in order to understand themselves and the world around them is to think about it. But unbridled fantasy, untethered by the limitations of reality, never foresaw the wonders of modern science. Without grasping the building blocks of nature, we will never be able to build structures of mind or matter that lie beyond imagination.

NOTES

1. Einstein's Nobel Prize for 1921 was given for his 1905 photon theory of the photoelectric effect and not relativity, which was still considered controversial at the time. No doubt, the Nobel committee still took relativity into account during their deliberations.

2. Of course, the speed of light in a medium is different from c. But this represents an average speed, as the photons follows a jagged path, scattering off the atoms of the medium. They still move at the speed of light between atoms.

3. One caveat here is that since the vacuum is not empty it has an effective index of refraction and so the speed of like in the "vacuum medium" is less than c. It also could have an energy dependence as a result. But the notion of an absolute speed of light in a "true vacuum" can be maintained.

4. I am not being inconsistent here in using the wave picture while arguing for a model of reality in which waves do not exist. I can still use waves for making physics calculations. Wavelength and frequency are theoretical quantities that have exact operational definitions in terms of specific measurement procedures that can be viewed in particulate terms.

5. In the 1920s, Einstein reportedly said that he misnamed the theory of relativity and instead should have called it the *theory of invariance*.

6. One of the many beauties of mathematics is that you can carry out calculations without having the foggiest notion what they mean, or even if they are meaningless.

7. This has not been recognized in the field of alternative medicine. See Stenger, 1999c.

GLOSSARY

ACTION. The average **Lagrangian** over a path from point A to point B times the time interval for the trip. Usually equivalent to **angular momentum**.

ADVANCED TIME. The time interval to a source of radiation when that radiation was emitted after it was detected, assuming the conventional **arrow of time**.

AETHER (OR ETHER). The invisible, frictionless field that, prior to the twentieth century, was thought to pervade all of space. Electricity, magnetism, and perhaps other forces were believed to propagate as waves in the aether.

ALPHA-RAYS. Nuclear radiation composed of doubly charged helium ions, helium atoms in which two electrons had been removed.

ANGULAR MOMENTUM. Rotational momentum. The product of the **linear momentum** and the perpendicular distance to the point about which the angular momentum is being measured.

ANGULAR MOMENTUM CONSERVATION. The principle which asserts that the total **angular momentum** of a system will be constant unless the system is acted on by an outside torque.

ANNIHILATION OPERATOR. In quantum mechanics, a mathematical operator that annihilates particles.

ANTHROPIC COINCIDENCES. Apparent delicate connections among certain physical constants, and between those constants and life.

ANTHROPIC PRINCIPLES. Hypothetical implications of the **anthropic coincidences.** *See* **FAP, SAP,** and **WAP.**

ARROW OF RADIATION. An arrow of time based on the apparent asymmetry of radiation.

ARROW OF TIME. Assumed direction of time.

AVOGADRO'S NUMBER. 6.022×10^{23}, the number of molecules in a mole of a substance. Essentially equal to the number of nucleons in one gram of matter.

BARYON. Member of the class of **hadrons** that have half-integer **spin** and are composed of **quarks.** The proton and neutron are baryons.

BARYON NUMBER. A quantity that identifies baryons and is conserved in all known interactions.

BETA DECAY. Nuclear decay with the emission of electrons.

BETA-RAYS. Nuclear radiation composed of electrons.

BIG BANG. Cosmological theory that our universe began in an explosion 13 to 15 billion years ago.

BILOCAL. An object is simultaneously at two places in space.

BLACK BODY RADIATION. The thermal radiation from a body.

BOSE CONDENSATE. A state of matter in which bosons are all collected in the lowest possible energy state.

BOSON. Particle with zero or integer spin. Not bound by **Pauli exclusion principle.**

C. The operation which changes a particle to its antiparticle. Also called charge conjugation.

CANONICALLY CONJUGATE. In mechanics, the connection between a spatial coordinate and its corresponding momentum.

CATHODE RAYS. Radiation from the negative electrode in a vacuum tube that is identified as a beam of electrons.

CHARGE CONSERVATION. The principle of physics which asserts that the total charge of a system is constant when no charge is added or subtracted from the system.

CHARM. A property of certain **hadrons** that is conserved in strong and electromagnetic interactions but not in weak interactions.

CHIRALITY. The handedness, left or right, of a physical body.

COLLAPSE. *See* reduction.

COLOR CHARGE. The "charge" of **QCD**.

COMMUTATOR. The difference between the product of two matrices AB and the product BA taken in the reverse order.

COMPLEMENTARITY. Principle enunciated by Bohr that quantum systems can be described equivalently in terms of one set of observables, such as coordinates, or another incompatible set, such as momenta, but never both simultaneously.

COMPTON EFFECT. The scattering of photons. off electrons.

CONJUGATE MOMENTUM. In mechanics, the momentum that corresponds to a particular coordinate.

COPERNICAN PRINCIPLE. The same laws of physics apply everywhere in the universe. No special place in the universe exists where the laws are different.

COSMIC MICROWAVE BACKGROUND. The 3K **black body radiation** left over from the big bang that fills all of space.

COSMOLOGICAL ARROW OF TIME. **Arrow of time** based on cosmology.

COSMOLOGICAL CONSTANT. A term in Einstein's **general relativity** that is present even in the absence of matter or radiation.

COSMOLOGICAL PRINCIPLE. *See* **Copernican principle**.

CPT THEOREM. A fundamental theorem in physics which says that every physical process is indistinguishable from the one obtained by exchanging particles with their antiparticles and viewing the process in a mirror and the opposite time direction.

CREATION OPERATOR. In quantum mechanics, a mathematical operator that creates particles.

CURIE TEMPERATURE. The temperature below which a material becomes ferromagnetic.

DARK ENERGY. The still-unidentified component of the universe responsible for the acceleration of the current expansion.

DARK MATTER. The still-unidentified component of massive matter in the universe.

DOUBLE SLIT INTERFERENCE EXPERIMENT. Experiment in which light is passed through two narrow slits giving an interference pattern on a screen. The same effect is seen with electrons and other particles.

DYNODE. An electrode in a **photomultiplier tube** that emits multiple electrons when struck by another electron.

EINSTEIN CAUSALITY. The hypothesis that causal relationships always occur in a single time direction, with cause preceding effect. Forbids **superluminal** motion or signals.

ELECTROMAGNETIC INDUCTION. Effect discovered by Faraday in which a time varying magnetic flux generates an electric current.

ELECTROWEAK UNIFICATION. The unification of the electromagnetic and weak interactions.

ENERGY CONSERVATION. Basic physics principle in which the total energy of an isolated system does not change with time.

ENERGY LEVEL. The discrete value of the energy of a quantum system such as an atom.

ENTROPY. A thermodynamic quantity that measures the disorder of a system.

ETHER. *See* **Aether.**

FALSE VACUUM. A region of space empty of matter and radiation but containing energy stored in the curvature of space.

FAP. *See* **Final Anthropic Principle.**

FERMAT'S PRINCIPLE. Law of optics in which the path that a light ray follows through a medium is that one along which the travel time is minimized.

FERMION. Particle with half-integer spin.

FEYNMAN DIAGRAM. Pictorial representation of fundamental particle interactions.

FEYNMAN PATH INTEGRAL. A method for making calculations in quantum mechanics developed by Richard Feynman.

FIELD. A mathematical function that has a value at every point in space.

FINAL ANTHROPIC PRINCIPLE (FAP). Intelligent, information-processing must come into evidence in the universe, and, once it comes into existence, it will never die out.

FINE STRUCTURE. Small splittings of atomic spectral lines.

FIRST LAW OF THERMODYNAMICS. Principle of physics in which the heat energy into a system must equal the increase in internal energy minus the work done by the system. Equivalent to **conservation of energy**.

FITZGERALD-LORENTZ CONTRACTION. The apparent contraction of the length of a body in the direction of its motion.

GALILEAN TRANSFORMATION. Procedure for transforming observables between two reference frames that assumes time is absolute.

GAMMA-RAYS. Photons with energies greater than **X-rays**.

GAUGE BOSONS. The generic name given to the elementary bosons exchanged in the standard model.

GAUGE INVARIANCE. In the **standard model**, a principle in which the equation do not change under certain phase changes of the **state vectors**.

GENERAL RELATIVITY. Einstein's 1916 theory of gravity.

GENERATOR. The mathematical operator than induces a transformation of observables between frames of reference.

GLUON. A boson exchanged in **strong interactions**.

GRAND UNIFICATION. The attempt to unify the **strong, weak,** and electromagnetic interactions.

GRAVITATIONAL MASS. The mass of a body that is acted on by gravity.

GRAVITON. The hypothesized quantum of gravity.

GUTS. **Grand unification** theories.

HADRON. Any strongly interacting particle.

HAMILTON'S PRINCIPLE. *See* **principle of least action**.

HEISENBERG PICTURE. In quantum mechanics, the **state vector**, or **wave function**, is assumed to be fixed while the **Hilbert space** coordinate axes evolve with time.

HIGGS FIELD. The **field** of **Higgs bosons**.

HIGGS BOSONS. In the **standard model, bosons** (not yet confirmed) that account for the masses of bodies.

HILBERT SPACE. An abstract space used in quantum mechanics to represent quantum states as vectors.

HUBBLE CONSTANT. The average ratio of the velocity at which a galaxies recede to their distance from us. Actually changes as the universe evolves.

INERTIAL FRAME OF REFERENCE. Reference frame in which **Newton's second law** of motion is observed.

INERTIAL MASS. The measure of the reluctance of a body to change its state of motion. It equals the mass m of a body that appears in **Newton's second law** of motion $F = ma$, where F is the force in a body and a is the acceleration.

INFLATION. The exponential expansion that the universe is believed to have undergone in its earliest moments.

INSEPARABLE. The feature of entangled quantum states in which the parts of a system can remain correlated even after they have becomes separated by spacelike distances.

INVARIANT. A quantity is invariant if it is unchanged when transformed to another frame of reference.

ISOSPIN. An abstract quantity mathematically like **spin** used in nuclear and particle physics to group different particles with similar properties in the same set.

KALUZA-KLEIN THEORIES. Theories in which space has more than three dimension, but the other dimensions are highly curved or "rolled" up so as to not be directly observable. *See also* **string theories** and **M-theories**.

KAON OR K-MESON. **Mesons** with the property of **strangeness**.

KET. In quantum mechanics, a symbol used to specify a quantum state vector.

LAGRANGIAN. A mathematical quantity, usually equal to the kinetic energy minus the potential energy.

LAMB SHIFT. A tiny shift in atomic spectral lines that was calculated in quantum electrodynamics.

LEAST ACTION. The basic law of classical mechanics, also known as **Hamilton's principle**, in which the path that a body follows through space is that one along which a quantity called the **action** is minimized.

LEPTON NUMBER. A quantity that identifies **leptons** and is conserved in al known interactions.

LEPTON. Generic term for fundamental particles such as electrons and neutrinos that do not interact strongly.

LEPTOQUARK. Hypothetical particle that is both quark and lepton. Also **quarton**.

LINEAR MOMENTUM. *See* **momentum**.

LINES OF FORCE. Imaginary lines used to map out fields.

LOCAL. Two events are local if they occur at the same place in some reference frame. They can still occur at different places in other references frames. That is, the event separation is **timelike**.

LORENTZ TRANSFORMATION. Procedure for transforming between two reference frames that assumes time is relative.

M-BRANES. The objects of **M-theories**.

M-THEORIES. Theories that extend **string theories** to higher dimensions.

MAGNETIC DIPOLE MOMENT. A measure of a particle's magnetic field strength.

MATRIX FORMULATION OF QUANTUM MECHANICS. Heisenberg's formulation of quantum mechanics which uses matrices to represent observables.

MAXWELL'S EQUATIONS. Equations discovered in the nineteenth century by Maxwell that allow for the calculation of classical electrical and magnetic fields from any given distribution of currents and charges. They successfully predicted the existence of electromagnetic waves travelling at the speed of light.

MESON. Generic term for particles originally intermediate in mass between the electron and proton, although many mesons since discovered have higher masses.

METRIC FIELD. The field describing curved space in general relativity.

MOMENTUM CONSERVATION. The principle which asserts that the total **linear momentum** of a system will be constant unless the system is acted on my an outside force.

MOMENTUM (LINEAR). The product of a body's inertial mass and velocity.

MULTILOCAL. An object is simultaneously at several places in space.

MUON. One of fundamental charged leptons. Like a heavy electron.

NATURALISM. The philosophy in which all phenomena can be explained in terms of material processes.

NEUTRINO OSCILLATION. The phenomenon of one neutrino type changing to another.

NEUTRINO. Fundamental neutral lepton.

NEWTON'S SECOND LAW OF MOTION. The time rate of change of the momentum of a body is equal to the total external force on the body. When the mass is constant, the force is equal to the **inertial mass** times the acceleration of the body, F = ma.

NON-INERTIAL FRAME OF REFERENCE. Reference frame in which **Newton's second law** of motion is not observed.

NONLOCAL. Two events are nonlocal if they occur at the different places in all reference frames. That is, the event separation is **spacelike**. Often mistakenly confused with **inseparable**.

NUCLEON. Generic terms for nuclear particle, either proton or neutron.

OLD QUANTUM THEORY. Quantum theory before 1925.

ORBITAL ANGULAR MOMENTUM. The angular momentum of a body in orbit around another body.

P. The **parity** operator, which changes a system to one viewed in a mirror.

PARITY. A property of the quantum state of particle which indicated whether or not the state changes sign under space reversal, as when viewed in a mirror.

PARTICLE-ANTIPARTICLE REFLECTION. The C operation in which a particle is changed to its antiparticle.

PAULI EXCLUSION PRINCIPLE. Only one **fermion** can be found in any given quantum mechanical state.

PHASE SPACE. An abstract space in which the axes correspond to the momenta and coordinates of all the degrees of freedom of a system.

PHOTOCATHODE. The metal surface of a **photomultiplier tube** that emits an electron when struck by a photon.

PHOTOELECTRIC EFFECT. The emission of an electron from a metal struck by a photon.

PHOTOMULTIPLIER TUBE. A very accurate photon detector that utilizes the **photoelectric effect**.

PHOTON. The particle, of light.

PION OR PI-MESON. Lightest mass meson.

PLANCK ENERGY. The rest energy of a body whose rest mass equals the **Planck mass**. Approximately 10^{-28} electron-Volts.

PLANCK LENGTH. The distance, approximately 10^{-35} meter, below which quantum gravitational effects become important.

PLANCK MASS. The mass, approximately 2×10^{-5} grams, of a sphere whose radius is equal to the Planck length.

PLANCK TIME. The time, approximately 10^{-43} second, that light takes to go the Planck length. It is the smallest time that can be measured by a clock.

PLANCK SCALE. A region of space-time of dimensions comparable to the Planck length and Planck time.

POINCARÉ GROUP. This mathematical group of operators of that contains all the space time translations and rotations.

POSITRON. Antielectron.

PRINCIPLE OF EQUIVALENCE. The principle that the inertial and gravitational masses of a body are equal.

PRINCIPLE OF SUPERPOSITION. Axiom of quantum mechanics which postulates that the state of a quantum system can be represented by a state vector in an abstract Hilbert space that contains one coordinate axis, and thus one dimension, for each possible outcome of a measurement of all the observables of the system.

PRINCIPLE OF RELATIVITY. Basic physics principle discovered by Galileo which states that no observation can distinguish between a body being at rest and being in motion at constant velocity. That is, there is no absolute motion.

PROBABILITY AMPLITUDE. A complex number whose absolute value squared gives a probability.

PROPAGATOR. The contribution made by an internal line (exchanged particle) or external line in a **Feynman diagram**.

PROPER MASS. The mass measured in the reference frame in which the body is at rest. Also **rest mass**.

PROPER TIME. The time measured in the reference frame in which the clock is at rest.

QCD. *See* **quantum chromodynamics**.

QED. *See* **quantum electrodynamics**.

QUANTUM. Generic term for a particle, such as the photon, that is associated with a quantum field. In the **standard model**, all particles are quanta of some field.

QUANTUM CHROMODYNAMICS (QCD). The theory of strong interactions in the **standard model**.

QUANTUM ELECTRODYNAMICS (QED). The study of photons, relativistic electrons, and electromagnetic radiation.

QUANTUM OF ACTION. The minimum increment, designed by ⍶, by which the **action** or **orbital angular momentum** of a particle may change in quantum mechanics.

QUANTUM POTENTIAL. Potential energy term introduced by Bohm to provide for quantum effects.

QUARK. Fundamental constituent of **hadrons**.

QUARTON. Hypothetical particle that is both quark and lepton. Also **leptoquark**.

QUINTESSENCE. The name given the repulsive **scalar field** that may be responsible for the apparent acceleration of the expansion of the universe.

R PROCESS. *See* **reduction**.

RED SHIFT. The shifting of the color of spectral lines to lower frequencies, or longer wavelengths.

REDUCTION. In conventional quantum mechanics, the process by which the **state vector** or **wave function** changes to a new form under the act of measurement. Also called state vector or wave function collapse.

RENORMALIZATION. In quantum field theory, the procedure by which infinities in calculations are subtracted by subsuming them in the self masses and charges of particles.

REST ENERGY. The energy of a particle at rest. It is equal to the particle **rest mass** multiplied by the speed of light squared.

REST MASS. *See* **proper mass**.

RETARDED TIME. The time interval to a source of radiation when that radiation was emitted before it was detected, assuming the conventional **arrow of time**.

ROTATIONAL SYMMETRY. When the observables of a system are **invariant** to a rotation of a the spatial axes, so that no special direction in space can be distinguished.

SAP. *See* **Strong Anthropic Principle**.

SCALAR FIELD. A field with no direction in space. Its **quanta** are spin zero **bosons**.

SCHRÖDINGER PICTURE. In quantum mechanics, the **state vector**, or **wave function**, is assumed to evolve with time. The **Hilbert space** coordinate axes remain fixed.

SECOND LAW OF THERMODYNAMICS. Basic physics principle in which the **entropy** of an isolated system either stays constant or increases with time.

SELF MASS. The mass of a body that results from its own internal interactions.

SINGLET. In quantum mechanics, the state in which the total **spin** of a set of particles is zero.

SNELL'S LAW. Law of optics that gives the path followed by a light ray as it moved from one medium to another.

SPACE TRANSLATION SYMMETRY. When the observables of a system are **invariant** to a translation of a spatial axis, so that no special position along that axis can be distinguished.

SPACE REFLECTION SYMMETRY. When the observables of a system are invariant to a reflection of a spatial axis, so that no special direction along that axis can be distinguished. Equivalent to mirror symmetry, where the **P** or **parity** is conserved.

SPACE REFLECTION. The **P** or **parity** operation in which a system is viewed in a mirror.

SPACELIKE. A separation of events in space-time in which no signal travelling at the speed of light or less can connect the events.

SPACE-TIME. The four-dimensional manifold of space and time.

SPECIAL THEORY OF RELATIVITY. Einstein's 1905 theory of space, time, mass, energy, and the motion of bodies.

SPIN. The intrinsic angular momentum of a particle.

SPONTANEOUS SYMMETRY BREAKING. The accidental breaking of underlying symmetries.

STANDARD MODEL. Current theory of fundamental particles and forces.

STATE VECTOR. A vector in abstract **Hilbert space** used to represent the state of a system in quantum mechanics.

STRANGENESS. A property of certain **hadrons** that is conserved in strong and electromagnetic interaction but not in **weak interactions**.

STRING THEORIES. Theories in which the elementary objects are one-dimensional strings. Also **Superstring theories**. *See* **M-theories**.

STRONG ANTHROPIC PRINCIPLE (SAP). The universe must have those properties which allow life to develop within it at some stage in its history.

STRONG INTERACTION. The strongest of the known forces, responsible for holding **quarks** together in **nucleons**.

SUPERLUMINAL. Faster than light.

SUPERSTRING THEORIES. Theories in which strings form the basic units of matter.

SUPERSYMMETRY (SUSY). Symmetry principle that unites bosons and fermions.

SUSY. *See* **supersymmetry**.

T. The operation which reverses time. *See* **time reflection**.

TACHYON. Hypothetical particle that always moves faster than light.

TAUON. One of fundamental charged leptons. Like a heavy electron, heavier than the muon.

TIME REFLECTION SYMMETRY. When the observables of a system are invariant to a reflection of the time axis, so that no special direction in time can be distinguished.

TIME TRANSLATION SYMMETRY. When the observables of a system are invariant to a translation of the time axis, so that no special moment in time can be distinguished.

TIME DILATION. The apparent slowing down of a moving clock.

TIME REFLECTION. The **T** operation in which a process is observed in reverse time, as in running a film backward through the projector.

TIME-DEPENDENT SCHRÖDINGER EQUATION. The basic dynamical equation of quantum mechanics that tells how a quantum state evolves with time.

TIME-INDEPENDENT SCHRÖDINGER EQUATION. The nonrelativistic equation used to calculate the energy levels of stationary quantum states.

TIMELIKE. A separation of events in space-time in which a signal travelling at the speed of light or less can connect the events.

TRIPLET. The quantum state in which a set of particles has a total **spin** one.

TRUE VACUUM. A region of space empty of matter and radiation and containing minimum energy.

U PROCESS. *See* unitary.

UNCERTAINTY PRINCIPLE. Principle discovered by Heisenberg which places limits on the simultaneous measurements of **canonically conjugate** observables.

UNITARY. In quantum mechanics, an operation in which either the **state vector** in **Hilbert space**, or the coordinate axes, is rotated without changing the length of the state vector. Assures that probabilities remain the same in the operation.

VECTOR FIELD. A field with direction in space (like the magnetic field). Its **quanta** are spin one **bosons**.

VIEW FROM NOWHEN. An imagined view from outside of **space-time**.

VIEW FROM NOWHERE. A view from outside of the space being viewed, such as when one looks down on a map.

VIRTUAL PARTICLES. Unobserved particles that seem to violate energy and momentum conservation but are allowed to exist for brief periods by the **uncertainty principle**.

W BOSON. A particle that is exchanged in **weak interactions**.

WAP. *See* **Weak Anthropic Principle**.

WAVE FUNCTION. A complex field specifying the state of a quantum system. When used for a single particle, its magnitude square gives the probability per unit volume for finding the particle at a particular position at a particular time.

WAVE PACKET. A localized superposition of waves.

WAVE-PARTICLE DUALITY. The notion that physical objects simultaneously possess both wave and particle properties.

WAVE MECHANICS. Schrödinger's formulation of quantum mechanics.

WEAK ANTHROPIC PRINCIPLE (WAP). The observed values of all physical and cosmological quantities are not equally probable but take on values restricted by the requirement that there exist sites where carbon-based life can evolve and by the requirement that the universe be old enough for it to have already done so.

WEAK INTERACTION. The nuclear force responsible for **beta decay** and the interaction of **neutrinos**.

WEAK ISOSPIN. A form of **isospin** used in the **standard model**.

WEINBERG ANGLE. A mixing parameter in **electroweak unification**.

WEYL CURVATURE. The tidal portion of the curvature of **space-time** in **general relativity**, the part of that curvature that is present even in an empty universe.

WORLDLINE. The path of a body through **space-time**.

X-RAYS. Photons with energies greater than ultraviolet photons but less than gamma-rays.

Z BOSON. A particle that is exchanged in **weak interactions**.

ZERO POINT ENERGY. The lowest energy any particle can have when confined to a finite region of space.

BIBLIOGRAPHY

Achinstein, Peter. 1991. *Particles and Waves: Historical Essays in the Philosophy of Science*. New York, Oxford: Oxford University Press.

Agrawal, V. , S. M. Barr, J. F. Donoghue, and D. Seckel. 1998. "Viable Range of the Mass Scale of the Standard Model." *Physical Review* D57: 5490–92.

Aharonov, Yakir, and Lev Vaidman. 1990. "Properties of a Quantum System During the Time Interval Between Two Measurements." *Physical Review* A 41: 11–20.

Anderson, Arlen. 1987. "EPR and Global Conservation of Angular Momentum." University of Maryland Reprint 88-096.

Anderson, Walter Truett. 1996. *The Truth About The Truth*. New York: Jeremy P. Tarcher/Putnam.

Aristotle. *Aristotle in 23 Volumes*. Vols. 17, 18. Translated by Hugh Tredennick. Cambridge, Mass.: Harvard University Press.

Aspect, Alain, Phillipe Grangier, and Roger Gerard. 1982. "Experimental Realization of the Einstein-Podolsky-Rosen *Gedankenexperiment*: A New Violation of Bell's Inequalities." *Physical Review Letters* 49, 91.

———. 1982. "Experimental Tests of Bell's Inequalities Using Time-Varying Analyzers." *Physical Review Letters* 49: 91–94.

Ayer, Alfred Jules. 1936. *Language, Truth, and Logic*. London: Victor Gollanncz; New York: Oxford University Press.

———. 1992. "Reply to Tscha Hung." In *The Philosophy of A. J. Ayer*. Edited by Lewis Edwin Hahn. La Salle, Ill.: Open Court.

Barrow, John D., and Frank J. Tipler. 1986. *The Anthropic Cosmological Principle.* Oxford: Oxford University Press.

Beard, David B. 1963. *Quantum Mechanics.* Boston: Allyn and Bacon.

Begley, Sharon. 1998. "Science Finds God." *Newsweek,* July 20: 46.

Bell, John S. 1964. "On the Einstein-Podolsky-Rosen Paradox." *Physics* 1: 195–200.

————. 1966. "On the Problem of Hidden Variables in Quantum Mechanics." *Reviews of Modern Physics* 38: 447–52.

————. 1987. *Speakable and Unspeakable in Quantum Mechanics.* Cambridge: Cambridge University Press.

Bergson, Henri. 1922. *Bull. Soc. Phil.* 22: 102.

Bernstein, Jeremy. 1993. *Cranks, Quarks, and the Cosmos.* New York: Basic Books.

Bethe, H. A., and E. Fermi. 1932. "Über die Wechselwirkung von Zwei Electronen." *Zeitschrift der Physik* 77: 296–306.

Bloom, Allan. 1987. *The Closing of the American Mind: How Higher Education Has Failed Democracy and Impoverished the Souls of Today's Students.* New York: Simon & Schuster.

Bohm, David. 1951. *Quantum Theory.* Englewood Cliffs, N.J.: Prentice-Hall.

————. 1952. "A Suggested Interpretation of Quantum Theory in Terms of 'Hidden Variables,' I and II." *Physical Review* 85: 166.

Bohm, D., and B. J. Hiley. 1993. *The Undivided Universe: An Ontological Interpretations of Quantum Mechanics.* London: Routledge.

Bondi, H. 1960. *Cosmology.* Cambridge: Cambridge University Press.

Born, Max. 1953. "Physical Reality." *Philosophical Quarterly* 3: 139–49.

Buechner, Ludwig 1870. *Force and Matter. Empirico-Philosophical Studies Intelligibly Rendered.* London: Truebner.

————. 1891. *Force and Matter. Or Principles of the Natural Order of the Universe. With a System of Morality Based Thereon. A Popular Exposition.* New York: Peter Eckler.

Bussey, P. J. 1982. " 'Superluminal Communication' in the EPR Experiments." *Physics Letters* 90A: 9–12.

Capra, Fritjof. 1975. *The Tao of Physics.* Boulder: Shambhala.

Carnap, Rudolf. 1936, 1937. "Testability and Meaning." *Philosophy of Science* B, 419 21; 4: 1–40.

Carter, Brandon. 1974. "Large Number Coincidences and the Anthropic Principle in Cosmology." In *Confrontation of Cosmological Theory with Astronomical Data.* Edited by M. S. Longair. Dordrecht: Reidel, pp. 291–98. Reprinted in Leslie 1990.

Castagnoli, Giuseppe. 1995. "Hypothetical Solution of the Problem of Measurement Through the Notions of Quantum Backward Causality." *International Journal of Theoretical Physics* 34 (8): 1283–87.

Costa de Beauregard, Olivier. 1953. "Une réponse à l'argument dirigè par Einstein, Podolsky et Rosen contre l'interpretation bohrienne de phénomènes quantiques." *Comptes Rendus* 236: 1632–34.

————. 1977. "Time Symmetry and the Einstein Paradox." *Il Nuovo Cimento* 42B (1): 41–63.

————. 1978. "S-Matrix, Feynman Zigzag, and Einstein Correlation." *Physics Letters* 67A, pp. 171–74.

———. 1979. "Time Symmetry and the Einstein Paradox—II." *Il Nuovo Cimento* 51B (2): 267–79.

Coveney, Peter, and Roger Highfield. 1991. *The Arrow of Time*. London: Flamingo.

Chopra, Deepak. 1989. *Quantum Healing: Exploring the Frontiers of Mind/Body Medicine*. New York: Bantam.

———. 1993. *Ageless Body, Timeless Mind: The Quantum Alternative to Growing Old*. New York: Random House.

Clauser, J. F., and M. A. Horne. 1974. "Experimental Consequences of Objective Local Theories." *Physical Review* D10, p. 526.

Clauser, John F., and Abner Shimony. 1978. "Bell's Theorem: Experimental Tests and Implication." *Rep. Prog. Phys.* 41: 1881–1927.

Craig, William Lane. 1979. *The Kalām Cosmological Argument*. Library of Philosophy and Religion. London: Macmillan.

Craig, William Lane, and Quentin Smith. 1993. *Theism, Atheism, and Big Bang Cosmology*. Oxford: Clarendon Press.

Crick, Francis. 1994. *The Astonishing Hypothesis: The Scientific Search for the Soul*. New York: Charles Scribner's Sons.

Davies, P. C. W. 1974. *The Physics of Time Asymmetry*. London: Surrey University Press.

———. 1977. *Space and Time in the Modern Universe*. Cambridge: Cambridge University Press.

———. 1983. "Inflation and Time Asymmetry in the Universe." *Nature* 301: 398–400.

———. 1992. *The Mind of God: The Scientific Basis for a Rational World*. New York: Simon and Schuster.

———. 1995. *About Time: Einstein's Unfinished Revolution*. London: Viking.

Davies, P. C. W., and J. Twamley. 1993. "Time-Symmetric Cosmology and the Opacity of the Future Light Cone." *Classical and Quantum Gravity* 10: 931.

Dawkins, Richard. 1999. "Snake Oil and Holy Water." *Forbes ASAP*, October 4, p. 235.

De la Mettrie, Julien. 1778. *Man a Machine*. La Salle, Ill.: Open Court, 1953.

Dembski, William A. 1998. *The Design Inference*. Cambridge: Cambridge University Press.

———. 1999. *Intelligent Design: The Bridge Between Science and Theology*. Downer's Gove, Ill.: Intervarsity Press.

Deutsch, David. 1997. *The Fabric of Reality*. New York: Allen Lane.

Dewdney, C., and B. J. Hiley. 1982. *Foundations of Physics* 12: 27–48.

DeWitt, Bryce, and Neill Graham, eds. 1973. *The Many-Worlds Interpretation of Quantum Mechanics*. Princeton: Princeton University Press.

De Jager, Cornelius. 1992. "Adventures in Science and Cyclosophy." *Skeptical Inquirer* 16 (2): 167–72.

D'Holbach, Paul Henri. 1853. *The System of Nature*. Boston: J. P. Mendum.

———. 1984. *The System of Nature*. New York: Garland.

Dicke, R. H. 1961. "Dirac's Cosmology and Mach's Principle." *Nature* 192: 440.

Dirac, P. A. M. 1927. "Quantum Theory of the Emission and Absorption of Radiation." *Proceedings of the Royal Society of London*, Series A 114: 243.

———. 1930. *The Principles of Quantum Mechanics*. Oxford: Oxford University Press. This book has had four editions and at least twelve separate printings. Page references here are to the 1989 paperback edition.

———. 1933. "The Lagrangian in Quantum Mechanics." *Phys. Zeits. Sowjetunion* 3: 64–72.

———. 1937. *Nature* 139: 323.

———. 1951. "Is There an Aether?" *Nature* 168: 906–907.

———. 1963. "The Evolution of the Physicist's Picture of Nature." *Scientific American* (May): 47.

Dobbs, Betty Jo Teeter, and Margaret C. Jacob. 1995. *Newton and the Culture of Newtonianism*. Atlantic Highlands, N.J.: Humanities Press International.

Dorit, Robert. 1997. *American Scientist*. September-October.

Dresden, Max. 1993. "Renormalization in Historical Perspective—The First Stage." In Brown 1993.

Dugas, René. 1955. *A History of Mechanics*. Neuchatel Switzerland: Éditions du Grifon. English translation by J. R. Maddox. New York: Central Book Company.

Dummett, M. A. R. 1954. "Can an Effect Precede Its Cause?" *Proceedings of the Aristotelian Societey, Supplementary Volume* 38: 27–44.

———. 1964. "Bringing About the Past." *Philosophical Review* 73: 338–59.

Dunbar, D. N. F., R. E. Pixley, W. A. Wenzel, and W. Whaling. "The 7.68-MEV State in C12." *Physical Review* 92: 649–50.

Durant, Will. 1953. *The Story of Philosophy*. New York: Simon and Schuster.

Dyson, F. J. 1949. "The Radiation Theories of Tomonaga, Schwinger, and Feynman." *Physical Review* 75: 1736.

———. 1953. "Field Theory." *Scientific American* 188T: 57–64.

———. 1979. *Disturbing the Universe*. New York: Harper and Row.

Earman, John. 1984. "Laws of Nature: The Empiricist Challenge." Quoted in Van Frassen 1989, p. 40.

———. 1987. *Philosophical Quarterly* 24 (4): 307–17.

Eberhard, P. 1978. *Nuovo Cimento* 46B: 392.

Eberhard, Phillippe H., and Ronald R. Ross. 1989. "Quantum Field Theory Cannot Provide Faster-Than-Light Communication." *Foundations of Physics Letters* 2: 127–79.

Eddington, Sir Arthur. 1923. *The Mathematical Theory of Relativity*. Cambridge: Cambridge University Press.

———. 1928. *The Nature of the Physical World*. Cambridge: Cambridge University Press.

Einstein, A., B. Podolsky, and N. Rosen. 1935. "Can the Quantum Mechanical Description of Physical Reality Be Considered Complete?" *Physical Review* 47: 777 .

Ellis, George. 1993. *Before the Beginning: Cosmology Explained*. London, New York: Boyars/Bowerdean.

Elvee, Richard Q., ed. 1982. *Mind in Nature*, San Francisco: Harper and Row.

Everett III, Hugh. 1957. " 'Relative State' Formulation of Quantum Mechanics." *Reviews of Modern Physics* 29: 454–62.

Ferguson, Kitty. 1994. *The Fire in the Equations*. London: Bantam

Feyerabend, Paul. 1975, 1988. *Against Method: Outline of an Anarchistic Theory of Knowledge.* London: Verso/New Left Books.

———. 1978. *Science in a Free Society.* London: Verso.

Feynman, R. P. 1942. "The Principle of Least Action in Quantum Mechanics." Ph.D. diss., Princeton University. Ann Arbor: University Microfilms Publication No. 2948.

———. 1948, "Space-Time Approach to Non-Relativistic Quantum Mechanics," *Reviews of Modern Physics* 20: 367–87.

———. 1949a. "The Theory of Positrons." *Physical Review* 76: 749–59.

———. 1949b. "Spacetime Approach to Quantum Electrodynamics." *Physical Review* 76: 769–89.

Feynman R. P., and A. R . Hibbs. 1965a. *Quantum Mechanics and Path Integrals.* New York: McGraw-Hill.

Feynman R. P. 1965b. "The Development of the Space-Time View of Quantum Electrodynamics." Nobel Prize in Physics Award Address, Stockholm, 11 December. In Les Pris Nobel en 1965 Stockholm: Nobel Foundation, 1966; *Physics Today*, August 1996: 31; *Science* 153: 599.

Feynman, Richard P. 1985. *QED: The Strange Theory of Light and Matter.* Princeton: Princeton University Press.

Fitzgerald, E. 1953. "The Rubôayôat of Omar Khayyôam." In *The Rubôayôat of Omar Khayyôam and Other Writings by Edward Fitzgerald.* London and Glasgow: Collins.

Fuchs, Christopher A., and Asher Peres. 2000. "Quantum Theory Needs No Interpretation." *Physics Today* (March): 70–71.

Gardner, Martin. 1979. *The Ambidextrous Universe.* New York: Charles Scribner's.

Gell-Mann, M., and J. B. Hartle. 1990. In *Complexity, Entropy and the Physics of Information.* Edited by W. Zurek. Reading, Pa.: Addison-Wesley, p. 425.

Gell-Mann, Murray, and James P. Hartle. 1992. "Time Symmetry and Asymmetry in Quantum Mechanics and Quantum Cosmology." In *Proceedings of the Nato Workshop on the Physical Origin of Time Asymmetry, Mazagon, Spain, September 30–October 4, 1991.* Edited by J. Haliwell, J. Perez-Mercader, and W. Zurek. Cambridge: Cambridge University Press.

Gell-Mann, Murray. 1994. *The Quark and the Jaguar: Adventures in the Simple and the Complex.* New York: W. H. Freeman.

Ghirardi, G. C., et al. 1980. "A General Argument Against Superluminal Transmission Through the Quantum Mechanical Measurement Process." *Lettre Al Nuovo Cimento* 27: 293–98.

Ghirardi, G. C., A. Rimini, and T. Weber. 1986. "Unified Dynamics for Microscopic and Macroscopic Systems." *Physical Review* 34: 470.

Gleick, James. 1993. *Genius: The Life and Science of Richard Feynman.* New York: Vintage, Random House.

Gossick, B. R. 1967. *Hamilton's Principle and Physical Systems.* New York: Academic Press.

Goswami, Amit. 1993. *The Self-Aware Universe: How Consciousness Creates the Material World.* New York: G.P. Putnam's Sons.

Greene, Brian. 1999. *The Elegant Universe: Superstrings, Hidden Dimensions, and the Quest for the Ultimate Theory.* New York: W. W. Norton.

Gregory, Bruce. 1988. *Inventing Reality: Physics as Language*. New York: John Wiley & Sons.

Griffiths, R. B. 1984. "Consistent Histories and the Interpretation of Quantum Mechanics." *Journal of Statistical Physics* 36: 219–72.

Griffiths, Robert B., and Roland Omnès. 1999. "Consistent Histories and Quantum Measurements." *Physics Today* 52 (8), Part 1, pp. 26–31.

Gross, Paul R., and Norman Levitt. 1994. *Higher Superstition: The Academic Left and Its Quarrels with Science*. Baltimore: Johns Hopkins Press.

Grünbaum, Adolf. 1964. "The Anisotropy of Time." *Monist* 48: 219. Amended version in *The Nature of Time*. Edited by T. Gold. Cornell University Press, 1967.

———. 1971. "The Meaning of Time." *Basic Issues in the Philosophy of Time*. Edited by Eugene Freeman and Wilfrid Sellars. La Salle, Ill.: Open Court.

Guth, A. 1981. "Inflationary Universe: A Possible Solution to the Horizon and Flatness Problems." *Physical Review* D23: 347–56.

Guth, Alan. 1997. *The Inflationary Universe*. New York: Addison-Wesley.

Harrison, Edward. 1985. *Masks of the Universe*. (New York: Collier Books, Macmillan.

Hawking, S. W. 1985. "Arrow of Time in Cosmology." *Physical Review* D32: 2489–95.

Hawking, Stephen W. 1988. *A Brief History of Time: From the Big Bang to Black Holes*. New York: Bantam.

Heisenberg, Werner. 1958. *Physics and Philosophy*. New York: Harper and Row.

Hesse, Mary B. 1961. *Forces and Fields: The Concept of Action at a Distance in the History of Physics*. London: Thomas Nelson and Sons.

Home, R. W., ed. 1992. *Electricity and Experimental Physics in Eighteenth-Century Europe*. Brookfield, Vermont and Hampshire, Great Britain: Ashgate Publishing.

Holton, Gerald, ed. 1972a. *The Twentieth-Century Sciences: Studies in the Biography of Ideas*. New York: W. W. Norton.

Holton, Gerald. 1972b. "Mach, Einstein, and the Search for Reality." In Holton 1974a, 344–81.

———. 1972c. "The Roots of Complementarity." In Holton 1972a, pp. 382–422.

———. 1993. *Science and Anti-Science*. Cambridge, Mass.: Harvard University Press.

Horgan, John. 1996. *The End of Science*. New York: Addison-Wesley.

Hoyle, F., D. N. F. Dunbar, W. A. Wensel, and W. Whaling. 1953. "A State in C^{12} Predicted from Astrophysical Evidence." *Physical Review* 92: 1095.

Hoyningen-Huene, Paul. 1993. *Thomas S. Kuhn's Philosophy of Science*. Chicago: University of Chicago Press.

Huby, Pamela, and Neal Gordon, eds. 1989. Translation (by committee) of *On the Kriterion and Hegemonikon by Claudius Ptolemaeus*. In *The Criterion of Truth*. Liverpool: Liverpool University Press, pp. 179–230.

Ikeda, Michael, and Bill Jefferys. 1997. "The Anthropic Principle Does Not Support Supernaturalism." <http://quasar.as.utexas.edu/anthropic. html>.

Jahn, Robert G., and Brenda J. Dunne. 1986. "On the Quantum Mechanics of Consciousness, with Application to Anomalous Phenomena." *Foundations of Physics* 16: 721–72.

———. 1987. *Margins of Reality: The Role of Consciousness in the Physical World*. New York: Harcourt Brace Jovanovich.

Jaki, Stanley L. 1966. *The Relevance of Physics*. Chicago: University of Chicago Press.

Jammer, Max. 1974. *The Philosophy of Quantum Mechanics: The Interpretations of Quantum Mechanics in Historical Perspective*. New York: John Wiley.

Jefferys, Bill. 1998. Private communication.

Jeltema, Tesla E., and Marc Sher. 1999. "The Triple-Alpha Process and the Anthropically Allowed Values of the Weak Scale." Submitted for publication. Preprint hep-ph/9905494.

Jordan, T. F. 1983. "Quantum Correlations Do Not Transmit Signals." *Physics Letters* 94A (6, 7): 264.

Kafatos, Menas, and Robert Nadeau. 1990. *The Conscious Universe: Part and Whole in Modern Physical Theory.* New York: Springer-Verlag.

Karakostas, Vassilios. 1996. "On the Brussels Schools' Arrow of Time in Quantum Theory." *Philosophy of Science* 63: 374–400.

Kauffman, Stuart. 1995. *At Home in the Universe: The Search for the Laws of Self-Organization and Complexity.* Oxford, New York: Oxford University Press.

Kazanas, D. 1980. *Astrophysical Journal* 241: L59–63.

Kennedy, J. B. 1995. "On the Empirical Foundations of the Quantum No-Signalling Proofs." *Philosophy of Science* 62: 543–60.

Kirk, G. S., J. E. Raven, and M. Schofield. 1995. *The Presocratic Philosophers.* 2d ed. Cambridge, New York & Melbourne: Cambridge University Press.

Kochen, S., and E. P. Specker. 1967. *Journal of Mathematical Mechanics* 17: 59–87.

Kuhn, Thomas. 1970. *The Structure of Scientific Revolutions.* Chicago: University of Chicago Press.

Laudan, Larry. 1996. *Beyond Positivism and Relativism.* Boulder, Colo., and Oxford: Westview Press.

Lederman, Leon, with Dick Teresi. 1993. *The God Particle: If the Universe is the Answer, What Is the Question?* New York: Houghton Mifflin, 1993.

Le Poidevin, Robin. 1996. *Arguing for Atheism: An Introduction to the Philosophy.* London: Routledge.

Leslie, John. 1990. *Physical Cosmology and Philosophy.* New York: Macmillan.

Levitt, Norman. 1999. *Prometheus Bedeviled: Science and the Contradictions of Contemporary Culture.* Piscataway, N.J.: Rutgers University Press.

Linde, Andre. 1987. "Particle Physics and Inflationary Cosmology." *Physics Today* 40: 61–68.

———. 1990. *Particle Physics and Inflationary Cosmology.* New York: Academic Press.

———. 1994. "The Self-Reproducing Inflationary Universe." *Scientific American* 271 (5): 48–55.

Lyotard, Jean-François. 1984. *The Postmodern Condition: A Report on Knowledge.* Manchester: Manchester University Press.

Mandelung, E. 1927. "Quantentheorie in Hydrodynamischer Form." *Zeitschrift für Physik* 43: 354–57.

McDonnell, John J. 1991. *The Concept of an Atom from Democritus to John Dalton.* San Francisco: The Edwin Mellen Press.

Mermin, N. David. 1985. "Is the Moon There When Nobody Looks? Reality and the Quantum Theory." *Physics Today* 38:38.

Milne, E. A. 1935. *Relativity, Gravitation and World Structure.* Oxford: Oxford University Press.

———. 1948. *Kinematic Relativity.* Oxford: Oxford University Press.

Mills, Robert. 1993. "Tutorial on Infinities in QED." In Brown 1993.

Moreland, J. P., ed. 1998. *The Creation Hypothesis*. Downers Grove, Ill.: InterVarsity Press.

Nagel, Thomas. 1986. *The View from Nowhere*. Oxford: Oxford University Press.

Nafe, J. E., E. B. Nelson, and I. I. Rabi. 1947. "Hyperfine Structure of Atomic Hydrogen and Deuterium." *Physical Review* 71: 914–15.

Neumaier. A. 1999. "On A Realistic Interpretation of Quantum Mechanics." quant-ph/9908071. Submitted to *Physical Review* A.

Okun, L. B., K. G. Selivanov, and V. L. Telegdi. 2000. "On the Interpretation of the Redshift in a Static Gravitational Field." *American Journal of Physics* 68 (2): 115–19.

Omnès R. J. 1994. *The Interpretation of Quantum Mechanics*. Princeton, N.J.: Princeton University Press.

Ortiz de Montellano, Bernard. 1997. "Post-Modern Multiculturalism." *Physics and Society* 24: 5.

Page, D. 1985. "Will Entropy Decrease if the Universe Recollapses?" *Physical Review* D32: 2496–99.

Pais, Abraham. 1982. *Subtle is the Lord: The Science and the Life of Albert Einstein*. New York, Oxford: Oxford University Press.

Parsons, Keith M. 1998. "Lively Answers to Theists." *Philo* 1 (1): 115–21.

Penrose, Roger. 1979. "Singularities and Time-Asymmetry." In *General Relativity: An Einstein Century Survey*. Edited by S. Hawking and W. Israel. Cambridge: Cambridge University Press.

———. 1989. *The Emperor's New Mind: Concerning Computers, Minds, and the Laws of Physics*. Oxford: Oxford University Press.

———. 1994. *Shadows of the Mind: A Search for the Missing Science of Consciousness*. Oxford: Oxford University Press.

Price, Huw. 1996. *Time's Arrow and Archimedes Point: New Directions for the Physics of Time*. Oxford: Oxford University Press.

Polkinghorne, John. 1994. *The Faith of a Physicist*. Princeton, N.J.: Princeton University Press.

Popper, Karl. 1934, 1959, 1958. *The Logic of Discovery*. English edition London: Hutchinson. New York: Basic Books, Harper & Row.

———. 1956. "The Arrow of Time." *Nature* 177: 538.

Prigogine, Ilya, and Isabella Stengers. 1984. *Order Out of Chaos*. New York: Bantam.

Press, W. H., and A. P. Lightman. 1983. *Philosophical Transactions of the Royal Society of London* A 310, p. 323.

Quine, W. V. 1969. *Ontological Relativity and Other Essays*. New York: Columbia University Press.

———. 1951. *The Two Dogmas of Empiricism*. In *From a Logical Point of View*. Cambridge, Mass.: Harvard University Press.

Redhead, M. 1987. *Incompleteness, Non-Locality, and Realism*. Oxford: Clarendon Press.

———. 1995. *From Physics to Metaphysics*. Cambridge: Cambridge University Press.

Rescher, Nicholas. 1967. *The Philosophy of Leibniz*. Englewood Cliffs, N.J.: Prentice Hall.

Ross, Hugh. 1995. *The Creator and the Cosmos: How the Greatest Scientific Discoveries of the Century Reveal God*. Colorado Springs: Navpress.

Russell, Bertrand. 1945. *A History of Western Philosophy.* New York: Simon and Schuster.

Schilpp, Paul Arthur, ed. 1949. *Albert Einstein: Philosopher-Scientist.* Evanston, Ill.: The Library of Living Philosophers.

Schweber, S. S. 1994. *QED and the Men Who Made it: Dyson, Feynman, Schwinger, and Tomonaga.* Princeton: Princeton University Press.

Schwinger, Julian. 1948a. "On Quantum Electrodynamics and the Magnetic Moment of the Electron." *Physical Review* 73: 416–17.

———. 1948b. "Quantum Electrodynamics I. A Covariant Formulation." *Physical Review* 74: 1439.

Schwinger, Julian, ed. 1958. *Selected Papers on Quantum Electrodynamics.* New York: Dover.

Seager, William. 1996. "A Note on the 'Quantum Eraser.'" *Philosophy of Science* 63: 81–90.

Sherer, H., and P. Busch. 1993. "Problem of Signal Transmission via Quantum Correlations and Einstein Incompleteness in Quantum Mechanics." *Physical Review* A47 (3): 1647–51.

Shimony, A. 1984. "Controllable and Uncontrollable Non-Locality." In *The Foundations of Quantum Mechanics: In the Light of New Technology.* Tokyo: Hitachi, Ltd.

Sokal, Alan, and Jean Bricmont. 1998. *Fashionable Nonsense: Postmodern Intellectuals' Abuse of Science.* New York: Picador (St. Martins Press). Originally published in French as *Impostures Intellectuelles.* Paris: Editions Odile Jacob, 1997.

Squires, Euan. 1990. *Conscious Mind in the Physical World.* New York: Adam Hilger.

Smith, Quentin. 1990. "A Natural Explanation of the Existence and Laws of Our Universe." *Australasian Journal of Philosophy* 68: 22–43.

Smolin, Lee. 1992. "Did the Universe Evolve?" *Classical and Quantum Gravity* 9: 173–91.

———. 1997. *The Life of the Cosmos.* Oxford and New York: The Oxford University Press.

Smyth, Piazzi. 1978. *The Great Pyramid: Its Secrets and Mysteries Revealed.* New York: Bell Publishing Company.

Stapp, Henry P. 1985. "Bell's Theorem and the Foundations of Quantum Mechanics," *American Journal of Physics* 53 (4): 306–17.

———. 1993. *Mind, Matter, and Quantum Mechanics* New York: Springer-Verlag.

Stenger, Victor J. 1988. *Not By Design: The Origin of the Universe.* Amherst, N.Y.: Prometheus Books.

———. 1990a. *Physics and Psychics: The Search for a World Beyond the Senses.* Amherst, N.Y.: Prometheus Books.

———. 1990b. "The Universe: The Ultimate Free Lunch." *European Journal of Physics* 11: 236–43.

———. 1995a. Review of *The Physics of Immortality* by Frank. J. Tipler. *Free Inquiry* 15: 54–55.

———. 1995b. *The Unconscious Quantum: Metaphysics in Modern Physics and Cosmology.* Amherst, N.Y.: Prometheus Books.

———. 1996. "Cosmythology: Was the Universe Designed to Produce Us?" *Skeptic* 4 (2): 36–40.

_____. 1998. "Anthropic Design and the Laws of Physics." *Reports of the National Center for Science Education* (May/June): 8–12.

_____. 1999a. "The Anthropic Coincidences: A Natural Explanation." *Skeptical Intelligencer* 3 (3): 2–17.

_____. 1999b. "Anthropic Design: Does the Cosmos Show Evidence of Purpose?" *Skeptical Inquirer* 23 (4): 40–63.

_____. 1999c. "Bioenergetic Fields." *The Scientific Review of Alternative Medicine* 3 (1): 16–21.

_____. 2000. "Natural Explanations for the Anthropic Coincidences." Submitted to *Philo*.

Story, Ronald. 1976. *The Space-Gods Revealed*. New York: Harper & Row.

Stückelberg, E. C. G. 1942. "La méchanique du point matériel en théorie de la relativité." *Helv. Phys. Acta* 15: 23–37.

Sutherland, Roderick I. 1983. "Bell's Theorem and Backwards-in-Time Causality." *International Journal of Theoretical Physics* 22 (4): 377–84.

Swinburne, Richard. 1990. "Argument from the Fine-Tuning of the Universe." In Leslie (1990): 154–73.

Taylor, Stuart Ross. 1998, *Destiny or Chance: Our Solar System and Its Place in the Cosmos*. Cambridge: Cambridge University Press.

Tegmark, Max. 1997. "Is 'the Theory of Everything' Merely the Ultimate Ensemble Theory?" *Annals of Physics* 270 (1998): 1–51.

_____. 1998. "The Interpretation of Quantum Mechanics: Many Worlds or Many Words?" *Fortschr. Phys.* 46: 855–62.

Thorne, Kip S. 1994. *Black Holes and Time Warps: Einstein's Outrageous Legacy*. New York" W. W. Norton.

Tipler, Frank J. 1994. *The Physics of Immortality: Modern Cosmology and the Resurrection of the Dead*. New York: Doubleday.

Tomonaga, S. 1946. "On a Relativistically Invariant Formulation of the Quantum Theory of Wave Fields." *Progress in Theoretical Physics* 1/2: 1–13.

_____. 1948. "On Infinite Field Reactions in Quantum Field Theory." *Physical Review* 76: 224.

Tryon, E. P. 1973. "Is the universe a quantum fluctuation?" *Nature* 246: 396–97.

Van Frassen, Bas C. 1989. *Laws and Symmetry*. New York, Oxford: Oxford University Press.

Vitzthum, Richard C. 1995. *Materialism: An Affirmative History and Definition*. Amherst, N.Y.: Prometheus Books.

Weinberg, Steven. 1989. *Annals of Physics (N.Y.)* 194: 336.

_____. 1972. *Gravitation and Cosmology: Principles and Applications of the Theory of Relativity*. New York: John Wiley.

_____. 1977. "The Search for Unity: Notes for a History of Quantum Field Theory." *Daedalus* 106 (2): 23–33.

_____. 1989. "The Cosmological Constant Problem." *Reviews of Modern Physics* 61: 1–23.

_____. 1992. *Dreams of a Final Theory*. 2 vols. Cambridge: University of Cambridge Press.

_____. 1995. *The Quantum Theory of Fields*. New York: Pantheon.

————. 1998. "The Revolution That Didn't Happen." *New York Review of Books*, October 8.

Weyl, H. 1919. *Annalen der Physik* 59: 101.

Wheeler J. A., and R. P. Feynman. 1945. "Interaction with the Absorber as the Mechanism of Radiation." *Reviews of Modern Physics* 17: 157–86.

Wigner, E. P. 1961. "Remarks on the mind-body question." In *The Scientist Speculates*. Edited by J. J. Good. London: Heinemann. Reprinted in E. Wigner 1967, *Symmetries and Reflections*, Bloomington: Indiana University Press; and in *Quantum Theory and Measurement*. Edited by J. A. Wheeler and W. H. Zurek. Princeton: Princeton University Press, 1983.

Will, Clifford M. 1986. *Was Einstein Right? Putting General Relativity to the Test.* New York: Basic Books.

Wilson, E. O. 1998. *Consilience.* New York: Alfred A. Knopf

Wittgenstein, Ludwig. 1922. *Tractatus Logico-Philosophicus.* Translated by C. K. Ogden. London: Routledge and Kegan Paul.

————. 1953. *Philosophical Investigations.* New York: McMillan.

Wootters, W. K., and W. H. Zurek. 1979. "Complementarity in the Double-Slit Experiment: Quantum Nonseparability and a Quantitative Statement of Bohr's Principle." *Physical Review* D19: 473–84.

Zeh, H. D. 1989, 1992 (2d ed.). *The Physical Basis of the Direction of Time.* Berlin: Springer Verlag.

Zellinger, Anton. 2000. "Quantum Teleportation." *Scientific American* (April): 50–54.

Zohar, Danah. 1990. *The Quantum Self: Human Nature and Consciousness Defined by the New Physics.* New York: Morrow.

Zukav, Gary. 1979. *The Dancing Wu Li Masters: An Overview of the New Physics.* New York: Morrow.

Zurek, Wojciech H. 1991. "Decoherence and the Transition from Quantum to Classical." *Physics Today* 36: 36–44.

Zurek, Wojciech H., et al. 1993. "Negotiating the Tricky Border Between Quantum and Classical." *Physics Today* (Letters to the Editor) 13–15 (April): 81–90.

INDEX

AMERICAN THEORISTS
OF THE NOVEL

The American theorists Henry James, Lionel Trilling, and Wayne C. Booth have revolutionized our understanding of narrative or storytelling, and have each championed the novel as an art form. Concepts from their work have become part of the fabric of novel criticism today, influencing theorists, authors, and readers alike.

Emphasizing the crucial relationship between the work of these three critics, Peter Rawlings explores their understanding of the novel form, and investigates their ideas on:

* realism and representation
* authors and narration
* point of view and centres of consciousness
* readers, reading, and interpretation
* moral intelligence.

Rawlings demonstrates the importance of James, Trilling, and Booth for contemporary literary theory and clearly introduces critical concepts that underlie any study of narrative. This book is invaluable reading for anyone with an interest in American critical theory, or the genre of the novel.

Peter Rawlings is Reader in English and American Literature and Head of English and Drama at the University of the West of England, Bristol (UK). He has published widely on Henry James, American theories of fiction in the nineteenth century, and the American reception of Shakespeare.

ROUTLEDGE CRITICAL THINKERS

Series Editor: Robert Eaglestone, Royal Holloway, University of London

Routledge Critical Thinkers is a series of accessible introductions to key figures in contemporary critical thought.

With a unique focus on historical and intellectual contexts, the volumes in this series examine important theorists':

- significance
- motivation
- key ideas and their sources
- impact on other thinkers

Concluding with extensively annotated guides to further reading, *Routledge Critical Thinkers* are the student's passport to today's most exciting critical thought.

Already available:

Louis Althusser by Luke Ferretter
Roland Barthes by Graham Allen
Jean Baudrillard by Richard J. Lane
Simone de Beauvoir by Ursula Tidd
Homi K. Bhabha by David Huddart
Maurice Blanchot by Ullrich Haase and William Large
Judith Butler by Sara Salih
Gilles Deleuze by Claire Colebrook
Jacques Derrida by Nicholas Royle
Michel Foucault by Sara Mills
Sigmund Freud by Pamela Thurschwell
Stuart Hall by James Procter
Martin Heidegger by Timothy Clark
Fredric Jameson by Adam Roberts
Jacques Lacan by Sean Homer
Julia Kristeva by Noëlle McAfee

Jean-François Lyotard by Simon Malpas
Paul de Man by Martin McQuillan
Friedrich Nietzsche by Lee Spinks
Paul Ricoeur by Karl Simms
Edward Said by Bill Ashcroft and Pal Ahluwalia
Gayatri Chakravorty Spivak by Stephen Morton
Slavoj Žižek by Tony Myers
Theorists of the Modernist Novel: James Joyce, Dorothy Richardson, and Virginia Woolf by Deborah Parsons
Theorists of Modernist Poetry: T. S. Eliot, T. E. Hulme, and Ezra Pound by Rebecca Beasley

For further details on this series, see www.routledge.com/literature/series.asp

AMERICAN THEORISTS OF THE NOVEL

HENRY JAMES, LIONEL TRILLING, WAYNE C. BOOTH

Peter Rawlings

Routledge
Taylor & Francis Group

LONDON AND NEW YORK

First published 2006
by Routledge
2 Park Square, Milton Park, Abingdon, Oxon OX14 4RN

Simultaneously published in the USA and Canada
by Routledge
270 Madison Avenue, New York, NY 10016

Routledge is an imprint of the Taylor & Francis Group, an informa business

© 2006 Peter Rawlings

Typeset in Perpetua by
Florence Production Ltd, Stoodleigh, Devon
Printed and bound in Great Britain by
TJ International Ltd, Padstow, Cornwall

British Library Cataloguing in Publication Data
A catalogue record for this book is available from
the British Library

Library of Congress Cataloging in Publication Data
Rawlings, Peter.
   American theorists of the novel: Henry James, Lionel Trilling,
and Wayne C. Booth/Peter Rawlings.
      p. cm. – (Routledge critical thinkers)
   Includes bibliographical references and index.
   1. Criticism–United States. 2. Fiction–History and criticism.
   I. Title. II. Series.
   PN99.U52R39 2006
   808.3–dc22   2005036198

ISBN10: 0–415–28544–5 (hbk)
ISBN10: 0–415–28545–3 (pbk)
ISBN10: 0–203–96947–2 (ebk)

ISBN13: 978–0–415–28544–5 (hbk)
ISBN13: 978–0–415–28545–2 (pbk)
ISBN13: 978–0–203–96947–2 (ebk)

SUCH AS IT IS,
IN MEMORY OF WAYNE C. BOOTH
(1921–2005)

CONTENTS

SERIES EDITOR'S PREFACE

The books in this series offer introductions to major critical thinkers who have influenced literary studies and the humanities. The *Routledge Critical Thinkers* series provides the books you can turn to first when a new name or concept appears in your studies.

Each book will equip you to approach these thinkers' original texts by explaining their key ideas, putting them into context and, perhaps most importantly, showing you why they are considered to be significant. The emphasis is on concise, clearly written guides that do not presuppose a specialist knowledge. Although the focus is on particular figures, the series stresses that no critical thinker ever existed in a vacuum but, instead, emerged from a broader intellectual, cultural and social history. Finally, these books will act as a bridge between you and their original texts: not replacing them but, rather, complementing what they wrote. In some cases, volumes consider small clusters of thinkers working in the same area, developing similar ideas or influencing each other.

These books are necessary for a number of reasons. In his 1997 autobiography, *Not Entitled*, the literary critic Frank Kermode wrote of a time in the 1960s:

On beautiful summer lawns, young people lay together all night, recovering from their daytime exertions and listening to a troupe of Balinese musicians.

> Under their blankets or their sleeping bags, they would chat drowsily about
> the gurus of the time ... What they repeated was largely hearsay; hence my
> lunchtime suggestion, quite impromptu, for a series of short, very cheap books
> offering authoritative but intelligible introductions to such figures.

There is still a need for 'authoritative and intelligible introductions', but this series reflects a different world from the 1960s. New thinkers have emerged and the reputations of others have risen and fallen, as new research has developed. New methodologies and challenging ideas have spread through the arts and humanities. The study of literature is no longer – if it ever was – simply the study and evaluation of poems, novels, and plays. It is also the study of the ideas, issues, and difficulties which arise in any literary text and in its interpretation. Other arts and humanities subjects have changed in analogous ways.

With these changes, new problems have emerged. The ideas and issues behind these radical changes in the humanities are often presented without reference to wider contexts or as theories that you can simply 'add on' to the texts you read. Certainly, there's nothing wrong with picking out selected ideas or using what comes to hand – indeed, some thinkers have argued that this is, in fact, all we can do. However, it is sometimes forgotten that each new idea comes from the pattern and development of somebody's thought and it is important to study the range and context of their ideas. Against theories 'floating in space', the *Routledge Critical Thinkers* series places key thinkers and their ideas firmly back in their contexts.

More than this, these books reflect the need to go back to the thinkers' own texts and ideas. Every interpretation of an idea, even the most seemingly innocent one, offers its own 'spin', implicitly or explicitly. To read only books on a thinker, rather than texts by that thinker, is to deny yourself a chance of making up your own mind. Sometimes what makes a significant figure's work hard to approach is not so much its style or content as the feeling of not knowing where to start. The purpose of these books is to give you a 'way in' by offering an accessible overview of these thinkers' ideas and works and by guiding your further reading, starting with each thinker's own texts. To use a metaphor from the philosopher Ludwig Wittgenstein (1889–1951), these books are ladders, to be thrown away after you have climbed to the next level. Not only, then, do they equip you to approach new ideas, but also they empower you, by leading you back to a theorist's own texts and encouraging you to develop your own informed opinions.

Finally, these books are necessary because, just as intellectual needs have changed, the education systems around the world – the contexts in which introductory books are usually read – have changed radically, too. What was suitable for the minority higher education system of the 1960s is not suitable for the larger, wider, more diverse, high technology education systems of the twenty-first century. These changes call not just for new, up-to-date introductions but new methods of presentation. The presentational aspects of *Routledge Critical Thinkers* have been developed with today's students in mind.

Each book in the series has a similar structure. They begin with a section offering an overview of the life and ideas of the featured thinkers and explaining why they are important. The central section of the books discusses the thinkers' key ideas, their context, evolution and reception: with the books that deal with more than one thinker, they also explain and explore the influence of each on each. The volumes conclude with a survey of the impact of the thinker or thinkers, outlining how their ideas have been taken up and developed by others. In addition, there is a detailed final section suggesting and describing books for further reading. This is not a 'tacked-on' section but an integral part of each volume. In the first part of this section you will find brief descriptions of the key works by the featured thinkers; then, following this, information on the most useful critical works and, in some cases, on relevant websites. This section will guide you in your reading, enabling you to follow your interests and develop your own projects. Throughout each book, references are given in what is known as the Harvard system (the author and the date of a work cited are given in the text and you can look up the full details in the bibliography at the back). This offers a lot of information in very little space. The books also explain technical terms and use boxes to describe events or ideas in more detail, away from the main emphasis of the discussion. Boxes are also used at times to highlight definitions of terms frequently used or coined by a thinker. In this way, the boxes serve as a kind of glossary, easily identified when flicking through the book.

The thinkers in the series are 'critical' for three reasons. First, they are examined in the light of subjects that involve criticism: principally, literary, studies or English and cultural studies, but also other disciplines that rely on the criticism of books, ideas, theories and unquestioned assumptions. Second, they are critical because studying their work will provide you with a 'tool kit' for your own informed critical reading and thought, which will make you critical. Third, these thinkers are critical because they are crucially important: they deal with ideas and questions that can overturn conventional understandings of the world, of texts, of everything we take for granted, leaving us with a deeper understanding of what we already knew and with new ideas.

No introduction can tell you everything. However, by offering a way into critical thinking, this series hopes to begin to engage you in an activity which is productive, constructive, and potentially life-changing.

WHY JAMES, TRILLING, AND BOOTH?

Why read James, Trilling, and Booth? The answer may not be immediately obvious. Writing from the 1860s and through to the early twentieth century, Henry James (1843–1916) is most widely renowned for works such as *The Wings of the Dove* (1902b), *The Golden Bowl* (1904), *The Portrait of a Lady* (1881), and his ghost story, 'The Turn of the Screw' (1898). But he also published ground-breaking prefaces to his own fiction and numerous critical essays. Lionel Trilling (1905–75) became well known as a literary critic in a 1950s academic scene dominated by, as we shall see, the 'New Criticism' of earlier decades. The academic career of Wayne C. Booth (1921–2005), on the other hand, has spanned the later twentieth-century transformation of literary 'criticism' into the myriad new approaches known as literary 'theory'.

So why read the texts of these three American critics, and why read them alongside one another? Because the landmark works of James, Trilling, and Booth have in just over a century revolutionized our understanding of what narrative, or story-telling is, and how prose fiction (novels and stories) functions. They are among the most widely cited theorists of the novel, and their work has had an enormous influence on the writing, reading, and criticism of fiction. Read by academics and the general reader alike, Trilling's *The Liberal Imagination* (1950) was a bestseller in the US and soon had a huge impact on

NEW CRITICISM

The focus of New Criticism is on literature itself and away from the lives and times (the context) of particular writers. The text is regarded as self-sufficient; and the task is to subject it to 'close reading'. In 'The Intentional Fallacy' (1946) and 'The Affective Fallacy' (1949), W. K. Wimsatt and Monroe C. Beardsley argued that neither the author's intention nor the reader's feelings were relevant to interpreting and judging works of literature. This movement held sway for much of the twentieth century. Although the New Historicism of Stephen Greenblatt and others has redirected attention to correspondences between texts and history, it remains unfashionable in many quarters to use biographical material to interpret literary texts.

critical thinking internationally. It has gone through many editions subsequently. Together with the rest of Trilling's work, *The Liberal Imagination* is attracting attention again now that literary theory has lost much of the ground it took in the later twentieth century (some critics refer to the current period as 'post-theory'). James's essays, and especially his prefaces to the New York edition of his work, continue to be a dominant force in discussions about fiction. Booth's *The Rhetoric of Fiction* has been indispensable to students of the novel ever since its first publication in 1961. Concepts from their work have become part of the fabric of novel criticism today: we have James's ideas on 'points of view' and 'centres of consciousness', Trilling's 'moral realism' and 'the liberal imagination', and Booth's 'implied author' and 'reliable/unreliable narration', to name but a few.

Their work has also had a huge effect on the status of the novel. In 1817, the Romantic poet and critic, Samuel Taylor Coleridge, was able to dismiss the reading of novels as a 'kill-time' rather than a 'pass-time', a 'species of *amusement*' akin to 'spitting over a bridge' (1817: 1: 34). Moreover, even at the end of a nineteenth century which had seen the achievements of novelists, (among many others) of Walter Scott, Charles Dickens, George Eliot, Gustave Flaubert, Ivan Turgenev, Leo Tolstoy, and Henry James himself, the minor American critic, George Clarke, was still comparing the effects of novel-reading with 'those of indulgence in opium and intoxicating liquors' (Clarke

1898: 362). At best, then, the novel was seen as a frivolous entertainment, and at worst, an immoral distraction from the practical world. Today, however, the novel is considered by a majority of critics to be a flexible form of art uniquely suited to the inspection of individual, social, and moral health. It has, as Trilling put it in *The Liberal Imagination,* a 'reconstitutive and renovating power' (1950: 253). To understand this new perspective, and the work from which it emerged, it is essential to engage with the writings of Henry James, Lionel Trilling, and Wayne C. Booth. This book provides a guide to their major work on theories of the novel and a companion for your own reading of the key texts.

DIFFERENT CONTEXTS, COMMON CONCERNS

Although the work of these three critics emerges from varied contexts, all three share a preoccupation with a set of ethical and moral questions about fiction that subsequent critics have been unable to ignore. Is it possible to have 'good' novels about 'bad' people? Should it be the function of the novel to make the reader a 'better', more socially responsible person? Do we, in any event, have common standards by which to assess such improvements? Should a novelist pass clear judgements on his characters? Is it morally dangerous for authors to multiply ambiguities or uncertainties about meaning?

The ethics of reading and writing and the moral consequences of formal and technical decisions are central concerns for these critics and, as a result of their influence, for theorists of the novel in general. On the basis of even a cursory glance at these concerns it is clear that James, Trilling, and Booth focus not only on what texts *are*, but also on *how they are put together*, or on what it is about their organization in language that makes them tick. In varying degrees, they are all interested in these matters of content, form, and technique; but they are even more preoccupied with what texts can *do*, with how they hook on to the world, and with the impact they can have on readers. As Trilling memorably expresses it, literary structures are not 'static and commemorative but mobile and aggressive, and one does not describe a quinquereme or a howitzer or a tank without estimating how much *damage* it can do' (1965: 11).

For these critics, communication, for good or for ill, is at the centre of the business of reading, writing, and grasping novels critically.

ETHICS AND MORALS

'Ethics' are the rules that regulate our behaviour in specific practical areas (such as medicine or literary criticism). 'Morals' are the underlying principles shaping these ethics.

Wayne Booth, the last in this theoretical genealogy, constructed a model of the communication process, making explicit many of the concepts that had been implicit in the work of the others. I shall turn to Booth's model shortly, as a slightly modified version of it provides the structure for this guide. At this point, however, we might consider a little more closely the lives and contexts of each of our three critics. As this guide examines aspects of their work, I shall necessarily return to the particular 'hooks' between the critics' own texts and their worlds, but it may be useful to set the scene with some background information, to which you might easily return later.

HENRY JAMES (1843–1916)

The American republic was less than seventy years old when Henry James was born in Greenwich Village, New York City, in 1843. By 1864, the family had settled in Boston, Massachusetts, after more than twenty years of moving between America and Europe. The family was of Irish and Scottish descent. Henry James's grandfather had made a considerable fortune in business, but the shrinking inheritance had eventually to be divided, in Henry's generation, between five children. For these five, then, there was no prospect of the life without work that had been enjoyed by their father, a devotee of the Swedish mystic, Emanuel Swedenborg (1688–1772). Henry's father had a relaxed, even rather a scattered, approach to child-rearing. As befitted a man whose youth had been somewhat dissipated, his emphasis was on 'being' rather than 'doing', and this resulted in a certain shiftlessness in his children. After dabbling in painting for a while, Henry's older brother, William James (1842–1910), became an eminent psychologist and philosopher and, as we shall see in Chapter 4, exercised a significant impact on James's theory and practice of fiction. Henry himself studied law at Harvard, fitfully, before turning in earnest to the writing of fiction.

Despite the influence of American writers on his fiction and criticism – especially that of Nathaniel Hawthorne (1804–64), an author most widely known today for his romance, *The Scarlet Letter* (1850) – James's attachment was to the culture of Europe, to the Old World rather than the New. In his 1879 book, *Hawthorne*, James protested that America lacked the 'complex social machinery' necessary to 'set a writer in motion' (1879: 320). After his unlikely year at Harvard (1862–3) and further trips to Europe, he settled in England in 1876, twelve years after the appearance of his first reviews and fiction. He returned to America only occasionally, and became a naturalized British citizen shortly before his death in 1916. Apart from *Hawthorne*, a series of prefaces to the New York edition of his fiction (1907–9), and the numerous reviews and essays he never collected, James produced four volumes of literary criticism and theory: *French Poets and Novelists* (1878), *Partial Portraits* (1888b), *Essays in London and Elsewhere* (1893a), and *Notes on Novelists* (1914). Most of this material had been published previously in journals such as the *Atlantic Monthly* and the *Nation*. James was a prolific writer of fiction as well as a critic: there are twenty-two novels (two were unfinished) and over a hundred short stories (and some are not so short). He also wrote a number of very bad and spectacularly unsuccessful plays such as *Guy Domville* (1894).

From his youth on, James read widely in the English and European novel traditions. His fiction and criticism attempt to reconcile the social and moral intensities of English novelists such as George Eliot (1819–80) with the formal self-consciousness of French writers who often seemed to disregard morality. French authors especially important to James were Honoré de Balzac (1799–1850), Gustave Flaubert (1821–80), and Émile Zola (1840–1902). When James started writing fiction in the 1860s, novels were tolerated by a good many influential reviewers only if they were heavily didactic; if they aimed, that is, to teach moral lessons. The legacy of Puritanism in America meant that the theme of adultery, which was especially prominent in the French novel, was often beyond the pale of what was acceptable there for most readers, critics, and writers. When the American writer Nathaniel Hawthorne (1804–64) tackled this theme in *The Scarlet Letter* (1850), it was described by one reviewer as having a 'running underside of filth' (Coxe 1851: 489). James found himself caught between admiring the technique, or what he considered the *art*, of many French

PURITANISM

The Puritans arose as a party within the Church of England during the Reformation, the Protestant rebellion against Catholicism, in the sixteenth and early seventeenth centuries. They were opposed to what they saw as the excessive ceremonies and rituals of the newly established Church of England and supported parliamentary government, rather than the monarchy, at the time of the English Civil War and its aftermath (1640–60). Puritans made up the majority of early European settlers in New England (America) in the early seventeenth century. The label 'Puritanism' became associated with strict and oppressively uncompromising moral attitudes.

novelists and condemning, with increasing reluctance, their 'off-limits' subject-matter.

The title of Henry James's major critical essay, 'The Art of Fiction' (1884), makes it clear that he considered the writing of novels and short stories as an art in its own right, and it is hard to imagine just how challenging this view was at the time. When James began to write, fiction was often regarded as dubious by narrow moralists because it tended towards the projection of escapist worlds of romance and fantasy. But as we have seen, writers who attempted to write more realistically by including glimpses of the adult bedroom (for example) were frequently condemned outright. James soon became known as a realist in two related senses. First, he dealt with the recognizable world of everyday reality, or at least the cultivated segment of it with which he was familiar. Second, he tackled morally complex situations in which the rules of conduct adhered to by conservative readers were unlikely to be universally helpful.

James was pulled in two directions: the morally intense world of his American context (especially that of Boston, with those powerful residues of Puritanism, in which he began to write), and the (mainly French) world of art with its increasing devotion to form and technique at the expense of morality and moralizing. The pressure in America and also in Britain, where James took up residence, was to produce a filtered version of reality, an ideal world full of messages promoting self-improvement. In France, the growing enthusiasm was for the representation of the world in all its lurid reality. Embedded